European-Russian Space Cooperation

From de Gaulle to ExoMars

Brian Harvey

European-Russian Space Cooperation

From de Gaulle to ExoMars

 Springer

Published in association with
Praxis Publishing
Chichester, UK

Brian Harvey
Templeogue
Dublin
Ireland

SPRINGER-PRAXIS BOOKS IN SPACE EXPLORATION

Springer Praxis Books
Space Exploration
ISBN 978-3-030-67684-1   ISBN 978-3-030-67686-5   (eBook)
https://doi.org/10.1007/978-3-030-67686-5

Cover image: Artist's impression of the ExoMars 2020 rover (foreground), surface science platform (background) and the Trace Gas Orbiter (top). Credit: ESA/ATG medialab, with additional elements furnished by cover designer Jim Wilkie.

Back cover image: ExoMars team with Proton Rocket. Credit: ESA.

Project Editor: Michael D. Shayler

This Springer imprint is published by the registered company Springer Nature Switzerland AG
The registered company address is: Gewerbestrasse 11, 6330 Cham, Switzerland

Contents

The original version of this book was revised. The correction is available at
https://doi.org/10.1007/978-3-030-67686-5_7

Author's introduction

The long story of space exploration has traditionally been dominated by a narrative of Russian-American competition and rivalry, now replaced by their cooperation together in the world's largest collaborative engineering project, the International Space Station (ISS). Often overlooked is the level of cooperation between Russia and the third power to reach space, Europe. The first collaborative ventures between the Soviet Union and European countries, primarily France, date to the early 1960s and such projects became increasingly structured as time passed. The cold war liaisons of the Soviet period gave way to joint enterprises in which European astronauts flew to Russian space stations, the Soyuz rocket found a new home in European territory in the South American jungle and science missions were flown to study the deep space universe. Their climax was the joint, two-part ExoMars mission to explore Mars, the detailed planning for which began in 2012 and whose first launch took place in 2016.

The story of European-Russian cooperation is little known and its importance undervalued. Because France was the initial and principal interlocutor for this venture, language barriers meant that this cooperation did not receive the attention it deserved in English language publications. This book looks at how that relationship evolved; what factors — scientific, political and industrial — most drove it; who gained most; which countries participated most and least — and why — and the scientific and other outcomes, climaxing in their joint exploration of Mars from 2016. Although the primary focus is on the technical aspects and outcomes of cooperation, the relationship is set within the wider diplomatic contexts of the cold war, the triangle of Russian-European-American interactions, the sanctions regime reintroduced from 2014 and the other elements that strained the relationships between the two sides. Cooperation was often controversial and even difficult at times. The analysis suggests that there were substantial gains in science and industrial efficiency and that the alternative of non-cooperation had its own costs.

To explore this long and complex narrative, a combination of a thematic and historical approach is applied. The first chapter, *Early days*, traces Russian-European cooperation to the 1966 agreement arising from the landmark visit to the Soviet Union by French President Charles de Gaulle. It examines the early outcomes, namely sounding rockets and the inclusion of French instruments on missions to the Moon and Mars. The chapter also sets such cooperation in the context of France's distinctive foreign policy in the 1960s. The second chapter, *Scientific cooperation*, looks at how the 1966 agreement matured through successful space science projects between France and the USSR in the 1970s, expanded to Germany and the European Space Agency in the 1980s and has continued right up to recent successful projects such as Spektr RG. The next logical area of cooperation was manned or piloted flight (*Chapter 3: Human spaceflight*), developed initially by France − though not without political tension − then Germany and then the European Space Agency countries and which endures to this day on the International Space Station. *Chapter 4: Industrial cooperation* looks at the industrial field, especially launchers, with the most visible form of cooperation being the use of the Soyuz rocket at the European rocket base in Guyana. Chapter 5 records the most integrated and contemporary of all cooperative projects, the two-part joint ExoMars project to send orbiters, landers and a rover to the planet Mars, a project that came under threat as a result of the political tensions that developed from 2014. Finally, chapter 6 comes to conclusions about the experience and what has been learned of the science, industry, diplomacy, politics and practicalities of cooperation.

For convenience, 'Europe' is defined here as what might be called 'western Europe' prior to 1991 and thereafter all of Europe from the Baltic states and Poland westward. This is normally synonymous with − but by no means limited to − European Union and European Space Agency countries. This is not to deny the European vocation of those European countries in the USSR and the socialist block, but they have been covered in the literature on the Soviet programme. Readers seeking information on pre-1991 cooperative Russian-European missions − including those of the socialist block − should read Zakutnyaya, Olga and this writer: *Russian space probes* (Praxis/Springer, 2011) and Burgess, Colin & Vis, Bert: *Interkosmos − the eastern block's early space programme* (Praxis/Springer, 2016).

Two principal currencies are used: the Russian rouble (R) and the European euro (€) or its predecessor currency used by the European Union, the European Currency Unit (ECU). For convenience, the word 'Russia' refers to both the USSR or Soviet Union (1922−91) or the Russian Federation (1992−) and the people living therein as 'Russians'. However, the specific political terms will be used when the context requires.

Brian Harvey,
Dublin, Ireland, 2020

Acknowledgements

I would like to thank those who assisted in this project through the provision of information, interviews and photographs, especially:

Jacqueline Myrrhe, *Go Taikonauts!* Germany
René Demets, ESA, the Netherlands
René Pischel, Head of ESA Permanent Representation in the Russian Federation, Russia
Olga Zakutnyaya, IKI, Russia
David Shayler, Britain
Dominic Phelan, Ireland
Marsha Freeman, United States
Gerry Webb, Britain
Bert Vis, Netherlands
Bart Hendrickx, Belgium
Christian Lardier, France
Massimo Cislaghi, ESA
Gerry Skinner, Britain
Romain Charles, France
Rainer Scharenberg, Germany
Jörg Feustel-Büechl, Germany
Davide Sivolella, Italy
Stanislav Klimov, Russian Academy of Sciences, Laboratory for Research into Electromagnetic Radiation
Alistair Scott, Britain
Bernard Tiedt, Germany
Jeanne Medvedeva, Exolaunch, Germany
Marco Siddi, Helsinki, Finnish Institute for International Affairs, Finland

Ruth McKenna, Department of Foreign Affairs, Ireland
Fernando Florindo, General Secretariat, Council of the European Union, Belgium
Gabriele Visentin, European External Action Service, European Union, Belgium
Rosine Lallement, Jean-Loup Bertaux, France.

Photographs are credited according to normal procedures, with CC for Creative Commons. Unmarked photographs are from the author's collection.

To Judith; Valerie and Alistair; Charlie and Robyn; and Justin

About the Author

Brian Harvey is a writer and broadcaster on space flight who lives in Dublin, Ireland. He has a degree in history and political science from Dublin University (Trinity College) and an MA from University College Dublin. His first book was *Race into space – the Soviet space programme* (Ellis Horwood, 1988), followed by further books on the Russian, Chinese, European, Indian and Japanese space programs. His books and chapters have been translated into Russian, Chinese and Korean.

1

Early days

European-Russian cooperation in spaceflight began on a hot summer's day, 20 June 1966, when a sleek French air force Caravelle passenger jet touched down at the newly built glass-and-steel terminal at Vnukovo airport, Moscow. The Caravelle was escorted in by a flight of MiG jets which quickly departed for their home airfields. Down the steps came the unmistakable, tall figure of the President of France, General Charles de Gaulle, accompanied by Madame Yvonne de Gaulle and their son Philippe, along with foreign minister Maurice Couve de Murville. They had scarcely been greeted by President Nikolai Podgorny, prime minister Alexei Kosygin and a guard of honour when a gun salute crashed out and a Russian military band struck up the *Marseillaise*. The visitors drove into the city in an open top black car, with the flags of each country on its wings. The glorious sunny weather was to last almost all of the next week. First there was a visit to the Bolshoi theatre and then Moscow Lomonosov state university, before the visitors travelled on 23 June to the capital of Siberia, Novosibirsk, Akademgorodok (the town for scientists) with its rows of apartments and riverside birch trees.

This was not the general's first visit. He had arrived in Moscow in 1944 as leader of the Free French by a long roundabout train journey via Baku. A hundred Free French pilots, the Normandie Niemen squadron, fought in the Red Air Force and would subsequently have 144 schools named after them. Earlier, when de Gaulle was a prisoner of war in Germany in 1917, a fellow prisoner was Mikhail Tukhachevsky, later Stalin's top marshal, so he had first-hand acquaintance with Russia [1]. Aviation links between Russia and France stretched back to the early 20th century [2].

The original version of this chapter was revised. The correction to this chapter is available at https://doi.org/10.1007/978-3-030-67686-5_7

The high point of the visit was on 25 June, when the President became the first westerner to visit the Soviet Union's main cosmodrome, Baikonour, in the heart of the Kazakhstan desert. The film record of his visit there is sparse, though that is hardly surprising given the officially secret nature of the base. Once off the plane, de Gaulle was driven in another black car, using his hat to shield himself from the bright sunshine. He was then taken to a viewing area, where he saw a rocket heading skyward amidst billowing smoke. This was later identified as a weather satellite, Cosmos 122, using a Vostok 2M rocket. His son Philippe saw the launch alongside his father, exclaiming *Collosal! Collosal!* (translatable into contemporary American English as *Awesome! Awesome!*). Accompanying them were his *aide de camp*, Admiral François Flohic and Maurice Couve de Murville.

President Charles de Gaulle. Bundesarchiv.

Next, the French president met a visiting delegation from Warsaw Pact countries and the legendary Soviet designer Mikhail Yangel, who was there for the launch of one of his missiles. One account says that de Gaulle was the only person to see the launch itself and that the rest of the French delegation (some of whom might have been able to assess the rocket) were kept busy at another function. They were then brought to the town of Leninsk, home of the workers on the base. It was renamed for the day as Zvezdograd ('Star Town'), which in reality was the name of the cosmonaut training and living centre in Moscow, though nobody outside the space programme knew that at the time. The town got a facelift, with roads resurfaced, paint applied and fences fixed. The *aide de camp* commented on how the local population was unusually young, masculine and short-haired, suggesting that the military there had been taken out of uniform to present themselves as a welcoming party. The whole visit was meticulously organized, with the Soviet authorities giving the operation the name 'Palm Tree': this was designated 'Palm Tree number 1' **[3]**.

The rest of the president's visit to the Soviet Union went equally well. His delegation flew to Leningrad, where their Soviet hosts thoughtfully arranged for them to attend mass at the Church of Notre Dame de Lourdes. Then it was on to Kiev and Volograd, with solemn moments to lay wreaths at the war memorials to the band music of Chopin's funeral march. There were meet-the-people walkabouts, a visit to a power station on the Volga and even a mock tank battle (de Gaulle had been a tank commander and they had shared experiences of taking on German panzers). The diplomatic high point of the visit came on 30 June back in Moscow: two agreements, one on space cooperation, the other on scientific, technical and economic cooperation, along with a permanent Franco-Soviet Commission and a hot line between the Kremlin and the Elysée palace, presumably to match that between the Kremlin and the White House. The agreements were signed in the Kremlin and co-signed by the two foreign ministers, Andrei Gromyko and Couve de Murville. De Gaulle returned to France on 1 July. Both sides agreed that it had been a triumph, the diplomatic story of the year. A 42-minute documentary was later made about the visit by Jean Lanzi for ina.fr, *Voyage en URSS*. In another footnote, a French company got the contract to build the landmark modern high-rise Kosmos hotel beside the space memorial. There is an imposing statue of the general outside.

Statue of President Charles de Gaulle at Hotel Kosmos

Fifty years later, in 2016, de Gaulle's visit was marked in Kazakhstan's then capital, Astana, at its National Space Centre, with an event organized by the country's Aerospace committee and attended by Kazakh cosmonaut Aydin Aimbetov and the first secretary of the Communist Party of Kazakhstan, Dinmukhamed Kunayev. There was an exhibition, which included postage stamps, a model of the first French satellite, Astérix, plus photographs of the 1966 visit. The French side involved the embassy, the Alliance française, President of the Fondation Charles de Gaulle, Jacques Godfrain and representatives of Space City Toulouse and Airbus.

Origins

While 1966 may have been the landmark year, the origins of European-Russian cooperation in space can be traced further back. From the start, the USSR participated in the International Astronautical Federation Congresses, whose first annual conference was held in Paris in 1950, so this offered an early opportunity for European space experts to meet and mingle with their Soviet counterparts. Indeed, the announcement of the launch of Sputnik was made just as delegates were gathering at the 1957 congress in Barcelona, Spain. Shortly afterwards, the Committee on Space Research (COSPAR) was formed as an international meeting place for space scientists, to enable them to meet outside the confines of government supervision. During the early days of what we now call the space race, there was a remarkably high level of cooperation that might seem surprising today. The two great, rival radiation scientists of their day, James Van Allen and Sergei Vernov, invited one another for lecture tours in their respective countries. As the cold war dragged on, however, such cooperation came to be regarded more suspiciously, certainly on the American side.

Soviet-French cooperation had an early personal and linguistic form in the character and life of Ari Sternfeld (1905–80). He was a Pole who lived his early professional life in France, where he was a mechanical engineer turned populariser of spaceflight. A supporter of the Communist movement, he wrote for the daily *L'Humanité* and, with shockingly bad timing, moved to Moscow in 1935. His main text was *Initiation à la cosmonautique (Initiation to cosmonautics)*, published in Russian in 1937, thereby inventing and introducing the word 'cosmonautics' as the distinctive path followed by the USSR. Presciently, he wrote *Artificial satellites of the Earth* in 1956 and his writing became enormously popular. In 1962, on the first anniversary of Yuri Gagarin's flight, the Soviet government ruled that a space 'man' was called a 'cosmonaut', a term also confirmed by the Academie Française in 1969 [4].

In March 1959 at COSPAR, the United States publicly offered to fly European payloads into orbit – principally using the small Scout rocket – an offer quickly taken up by Britain (Ariel, 1962), France and Italy. While this was a generous gesture in the heady, early days of space exploration, retrospective analysis suggests that the US was also motivated by the desire to keep any moves into the space field by Europe within the American, rather than the Soviet orbit [5].

At the meeting of COSPAR the following year, 1960, French space scientist Jacques Blamont cheekily asked the President of the Soviet Academy of Sciences, former artillery general Anatoli Blagonravov, whether the USSR would give France a place aboard one of its rockets. Nothing came of this immediately, but on 16 January 1961, Blagonravov made just such an offer. Nobody is really sure where, but it was picked up by *Le Monde* newspaper, so it was something that the government could not ignore. The timing was good, because April saw the flight of Yuri Gagarin which, as in the rest of the world, was a sensation in France. A sports centre in Marseille was named after Gagarin only two weeks later, followed by streets and schools, especially in those towns where the Communist Party was strong. Despite its image of secrecy, the Soviet Union attempted to make several gestures of openness, for example in giving two Australian journalists, Anthony Purdy and Wilfred Burchitt, a bird's eye view of its programme in 1961 **[6]**. On 27 September 1963, Gagarin arrived in France in a Tupolev 104 jetliner for the International Astronautical Congress in Paris and subsequently toured the country. He made another visit two years later for the Paris Air Show, where he met the prime minister, Georges Pompidou and flew a Caravelle jetliner into Toulouse Blagnac airfield. During the air show, French aviation journalists managed to organize an invitation for a tour of Soviet facilities, which duly took place in 1966. Gagarin made a third trip in September 1967, as part of worldwide ceremonies to mark the 50th anniversary of the 1917 revolution. He was followed to Paris by the first woman in space, Valentina Tereshkova, who was likewise given a rapturous welcome.

Anatoli Blagonravov

At this stage, political and diplomatic factors came into play which set the context for cooperation in spaceflight. A diplomatic rapprochement between France and the USSR began, with respective foreign minister visits by Couve de Murville and Andrei Gromyko. There was also a brief moment of opportunity for détente in 1963, in the aftermath of the Cuban missile crisis. The United States and Soviet Union signed the nuclear test ban treaty that year and on 10 June, US President John F. Kennedy made what many consider to be his least well-known but most remarkable public address, *Strategy for peace*, a text that would make salutary reading to this day. He proposed a reconciliation with the USSR in a speech which so moved his rival, General Secretary Nikita Khrushchev, that he asked for the entire text, unedited, to be circulated across the Soviet media. That October, the opening rounds of discussions began on America's Moon project becoming a joint venture with the USSR and progress was being made at the very moment Kennedy was assassinated. His successor, Lyndon B. Johnson, was much less interested in détente and the Vietnam war caused a hardening of positions.

De Gaulle in effect resumed where John F. Kennedy (whom he greatly respected and was terribly shaken by his death) left off. De Gaulle had his own reasons too, for he saw his role as rebuilding France as a great European nation in the aftermath of the war and the political uncertainty of the Fourth Republic. De Gaulle designed a policy of independence and equilibrium: independence from the Atlantic world of Britain and the United States; promoting equilibrium between east and west. He spoke of the importance of a less bipolar world and of France being less dependent on the US, with an 'opening to the east'. In 1965, he was re-elected president and on 7 March 1966, to demonstrate independence and equilibrium, he withdrew France from the NATO command. France also took its own line with China. Although the government of the Fourth Republic did not formally recognize the government of the People's Republic of China (PRC) in 1949, it nevertheless maintained lines of communication. In 1962, de Gaulle initiated the process that would lead to formal recognition of the PRC – with the exchange of full ambassadors – on 27 January 1964. France wanted to be seen to be even-handed in dealing with the space superpowers and prepared to demonstrate its independence, recognizing that the Soviet space programme offered opportunities not available in the United States. De Gaulle was very clear that he did not want the kind of 'special relationship' that the United States had with Britain, which he regarded as subservience by the latter.

The withdrawal of France from the NATO command was criticized both at home and abroad, but de Gaulle was by no means uncritical of the USSR. He declined to sign a much closer 'treaty of assistance and friendship' (a form of 'best friend' or even ally status), refused to recognize the German Democratic Republic (GDR) and, in Gdansk in September 1967, encouraged Poland to shake off Soviet domination. He spoke of a 'greater Europe from the Atlantic to the Urals', an idea re-invented, equally unsuccessfully, as 'the common European home' by Mikhail Gorbachev in the 1980s [7].

While de Gaulle's policies, especially the withdrawal from NATO, infuriated the Americans and British and created fears that the USSR would use the French connection to weaken NATO, de Gaulle always made clear his affirmation of 'France first', not least through the independent French nuclear deterrent which, unlike the British one, did not operate a dual control system with the Americans. Neither de Gaulle nor his successor Georges Pompidou were ever going to weaken NATO fatally – France found its way back in – nor the European Communities, which France led.

This was the backdrop to a visit to the Soviet Academy of Sciences in 1964 by Gaston Palewski, Minister of State for Scientific Research and Atomic and Space Questions, which included a meeting with Nikita Khrushchev. Space cooperation was also discussed in Paris when General de Gaulle received Soviet foreign minister Andrei Gromyko on 27 April 1965. On 1 July, Gromyko sent a memo to the French ambassador in Moscow, Philippe Baudet, which began the paperwork for formalizing some kind of relationship in the space field. In the meantime, France and the USSR signed an agreement for cooperation in nuclear energy.

In October 1965, a Soviet delegation led by Leonid Sedov, a leading member of the International Astronautical Federation, visited the French space agency, the Centre Nationale d'Études Spatiales (CNES) in Paris to see what might be possible. The French felt that they were working in the dark regarding both the preparedness of the USSR to cooperate and what the practical limits were, so they decided to bring matters to a head at a meeting between Sedov and the head of the French space agency, Robert Aubinière. The French foreign ministry supplied a Russian aristocrat, Prince Konstantin Andronikov, as interpreter, who combined elegance with a perfect knowledge of the two languages. Sedov told the French that they could ask for any mission: there were no limits. This was welcome, because Soviet rockets were powerful whereas both the small French Diamant and the American Scout launchers limited their satellites to only 150 kg.

The French were astounded when the USSR suggested launching for France a high-altitude (40,000 km) scientific satellite, a communications satellite and a lunar orbiting probe, something that had not yet been done. This was far beyond France's technical capacities, not to mention budget. The Quai d'Orsay, the French Ministry for Foreign Affairs, described this as 'extremely ambitious'. Prime Minister Georges Pompidou recoiled, telling his minister for space affairs, Alan Peyrefite, that they were getting carried away if they thought they could do this.

A formal text was sent to Paris by Andrei Gromyko on 17 March 1966. This was followed by a delegation of CNES president Jean Coulomb and his scientific advisor Jacques Blamont (scientific and technical director, CNES, 1962–76) heading to the USSR the following month. In advance, the French government advised them to limit themselves to 'a little science, some meteorology and no communications'. When they arrived in Moscow, the CNES delegation was accommodated in the skyscraper hotel Ukrainia, with its infrequent lifts, *dezhurniyas* (floor managers) and slow service in the crystalline restaurant. They were taken to meet Mstislav Keldysh (1911–1978), later to become President of the Academy of Sciences, together with all the key figures of the Soviet space science programme, in a big hall under the watchful eye of a portrait of Peter the Great. Keldysh not only came from a highly educated family of engineers and mathematicians, he also spoke perfect French. He spoke to the delegates about how he would like to undertake a joint mission to Mars, carrying an instrument for detecting life. The French were taken aback a second time, not with any offer but with an invitation for a state visit to the Baikonour cosmodrome.

Hotel Ukrainia. CC Gennadiy Grachev

The idea of cooperation in spaceflight was probably something that appealed to General de Gaulle, not out of any rosy notions of exploring the cosmos, but because he saw it as part of his country's process of national modernization. Ever since 1939, when the Centre National de la Recherche Scientifique (CNRS) was established by President Albert Lebrun, the French state had taken a directing role in science in a way unimaginable in the Atlantic world. CNRS became an empire of public service institutes and laboratories. It now boasts over 30,000 scientists and engineers and is adjudged one of the most prestigious, highest-performing scientific bodies in the world. De Gaulle was keen to complete the process of recovery from the scarring effects of the war on the economy, infrastructure and politics. He was an old man in a hurry, who desired a new republic in which France would have a leadership position in Europe, build a strong defence and modernize, reconstructing its railways, aircraft industry, airlines and its scientific and technological capacity. Although an economic conservative − his electoral foe was the socialist party candidate François Mitterrand – de Gaulle was a strong believer in the state directing investment into key industrial and scientific areas and locations, aided by state agencies and enterprises; what we call *dirigisme*. The government directed that its industrial base should be the capital of the backwater south-west, a deliberate − and successful − attempt to reinvigorate provincial areas. In 1964, his prime minister, Georges Pompidou, approved the construction of a French launch base in Guyana, which General de Gaulle visited the following year and announced that it would be 'the site of a great French undertaking, one that would be recognized throughout the world'. The following year, France launched its first satellite. So, space fitted in well with de Gaulle's ideal of a resurgent, modern France.

The French decision to set up a space programme was made at a remarkably short meeting of the government of prime minister Michel Debré in 1961, which de Gaulle concluded by saying 'just do it!' before walking out. The French space agency, CNES, was set up by year's end. France had a three-pronged strategy of developing its own national space industry (CNES), working with 'Les Grands' (the USA and USSR) and building a European space industry. The latter turned out to be the most difficult part, not least because of low spending by Germany and Britain's lack of interest (Britain did not set up a space agency until 2010). At CNES, Jacques Blamont pondered the reasons for the Soviet advances. They wanted western technology and they could offer their powerful rockets. They would not talk to the Germans, while the British were hostile and in the American camp, so this left only the French. Long-standing links between the French and Russian intelligentsia helped.

CNES

Despite political backing from the very top, French space scientists faced an on-going challenge to win political support, for space activities in general and cooperative efforts in particular later. Prime minister Georges Pompidou made it clear that, in his opinion, it would all cost too much. 'We can do something, but around the Earth. Missions further afield: that's for the big powers. We have to look after the [war] veterans', he warned.

In the meantime, the story of Soviet-French cooperation is interrupted by a detour into what could have been an important strand of Soviet-British cooperation. Britain had one great asset: the most powerful radio telescope in the world.

Soviet-British cooperation: Bernard Lovell, Alla Masevich

Given France's pre-eminence, it is now difficult to imagine that Britain had long been seen as the country most likely to lead European space cooperation. Britain had been the world leader in what we would now call electronics and its radar systems had played a key role in winning the Battle of Britain in 1940. Determined to maintain this lead, its scientists constructed what for a long time was the world's largest radio telescope, Jodrell Bank near Manchester, directed by Bernard Lovell (1913–2012). Jodrell Bank had tried to follow the first Sputnik, eventually picking up its signals as it passed over the Lake District on 12 October 1957. This led to a request for the tapes from Russia on 25 October. Two years later and tired of international allegations that its Moon flights were a hoax, the Soviet Union asked Jodrell Bank to track the Second Cosmic ship (Luna 2) as it approached the Moon. Jodrell Bank would be able to provide independent verification that they had actually reached the Moon, which it duly did. Jodrell Bank also assisted the Americans with tracking their Moon shots, but less well known was its role in the American-British early warning system against Soviet missile attack.

Jodrell Bank. CC Mattbuck.

When the first Soviet spacecraft to Venus broke down after only a few days in February 1961, the USSR dispatched one of its top scientists to attempt to regain contact. She was Dr Alla Masevich (1918–2008). Born in Tiflis, Georgia, to a wealthy family with a Polish and French background, at school Masevich was inspired by the space writings of Yakov Perelman, with whom she had a lengthy correspondence. She graduated in physics from Moscow University in 1941, going into metallurgy during wartime but beginning the astrophysics career she always hoped for straight after. She became Professor of Astrophysics at Moscow University in 1956 and was charged the following year with devising the tracking system for the first Sputnik, which used a combination of optical and radio tracking. She could speak four languages and became involved in developing scientific contacts abroad, so she was well qualified, both linguistically and technically, for the Jodrell Bank assignment. Later, she became professor of space geodesy, authored key texts on stellar evolution and was rated one of the ten most influential Russian women of her time. Despite her tsarist-period background, Masevich was a party member and trusted by the authorities to travel abroad, journeying to the International Astronomical Union in Rome in 1952. In 1953, she made waves when she denounced the steady state theory of the universe **[8]**.

Jodrell Bank was immediately asked to assist in finding the Venus probe and listened in for signals on 4 and 5 March, once more for a Venus flyby on 19 May and again thereafter. Alla Masevich travelled to Manchester with a colleague, Jouli Khodarev, on 9 June 1961, spending several weeks trying to recover signals. They did indeed pick up weak signals on 11–12 June from the expected location, but they were not from the spacecraft. Their final attempts were on 20 June, after which the two Russians returned home.

Alla Masevich. CC Josef Blažej

The USSR reciprocated by inviting Bernard Lovell to visit Russia. This was the highest profile visit by any European scientist. It was directed by Jouli Khodarev and guided by Alla Masevich, who also acted as his interpreter. Lovell arrived on 25 June and spent the first two days at Moscow University and the Academy of Sciences. On the third day, he flew to Crimea, where he spent four days visiting its tracking facilities, notably the deep space tracking antennae. The main one was called 'the battleship' because it was built from left-over ship parts (the same was true of parts of Jodrell Bank). Lovell was the first westerner to see the facilities, with no one else getting such an opportunity for at least 20 years. He was not allowed to take photographs and was asked not to disclose its exact location (though the Americans already knew it). While there, Lovell made an agreement for cooperation between Jodrell Bank and the Crimean Astrophysical Observatory over radio observations of the stars and planets.

The Bernard Lovell visit had a number of postscripts. Firstly, he came back with the story that the USSR was not in the Moon race, which was true (the USSR did not commit itself to a Moon landing until August 1964). Secondly, the claim was made – not by him – that the Russians had tried to persuade him to defect. It is certainly possible that they offered to build him a radio telescope, with no expense spared, which he could direct, but there did not appear to be a condition that he would forever have to live there, so the term 'defect' may not be appropriate. Thirdly, there was an allegation that he had been radiation-poisoned, in what was speculatively claimed was an attempt to wipe the memory of the Yevpatoria visit from his brain. Certainly, Lovell was unwell after his return, but his son later explained this as his tiredness after such a hectic visit **[9]**. The story was revisited after Lovell's death when the *Daily Express* claimed that Alla Masevich had used her charms to front a KGB operation to persuade him to defect, but that he proclaimed himself to be an Englishman always loyal to his country (and the game of cricket, he might have added). The historical records have no evidence that the connection between the two was anything other than a professional friendship **[10]**.

Bernard Lovell. CC Jodrell Bank.

However, the 1963 visit was the basis for starting a British-USSR axis of cooperation. But just as the French-Russian romance was reaching a critical stage, the British one ended abruptly. In February 1966, Jodrell Bank picked up Luna 9's signals that sent the first photographs from the surface of the Moon and published them the next morning in the *Daily Express* before the USSR had the chance to do so. The Russians were furious and all their subsequent spacecraft sent encoded signals. Either way, a promising line of British-Russian cooperation had run its course.

France and USSR move on

President de Gaulle's visit was a big event in European, cold war and diplomatic politics, even if it did not attract much attention in the anglophone media. This was a high-visibility break in the bipolar world, one that marked the end of the almost complete isolation of the USSR. De Gaulle returned from Moscow with his stature at home greatly enhanced. He even attracted unexpected bonus support from the political left and the intelligentsia – not his natural allies, but both sympathetic to the Soviet Union and critical of American foreign policy. As for the agreement, its precise terms were:

The governments of France and the Soviet Union:

- *Recognizing the importance of the study and exploration of outer space;*
- *Considering that cooperation between France and the USSR in this field will enable the extension of cooperation between the two countries and will be an expression of the traditional friendship between French and Soviet peoples;*

have decided to prepare and implement a programme of scientific and technical cooperation between France and the USSR for the peaceful study and exploration of outer space.

The agreement specified annual conferences, alternating between the two countries and organized by CNES and the Soviet space cooperation body, Interkosmos, respectively and, in particular:

- The study of space, with the USSR to launch a French satellite
- Cooperation in three fields: space science; meteorology and aeronomy; telecommunications
- Exchange of information, conferences, studies and exchanges of students
- Scientific information shared equally between the two parties.

Further details were agreed when Palewski's deputy, Alain Peyrefitte, visited Moscow in October 1966. There was a 'Grande Commission' of the President of CNES and the President of the Advisory Board of Interkosmos, meeting annually to review past, present and future projects. They were assisted by four working groups: science; meteorology; medicine and biology; and communications and annual reunions. The 1966 agreement between France and the USSR was later formally renewed on 4 July 1989, specifically adding new areas of cooperation: the terrestrial environment; space vehicles (including new flights to Mir and the development of shuttles); telecommunications; and Mars. The accord was formally dated as decree 90-79 in French law on 17 January 1990. The 1966 agreement was legally replaced by a new one 30 years later on 26 November 1996, recognizing the extinction of the Soviet Union and adding industrial and commercial cooperation. There were other supplementary agreements en route (e.g. piloted flight 1979), but the original 1966 agreement is still legally on the books [11].

The annual, week-long reunions became a key part of the process of cooperation. It was agreed from the start never to meet in Moscow or Paris, but in interesting provincial locations. They chose places intended to be both suitable for meetings and scenic, visiting a new place very time (Yerevan and Ajaccio both hosted two meetings, but they were the exceptions). Two turned out to be wet: Rodez and Kishinev, while some were not well known to the other side. The French travelled not just to Leningrad or Kiev, but to Samarkand, Tbilisi, Baku, Minsk, Pitsunda, Talinn and Kaliningrad; while Soviet visitors travelled to Marseille, La Grande-Motte, Toulouse, Rambouillet, Trouville, Tours and Cannes. About 60 scientists, engineers and technicians attended from each country to review present and future programmes. Typewriters clacked throughout the night before the last day to agree the final text, which would be the bible for the next year's work. Any changes and it all had to be retyped. There was also a social programme: football matches, cruises on the Dnepr and, in Russia, concerts, ballet and opera. The Soviet side offered tours of the golden ring, Vladimir and Suzdal, while the French offered Versailles and the chateaux of the Loire. Many records of the annual reunions have photographs of happy picnics, with stories of late nights, toasts and vodka. They were considered important events, being reported in the French press. The host country paid for the costs of the visitors and interpreters, with all participants given daily allowances (*per diems*).

The working methods were simply to agree projects at an annual meeting alternating between the two countries, an arrangement that is still in effect. All scientific results of cooperation would be shared and the principle of 'no exchange of funds' adhered to (human spaceflight later became an exception). CNES dealt with Interkosmos, the body attached to the Academy of Sciences that dealt with all cooperating countries, both in the socialist block and further afield. It was originally headed by Boris Petrov (1966–80), then by Vladimir Kotelnikov, the Director of Radio and Electronic Engineering and an expert in planetary radars. With an initial 12 or so administrative staff, it had the full confidence of the government which did not interfere in its operations. At government level, the French found themselves dealing with Vladlen Vereschetin, his first name being a concatenation of 'Vladimir' and 'Lenin'. Neither a scientist nor engineer, this brilliant, omniscient administrator was able to get all the paperwork through the highest levels of government.

The French found some of the Soviet working methods challenging, especially compared to the Americans. They could only meet with scientists and did not have access to engineers, the space industry, or industrial or launch facilities. The exact status and timetable of a joint project was difficult for the French to ascertain, as their Soviet colleagues were frustratingly vague despite being full of goodwill. On the positive side, there were never any arguments about costs or funding and the programme clearly had support at the highest levels of both governments. Procedures could be slow, however and by the 1990s, other problems began to surface as the Russian programme, though more open, began to contract in the growing economic chaos and financial issues became dominant. The French were always struck by the brilliance of the Soviet scientists, which they contrasted with the quite challenging environments – organizational, political and working conditions – in which they found themselves. There were surprises too. One western scientist recalls how discussions about flying his equipment on a Soviet spacecraft finally concluded in mid-morning. The Russians organized a celebration at which there were two ground rules: the first was that once a bottle of vodka was opened, it must be finished, which it was; the second was that vodka should not be drunk without food, so bread and pork sausage were provided.

Jacques Blamont admitted that he had no idea how to deal with the nebulous Soviet space programme. 'They refused to explain their organization, their working methods, their plans. The function of the people we met was unknown, they were not allowed to speak of future projects and they could say little about their earlier experiments. We had to guess who was who and who did what. When we presented our ideas, we had no idea if they interested them or not. We got a yes or a no or a counter proposal.' Later, as the annual reunions allowed each side to get to know one another better, the French pressed the Russians to explain their organization, only to be told that they did not understand it themselves. Western scientists also dealt with the institute for space research, IKI, but although it was a civilian institute it was not well-known domestically and visits were under escort. Over time, foreigners came to learn the interplay of roles between the Academy of Sciences (which included IKI) and its space committee, the ministry responsible for the space industry (the Ministry of General Machine Building), the various design bureaux responsible for spacecraft design and testing (OKBs in Russian) and the production centres (*zavod* in Russian). There was an invisible, never described, but real and complex decision-making cycle.

Over time, cooperation with France became so well developed that quite a number of the Soviet scientists had learned French (and some of the French learned Russian). Although interpreters were provided by the host countries, some western countries were short of Russian interpreters, so exceptions were made to allow some Russian interpreters to travel to western countries accompanying their delegation, a much-prized opportunity for them.

Then there was the cold war. The French had their own (intelligence) 'service' present and participants departing for the USSR were briefed about not taking technical documents with them, using only fresh notebooks and assuming that rooms would be bugged. French visitors were sometimes asked to bring in medicines not available in the USSR, which they did, on one occasion even saving the life of an astronomer who was suffering from cancer. Later, others brought antidepressants not otherwise readily available. Both sides provided cash *per diems* for their guests, but for westerners there was little on which to spend money in the USSR, while the Russians saved their cash to spend in the *beriozhkas*, (foreign currency shops) on their return. In the case of one meeting in Moscow, the western scientists and engineers in the *Akademicheskaya* hotel were given a bundle of rouble notes as *per diems*: 'It was difficult to find much on which to spend the notes, illegal to leave with them and impossible to change [them] into western currency [for which one would need documents to show how they had been acquired]. So any excess was pushed into the hands of one of the locals before leaving.' French visitors liked to bring back caviar from the kitchen of the *Ukrainia* hotel, typically smuggled inside Lenin statues that were missed by the x-ray machines and customs officials. There was the occasional attempted honey trap of French visitors by friendly 'students'. Some genuine romances took place, but on the Russian side they knew they could never leave the USSR legally and that defecting during one of the reunions would have bad consequences for their families. Rooms were searched when participants were attending conference sessions, or, as one French participant put it, the KGB was never very far away. When the Russians came to Esrange, Sweden, in the 1970s, the Swedes quickly noticed a nervous man who did not really fit in: his job was to anticipate and prevent defections.

One of the practical problems that they had to address was data: the Russians simply handed the French magnetic tapes, but as the volume grew, this became problematical. Accordingly, a 10,700 computer was installed in Toulouse to take data direct from IKI in Moscow, with a data officer available day and night via a hotline.

How did Russians react to the arrival of the French? Those who were involved in formal cooperation entered a new world of people, ideas and travel – even fashion, for they used to compare the different French fashions afterwards. Those not involved were jealous of those who were. They had to be wary of any in their midst who might be professional stool pigeons, ready to report on those too free with their political opinions, especially under the influence of vodka or wine. Translators had to supply a report afterwards indicating any political deviations or anyone 'at risk' of defecting, with one mischievously but correctly identifying all the spies in her group and declaring them to be the most 'at risk'. For Soviet visitors to France, there were unexpected freedoms. Roald Sagdeev stayed on two weeks after the end of a meeting to 'edit his notes' (at least, that was what the embassy was told). In fact, he rented a car and explored around France on his own.

The two sides developed their own circles of who-knew-what. In a discussion on the first proposed satellite, the Soviet side accidentally let it slip that the upper stage could not spin satellites before releasing them, their embarrassment being immediately and obviously visible because that was officially secret information. The information stayed tightly within the group, as 'our secret'. IZMIRAN, the centre for terrestrial magnetism with whom they were then dealing, had no photocopier, because organizations were not permitted them in case they were used to copy documents that should not be copied. On the western side, photocopiers were on the list of prohibited exports, but still the French found a roundabout way of sending 'repair' parts. Over time, the French and Russian scientists became aware of who in their network knew what and what they were supposed to know and not know – a trust that was never broken.

Following the 1966 agreement, CNES proceeded to plan both a 150 kg satellite in an eccentric orbit (180,000 km), ROSEAU (Radio Observation par Satellite Excentrique à Automatisme Unique, normally written Roseau) and a lunar probe, appointing principal investigators to each. The idea of the lunar probe reached France even before de Gaulle had returned from Russia **[12]**. Roseau was first in the queue and the design was signed off by April 1968. There were seven instruments, of which five were called *Sondeur, Champs électriques, Radioastronomie, Particules* and *Rayons cosmiques.* They offered a leap forward for France, as the Diamant launcher could only put quite small payloads into low Earth orbit. The only hiccup was that the Russians refused to provide any details about their launcher, which was secret, nor would they permit the French to integrate their satellite with the Russian launcher at the pad. The embarrassed Russian scientists explained that they could not integrate their own satellites at the pad either and they appeared to know as little about their own launchers as the French. They always handed over their satellites to the military in Moscow, who took over from there. That was the last they heard until they got the good news that the satellite was in orbit.

Then came the May 1968 political crisis in France. Georges Pompidou had always been sceptical of space projects and the government blamed scientists, thinkers, researchers and intelligentsia for causing them all the political trouble that year. Roseau was cancelled in the middle of the political upheaval, along with the lunar probe. Those involved in the project were taken aback by the sudden, brutal nature of the decision, the lack of explanation and how it was communicated. Moreover, the CNES budget, which had been one million French francs

(FR) annually when it was set up and reached FR2,730 million by 1968, was frozen and flatlined until 1980. The young space enthusiasts of the 1960s were thoroughly demoralized. 'Our government had no interest in space science,' declared Blamont. The Russians retrospectively believed that the decision was a form of disapproval of their country's handling of Czechoslovakia and could not believe that it was a budgetary decision, a problem which never affected them. The mission held out the promise of a significant French scientific success and its cancellation broke up some highly successful teamwork among laboratories across France. Not all was lost, however, as the ideas that inspired Roseau later found their way into Aureole, Prognoz and Interball.

It was a miracle that the Franco-Russian cooperation survived this test, the scientists said afterward. Although they did not realize it at the time, the French had already constructed some important institutional protections for cooperation. Firstly, there were the annual reunions, which built good personal relationships, even friendships, between the two sides. For the 1968 meeting in Tarbes, Keldysh sent one of his best young scientists, Roald Sagdeev, who was personally highly committed to international cooperation. Secondly, the French side was not just CNES but several key laboratories: the Service d'Électronique Physique, the Service Radioastronomie Spatiale, the Centre d'Étude Spatiale des Rayonnements (CESR) and the Groupe de Recherche Ionosphérique, all of which built the breadth of cooperation. On the Soviet side, their key institutes were brought into the process too: the Institute for Space Research (IKI), the institute for terrestrial magnetism (IZMIRAN), the Lebedev Institute, the Crimean Observatory and the Vernadsky Institute.

First outcomes of the 1966 agreement: small rockets and balloons

Despite the disaster with Roseau and the lunar orbiter, modest cooperation was already underway. The first was when researchers of the USSR Hydrometeorological Services and the Aeronomy Service of CNES travelled to Heiss Island in Franz Joseph Lund, to make polar, thermosphere and magnetic studies from the firing of two MR-12 meteorological rockets which released an artificial sodium cloud. The first two launches were on 9 and 10 October 1967 and 50 rockets were eventually launched under this programme [13].

The second was the OMEGA programme in March-April 1968. This brought together IZMIRAN in Troitsk with the French centre for the study of cosmic rays, CESR in Toulouse, for studies on the French Indian Ocean island of Kerguelen, coordinated with Soviet observations in Oboziorsky, Archangelsk, two magnetically conjugated points. Radio links were set up between the two stations. At the Indian ocean station, amidst the penguin colonies, they set up under a big tent marked 'CNES Interkosmos'. In the first round, France launched seven and the USSR two balloons to measure aerial currents and x-rays during periods of magnetic disturbance.

Kerguelen —French ship arriving. CNES.

Kerguelen became an important part of the Soviet-French story. These windswept, glacial islands were named after their French discoverer, Yves-Joseph de Kerguelen-Tremarec, who found them in 1772. British explorer Captain James Cook preferred to call them the 'islands of desolation' and they are sometimes so called in French (Îles de la Désolation). They are almost 2000 km from the nearest landfall. The Ross Antarctic expedition set up a magnetic station there during a 68-day stay as part of the great *Erebus* and *Terror* expedition in May 1840 [14]. Nowadays, the islands are reached by a quarterly ship from Réunion. Apart from penguins and other wildlife, the only other inhabitants are scientists. An agreement for the use of Kerguelen for space cooperation was made in Moscow in October 1969, formalized as an 'exchange of notes' between the two countries and counter-signed by Boris Petrov for Interkosmos and Jean-François Denisse for CNES. The French were not overly keen on northern Russia: frozen in winter, mud and mosquitos in summer.

For the second round, in January-February 1969, Franco-Soviet teams released balloons from Archangelsk, extending their studies to chemical analysis of the upper atmosphere between 100 km and 430 km, with France simultaneously firing Dragon sounding rockets from its land base of Landes. A third OMEGA campaign took place in February-March 1970 from Archangelsk and a fourth from the two locations in February-March 1971. Under the COLOME programme of 1968–9, there were coordinated weather observations by Cosmos 226, Meteor and French balloons launched to 30 km.

These experiments continued with the launch of three MR-12 rockets from the oceanographic ship *Professor Zubov* in December 1971, to study the ion composition of the atmosphere, electronic density and temperature from 100 km to 230 km. A French team boarded the all-white painted ship *Professor Zubov* at Cherbourg for the journey to the launch site off Kourou in French Guyana, where it dropped anchor in the blue waters off Devil's Island, its arrival causing a sensation. The Soviet sailors went ashore to visit the islands and then, despite their lack of fitness from being at sea, agreed to a football match with local residents. The Russian rockets carried a French-made magnetic mass spectrometer. These firings were synchronized with launches of French Véronique rockets (with Soviet mass spectrometers) from their land launch base at Kourou. From 20 September to 3 October 1973, 14 M-100 rockets were fired off Kourou by the *Akademik Korolev*, with a further 14 from 3–31 October 1977 **[15]**. From a western point of view, these launch campaigns provided a first-hand view of Soviet meteorological rockets at work and their very fine ocean-going scientific ships.

Akademik Korolev

Akademik Korolev preparing rocket

This programme was then extended to Kerguelen. In February-March 1973, there were 20 launchings of Soviet M-100 meteorological rockets, which took place off Kerguelen from the *Professor Zubov* and *Navarin* to measure the atmosphere, wind direction and the concentration of atomic oxygen. The ship *Borovichi* later anchored in Kerguelen to pick up signals from the Russian Spektr instrument on the French Eridan rocket. The last joint programme in Kerguelen took place in May-June 1980 with the firing of M-100Bs to examine the parameters of the atmosphere at middle latitudes. Launches then moved to Heiss Island in December 1980 with the MR-12. Rocket launchings on MR-12s were also made from Krenkel, Franz Joseph Land, to study ion concentrations in the atmosphere in four campaigns (February-March 1977, March 1979, September 1983 and July 1987). Outcomes were discussed at a meeting in Marseilles in 1985.

Kerguelen — Eridan launch site. CNES.

Balloons were an important part of the cooperation. France was the world pio-neer of balloons for meteorology and space research, hardly surprising consider-ing the first-ever balloon ascended from Paris in 1783. More than any other country, France invested in balloons as a low-cost means of understanding the atmospheric model, from the global to the local. Indeed, this was celebrated in 2016, when the Centre Spatial de Toulouse exhibited *Balloons for science at the edge of space*. Modern balloons were supported by CNRS, the Centre National de la Recherche Scientifique (National Centre for Scientific Research) and were a high priority for CNES from its foundation. They had the advantage of low cost and the ability to fly in the high atmosphere for long periods, providing a stable platform for observations of the Earth, atmosphere and sky, as well as for testing out and calibrating instruments and materials for satellites. CNES developed a fleet of balloons able to fly up to 45 km high, with one even circling the globe three times in one mission over 71 days. By 1971, 500 Éole balloons had been launched. That year, the Americans launched the Éole satellite for France, its function being to interrogate and transmit to the ground the data from a fleet of balloons. As early as 1967, Jacques Blamont identified the balloon as an ideal means of exploring the atmosphere of Venus.

Éole balloon. CNES.

This would be a good place to say more about Jacques Blamont (1926–2020), a key personality in Soviet-French cooperation. An astrophysicist, Blamont is credited with original discoveries in the turbopause, interstellar wind, the hydrogen envelope of comets and noctilucent clouds, while contributing to the development of the Véronique rocket and spacecraft instruments. He rose rapidly through the institutes in French science and spaceflight, while also helping to establish the Indian space programme. Here, he is best known for his unrelenting campaign to explore the planets, especially Venus, by balloon. A man of strong convictions and abundant energy, Blamont led the Service d'Aéronomie for 25 years and was Scientific and Technical Director to CNES – in practice, advisor to all eleven of its Presidents – and was still working in his office there at the age of 94.

Jacques Blamont. Government of India.

Balloons were an important part of the follow-up to OMEGA, called SAMBO (Synchronized Auroral Multiple Balloon Observations). SAMBO 1, of 20 balloons, ran between January-March 1974; SAMBO 2 was November-December 1976; and SAMBO 79 was January-March 1979. SAMBO was a three-sided venture, hosted by the new (1972) Swedish Space Corporation (SSC), the idea being to launch balloons from its base at Esrange in northern Sweden, from where they drifted toward the Ural Mountains. The SSC had an ulterior motive, hoping to impress its own government with Esrange's high-profile international role to the point that it would expand the sounding rocket centre there (Esrange later did become a premier space facility) **[16]**. The three-sided campaign between SSC, CNES and IZMIRAN, the centre for terrestrial magnetism in Troitsk, began badly, with the Soviet crew failing to consider the alarm caused by its arrival in a fleet of red-starred military trucks, prompting reports of a sneak invasion. Things improved after that, with the Russian crews teaching the Swedes new techniques of ice fishing and fish recipes, improving the cuisine at the Esrange hotel. There were many late-night parties.

SAMBO. CNES.

The final part of this series was ARAKS, led by Yuri Galperin (1932–2001), one of the Soviet Union's most prominent scientists, who was to become the leading personality of French-Russian cooperation. Galperin began his life making scientific auroral observations from the polar base of Loparskaya. He came from a multilingual family, his father being a professor of English and his mother a French teacher, so he spoke both languages. From 1962, he built instruments for the Cosmos programme of scientific satellites and later became involved with the first satellites of the Interkosmos programme with the socialist countries (Cosmos 261, 1968). From there, it was a short step to his involvement in programmes with France. The ARAKS programme was agreed at the Yerevan reunion in 1970.

ARAKS stood for Artificial Radiation and Aurora between Kerguelen and the Soviet Union – though it was also the name of a river between Turkey and Armenia where the first planning meeting for the series took place in 1970. The idea was to match Soviet sounding rocket experiments from 70°N, near the Arctic, with equivalent latitudes in the southern hemisphere, so that these two locations were linked with one and the same magnetic flux tube. Not only that, but active experiments would be carried out – injecting charged particles into the ionosphere and magnetosphere. Under the Zarnitsa programme, the USSR had developed an electron cannon able to fire an energy of 10 keV from sounding rockets.

ARAKS. CNES.

This was what we would now call 'exotic' science. The electron beams were an example of cutting-edge science developed in the Soviet period, in this case by the EO Paton Electric Welding Institute in Kiev, the world leader in welding technology. Its electron guns could inject plasma blobs and powerful electron beams up to 30 kW into the ionosphere and magnetosphere. In October 1969, it was agreed to conduct these experiments from Kerguelen, where Galperin and his colleagues encamped for several months, escorted by the colonies of penguins. Soviet tracking cameras, AFU-75s, electron guns and other instruments to measure particles and waves were installed there. The first ARAKS launches were of Eridan rockets from Kerguelen on 26 January and 15 February 1975. The beams were fired between 160 km and 185 km high. The electrons and plasma clouds followed the force line of the magnetosphere in a high arc and duly descended over Sogra, Archangelsk, but the brightness of the artificial aurora was no higher than a seventh magnitude star. ARAKS thus succeeded in using artificial low frequency emission to stimulate electron precipitation from the radiation belts for the first time. Radio waves and whistlers generated by the beams were also studied.

ARAKS involved a substantial number of scientists and was a notable upward shift in the level of cooperation. The outcomes were discussed at a joint conference organized by CESR in May 1976. The ARAKS report was a rollcall of the most senior scientists of France and the Soviet Union from IZMIRAN, IKI, the Polar Geophysical Institute, the Paton Institute, the Kurchatov Institute, CESR and French universities (Paul Sabatier in Toulouse, Paris VI). There was also American participation, via a team from the physics department of the University of Houston [17].

In the meantime, meteorological and communications experiments set down in the 1966 agreement were developed. Both countries undertook the COSCOL programme of observations of cloud cover of the Earth over 1971–2, using French *Colombe* balloons drifting 30 km high plus Soviet Meteor weather satellites and Cosmos 226. Communications experiments were conducted between France and the USSR from 1966 to 1975, using the Soviet Molniya communications satellites, the Franco-German Symphonie satellite, the European Orbital Test Satellite (OTS) and later Luch (1983) for relays between the two countries.

These balloon, sounding rocket and meteorological rocket campaigns are not well known and do not easily fit into the anglophone narrative of space history, where they may be considered low-tech. They are noted here in some detail because they were an important successor to the research begun during the International Geophysical Year (IGY; 1957–8), yielded important research results, used the assets and locations of both France and the USSR to greatest advantage and were an important field of their cooperation. The next stage was when the Soviet Union invited France to provide instruments on its Moon, Venus and Mars probes.

Second outcome: instruments to the Moon, Venus and Mars

Before the demise of the lunar orbiter project, CNES and the Academy of Sciences signed an agreement in 1967 for France to make a laser reflector for landing on the Moon. The proposal originated from the Lebedev Institute directed by Nikolai Basov (1922–2001), one of the country's most original scientists and developer of the molecular oscillator. Each laser reflector was made of 14 prisms of homosil glass and two were delivered to Russia, but the French were not told on what type of spacecraft their laser would be placed. At the time, they did not know that the USSR was developing large lunar roving vehicles. Two ground stations were calibrated to work with the laser and thus measure the precise distance between the Earth and the Moon: the 2.6m telescope at the Crimean Observatory and the 1.2m telescope in Pic du Midi.

Although they were likely not told so at the time, the first reflector was probably lost when the Proton rocket for what would have been the first Moon rover crashed on 23 February 1969. However, the second attempt was successful when Luna 17 landed the first Moon rover, Lunokhod, in the Sea of Rains on 17 November 1970, with the French laser, called the Télémétrie Laser Terelune (TL2), placed on the front. The French did have some weeks advance warning when Georgi Babakin, the head of the Lavochkin design bureau which made Lunokhod, told CNES in the greatest confidence that the next Moon probe would be a rover, carrying their laser. Laser signals were first sent from Pic du Midi on 5 December 1970 and the last during surface operations were on 17 June 1971. In the event that it failed to awaken after lunar night, Lunokhod was always parked at night with its laser pointing at Earth. The rover fell silent that autumn, but it was possible to continue laser ranging over 1974–5, which was used to fix the Earth-Moon distance with great accuracy and also to pinpoint Lunokhod's own final position (for the record, 38.3689°N, 35.1537°W). To raise funds for the ailing space program in the 1990s, the rover was sold at auction to a private collector, so pinpointing its position was important information to that collector.

Lunokhod and its laser.

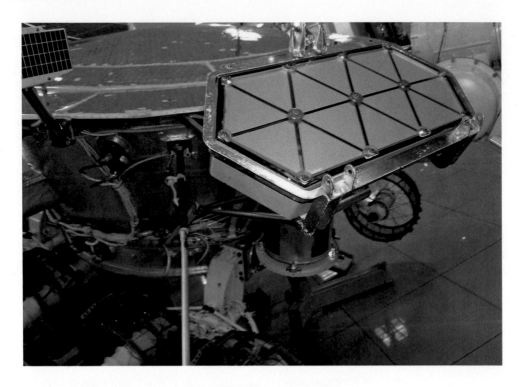

Lunokhod laser close up.

The story has a sequel, as Lunokhod was observed again on the Moon's surface by the powerful cameras of the American Lunar Reconnaissance Orbiter (LRO). So informed, astronomers at Apache Point Observatory in New Mexico re-activated Lunokhod's laser on 22 April 2010 with their 3.5 m telescope, getting a surprisingly strong signal of 2000 photons and making possible the first-ever day-time ranging. It was determined that the Moon is slowing down by 25.88 seconds per century and its distance from Earth increasing by 3.83 m per century.

Based on the French experience, the Russians decided to build their own laser for the next Lunokhod, or so they said. Lacking a laser-testing facility themselves, they sent a delegation of six to test it at the French facility in Cannes on 13 October 1971, also taking the opportunity to visit the rest of the factory, which by chance happened to make French missiles for its strike force. In the end, they decided to opt for a French laser for Lunokhod 2 which soft-landed in 1973. The laser was fired 4000 times to determine the Earth-Moon distance to an accuracy of 25 cm. In recent times, Lunokhod 2 was also pinpointed by LRO, with astronomers finding its laser much dimmer than that of its predecessor (750 photons).

The Soviet lunar programme concluded with Luna 24 in 1976. The USSR gave France samples of Moon rock brought back by Luna 16, 20 and 24, the first western country to be given such an opportunity. France also had the benefit of Apollo samples, so was able to scan a wide range of lunar locations with an electron microscope. This provided information on the nature of the soil and the effects of solar wind, enabling its scientists to form their own histories of the origin and development of the Earth and the Moon.

Venus

France also became the first European country to reach the planets. When Venera 4 entered the Venusian atmosphere in October 1967, there was disagreement in the scientific world as to its findings. The Soviet Union claimed that Venera 4 had reached the surface, but its pressure and temperature readings did not tally with simultaneous American measurements by Mariner 5, which indicated much higher figures. Soviet scientists then modified their view, suggesting that Venera 4 had landed on and broadcast from a high mountain, an explanation not accepted internationally. France was then invited to supply thermometers and barometers that might give independently verifiable estimates for the next set of Venus missions, Venera 5 and 6, set for January 1969. Completion of the instruments was delayed by the uprising in France in May 1968 and they were not delivered until September, too late for installation [18]. Venera 5 and 6 suffered the same fate as their predecessor and, the scientists realized, were crushed some time before reaching the surface. France was not invited to participate in the next mission, Venera 7, which achieved a successful touchdown the following year, but did have an instrument on Venera 8. However, although the instrument worked, the data capture failed.

Venera early landers.

In the meantime, the USSR was preparing the launch of a new generation of Venus probes, which, with the benefit of the Proton rocket, could be up to four tonnes in weight. The normal pattern was to launch double identical missions to the planet. In 1975, Venera 9 and 10 became the first probes to orbit Venus, sending down landers which successfully photographed the surface. Following an invitation, Jacques Blamont and Jean-Loup Bertaux provided a Lyman α instrument to measure the temperature and distribution of hydrogen in the Venusian atmosphere. As was normal practice, they knew nothing of the progress of their instruments from the point of handover until the first five orbits had been completed, when they were given the data.

For Venera 11 and 12 in 1978, France supplied gamma ray detectors called Signe 2MS3. They combined an omnidirectional gamma and x-ray radiation detector which picked up dozens of gamma bursts and determined their precise origins. Jacques Blamont and Jean-Loup Bertaux supplied an extreme ultraviolet spectrometer to detect hydrogen, helium and ionized helium and oxygen. This duly found two populations of atomic hydrogen, a normal one of 275 to 300 K and a suprathermal one 1000 to 40,000 km over the planet.

In 1982, Venera 13 and 14 carried a common programme of experiments developed in the USSR, France and Austria, in the areas of x- and gamma rays, magnetic fields, the solar wind, cosmic rays and interplanetary plasma. Blamont provided an ultraviolet spectrometer. France helped with the Soviet instruments by launching a stratospheric balloon on 10 September 1981 to test their spectrometer PS-17B designed to observe x-rays in the 20 keV and 800 keV range.

The Venera 9–14 missions attracted the most attention for their landings on the hostile Venusian surface, but they were also important for the scientific instrumentation installed on the mother spacecraft, which orbited Venus in the case of Venera 9 and 10 and flew past while dropping landers in the case of the others. They provided important platforms for instrumentation to study stellar events from the vantage point of the inner solar system. Altogether, 17 Signe detectors flew on board Soviet planetary spacecraft, the summary results being presented by Jean-Luc Atteia in 1987. They recorded 200 x-ray bursts, ranging from many keV to MeV, which he mapped according to duration and location. He was of the view that they were traces of a new population of neutron stars. These missions carried instruments, but discussion began during this period regarding a much more ambitious joint project to send balloons into the atmosphere of Venus (see *Chapter 2: Space science*).

Mars

France had further planetary success with Mars. Despite its image of running a closed, secret programme, the USSR invited international participation on its planned three launchings of an orbiter and two lander-orbiters to Mars in 1971.

France was the first and only country to respond, providing two instruments called Stereo, a solar radio-astronomy experiment designed to study solar radiation at 169 MHz and make three-dimensional images from the data. Built in the University of Meudon, it took the form of a T-shaped aerial installed on the solar panel. The French simply handed the experiment over, so not only were they uninvolved in integrating their experiments with the spacecraft, but they were not even permitted to see drawings of how this was done. The French were simply told it had been installed.

The spacecraft was duly launched and named Mars 2 and although it was stated that everything was working perfectly (which it was), the French were puzzled that they were getting no data from their solar radiation instruments. This obliged the Russians to explain that there was an electronic fault and that sadly, data could not now be returned. Years later, the French found out that the first Stereo had been placed on the orbiter spacecraft, Mars 71S ('S' for Sputnik) which had been launched earlier. When it failed to leave Earth orbit, it was redesignated Cosmos 419 and was not announced as a Mars probe. It quickly fell from orbit, bringing with it the first Stereo instrument, but the French knew nothing about this. The second Stereo was placed aboard the next probe, Mars 3. The French got their data as soon as it was on its way and were able to able to obtain information on solar rays and solar ray bursts which were matched against ground observations in Troitsk and Nancy. Mars 3 made the first soft-landing on Mars that December.

Two years later, the USSR launched a Mars fleet of four spacecraft. Mars 5, an orbiter, carried Stereo 3, a successor instrument which measured radio-electric rays from the Sun, their direction and solar bursts from 30 to 60 MHz. Mars 6 and 7 carried Stereo 5 to study solar emissions and Zhemo to study solar protons and electrons. Mars 6 and 7 also carried Gémeaux S and T, made in the Centre for Nuclear Studies in Saclay (S) and the Centre for Cosmic Ray Studies in Toulouse (T), respectively. Gémeaux was designed to record protons from 4 to 330 MeV, medium electrons from 2 to 130 MeV and alpha particles in the range 180–500 MeV. It also had two spectrometers to study protons and helium ions from 100 eV to 10 keV. Gémeaux S was developed in a CNES-Interkosmos working group set up in March 1971, with a target date of July 1973 for the experiment to fly to Mars. The group was assigned a translator, Maria Fedossova, whose accentless French had been achieved without ever having set foot there. The prototype was built in France, tested in IKI's huge four-floor integration hall and duly left for Mars on 9 August 1973. The results enabled the scientists to present the filamentary structure of the magnetic field around the Sun, for which Alain Raviart received an award from Academician Petrov, presented to him in a ceremony aboard a cruiser on the Dnepr near Kiev the following year.

Gémeaux. CNES.

For France, the sounding rocket, meteorology and balloon campaigns, ARAKS, Lunokhod lasers and instruments to Venus and Mars were the first outcomes of the 1966 agreement, with more missions in the pipeline. The story of the early days ends with a second presidential visit to Baikonour, that of de Gaulle's successor, President Georges Pompidou (1969–74).

Pompidou

Despite the disappointment over Roseau and the lunar orbiter, the Soviet Union was keen to maintain a similar type of relationship with General de Gaulle's successor, inviting him for a week-long visit in October 1970 during the second year of his presidency, following a similar schedule.

President Pompidou and his wife Claude were welcomed in a manner similar to that of his predecessor, with General Secretary Leonid Brezhnev greeting him at Vnukovo, the first such honour bestowed on a visitor. Pompidou was also given a room in the Kremlin. This time, more agreements were signed, covering coal, coke and metals and related technology. The visit attracted little western attention, with the typical and honourable exception of the *New York Times* [19].

In advance of the visit, Pompidou expressed an interest in seeing a rocket launch and the Soyuz spacecraft. He arrived at Baikonour on 8 October with President Podgorny and met again with General Secretary Leonid Brezhnev and prime minister Alexei Kosygin there the next day. The entire day may not have been spent on cosmic matters, with some of it likely spent on political talks. Pompidou saw the launch of the Cosmos 368 biological mission and he may also have seen Cosmos 369, launched the same day. His visit was called 'Palm Tree 4'.

Information about his visit was as scarce as that of General de Gaulle. Unlike his predecessor though, Pompidou found himself hosting a press conference afterwards, on 11 October in Tashkent. Agence France Presse duly questioned him on the visit and he told reporters that he was present for the launching of a satellite and 'also for another experiment, let's say military', suggesting that he might well have seen Cosmos 369. He stated that he had seen Yuri Gagarin's return cabin (doubtful, because that was most likely in a Moscow museum, but could well have been another Vostok cabin) and the large rocket that had sent Luna 16 to the Moon, so he probably saw the Proton, photographs of which were not released until 1988. Pompidou had earlier visited Cape Canaveral and when asked to compare the two, he volunteered that their working methods were quite different. The Russian system was 'more simple and effective, as its results proved' [20].

Prime Minister Georges Pompidou (standing, right), seen here at the 1965 Paris air show with (left to right) Yuri Gagarin, US Vice President Hubert Humphrey and Gemini 4 astronauts Ed White and Jim McDivitt. NASA.

Early days: conclusions

We have a precise starting point to mark this story: 20th June 1966, when General de Gaulle touched down in Vnukovo for his ten-day visit to the USSR, including a trip to the Baikonour cosmodrome and concluding, before he departed, with the signing of a treaty for space cooperation.

Likewise, the second French presidential visit, of Georges Pompidou, brought the early years of Russian-European cooperation to an end.

After hesitant beginnings, an accord was signed in 1966 which led to cooperation in three distinct areas: sounding rockets (and related areas of meteorological rockets and balloons), lunar exploration and planetary exploration. The 1966 agreement also set in motion a much wider set of cooperative activity that was not to come to fruition until the 1980s, both in the field of space science (see chapter 2) and human spaceflight (see chapter 3).

At this point, France was the only European country with such privileged access to the Soviet Union. European-Russian cooperation in space had its origins in the political circumstances of France's role in Europe in the 1960s. This was very much due to France's particular position in Europe and the vision of General de Gaulle, which emphasized independence from the Atlantic alliance and equilibrium in dealing with the Soviet Union and United States. The story is inseparable from the personal history, character and politics of the President of France, Charles de Gaulle. A prisoner with Russians in a German first world war camp, a visitor to the USSR during the second world war, de Gaulle was better disposed to Russia than many of his contemporaries. Once his presidency was under way, he believed in a strongly independent role for France and would have neither France nor Europe a prisoner of American foreign policy, desiring France to be free to build its own relationship with Russia, or China, as it wished. De Gaulle seems to have had no special interest in spaceflight itself, but he did see it as an instrument of post-war recovery, modernization and development of a technologically advanced economy and was prepared to direct government resources to ensure that outcome (*dirigisme*). It was also clear from the start that the French programme would be led from the top. When France launched its first satellite in 1965, General de Gaulle invited the scientific and technical director to dinner to say 'La France vous remercie' (France thanks you) − hard to imagine in some other countries − and most subsequent presidents lost no time in associating themselves with the space programme. Nicholas Sarkozy, who was elected in 2007, visited the Kourou launch base as early as February 2008.

On the Soviet side, there was an openness to cooperation. Boxed in by NATO and the demarcating lines of the cold war, the USSR was always interested in avenues that could break its isolation, though this would inevitably attract the criticism that doing so was a subtle attempt to undermine and divide the western alliance. The experience with Jodrell Bank was an example of an attempt to reach out to Britain and its leading scientist.

According to the analysis by Dupas, the agreement opened the door to France for opportunities to fly experiments complementary to national, European and American opportunities; enabled the USSR to fly experiments which it had not yet devised itself; and gave access by the Soviet space community to both France and, through French colleagues, the rest of Europe [21]. The USSR proved to be a reliable partner, always carrying out its side and timetable of the agreement until financial problems emerged in the 1990s. The ingredients for this successful cooperation, Dupas said, were long-term political support at the highest level and finding the right areas to work together, with patience and goodwill on both sides to overcome and survive any problems arising. The fact that the USSR's most senior scientist (and later President of the Academy of Sciences) Mstislav Keldysh and its leading magnetospheric scientist, Yuri Galperin, were French-speaking probably helped too. It is notable that the terms of the 1966 France-USSR agreement were restated in the 1996 France-Russia agreement, so it has proved an exceptionally long-lasting partnership. The idea of regular, alternating annual meetings appears to have been one of its strongest points, for it meant that progress was under constant bilateral review and provided a place where new projects could be floated, discussed and initiated. Not only that, but these week-long meetings cemented personal ties that endured setbacks, disappointments and political changes, an important human chemistry. The USSR must have been disappointed by the French government's decision to pull out of Roseau and the lunar orbiter, but they kept their feelings to themselves. This must have raised questions about France being a reliable partner with which to do business and there were more disappointments to come (see Chapter 2).

Both sides required a period to acclimatize to the cultures and working methods of the other, so it was probably as well that they started small, beginning with meteorological and sounding rockets (1967), the OMEGA programme (1968) and then joint firings of rockets from ships off Guyana and Kerguelen. Small rockets might appear nowadays to be a relatively unambitious form of cooperation, but many countries still fire such small suborbital rockets to make important contributions to science and the development of instruments. Europe still sustains a sounding rocket base in Kiruna, Sweden, which is often used by German scientists. ARAKS was in effect the first joint campaign by the two teams of scientists working together, which formed the basis of personal working relationships that lasted for many years. ARAKS was more ambitious, exploring as it did the cutting edge of active magnetospheric science.

If this was the low-profile end of space cooperation, then transporting French equipment to the Moon, Venus and Mars was the high-profile end. France was given the earliest and first opportunity to fly equipment on Soviet spacecraft to these distant destinations, with the nation able to offer instruments that, one surmises, were less well developed in the USSR, like lasers and Stereo. The French, though, had to endure Soviet procedures, whereby they simply handed over their equipment and that was the last they heard of it until it was successfully launched (or not).

The paths of cooperation appeared to be working well, but the Russians took nothing for granted and took the precaution of similar red carpet treatment for de Gaulle's successor, Georges Pompidou, including a return visit to the Baikonour cosmodrome, reinforcing the progress made in the 1960s decade. By this stage, the relationship had delivered results in three core areas: for the Russians, a reduction in their cold war isolation; for the French, a highly visible form of cooperation that exemplified France's independent, leadership role in Europe; and for both, practical scientific results from the sounding rocket campaigns, Lunokhod and the Venus and Mars probes.

An American assessment of cooperation is available from the congress [22]. This explained Soviet-French cooperation as the Soviet need for sophisticated western technology, especially in computers; the French need for rocket lifting power; and together, the desire to build French-Soviet political interests. This is broadly accurate, but overstates the weakness of Soviet technology. Europe may have been ahead in some areas of instrumentation, but there was nothing wrong with Soviet computers, as demonstrated by automated Earth orbital rendezvous and docking in 1967, unmatched by the west at the time and the system still used for re-supplying the International Space Station to this day; nor with its automated deep space astro-navigation, used by its Mars probes from 1971.

One of the important outcomes of Soviet-French cooperation was what did *not* happen. Although, intuitively, one might have expected the Americans to have backed away from their cooperation with France as a result, this was not the case and good levels of cooperation between France and the United States were unaffected. On their side, the French made it clear that cooperating with the USSR should not be mistaken as political alignment. Jacques Blamont, the scientific and technical director of CNES from 1962–72, insisted that they were 'never trying to be pro-communist'.

Some historians challenge us not to view this period through the sole narrative of competition, the space race and the cold war [23]. A strong thread of cooperation, most evident during the earlier International Geophysical Year in the 1950s, ran alongside the prevailing 1960s discourse of rivalry. In 1963, this pathway came close to dominant, with the opening moves by President Kennedy to turn the Moon landing project into a joint US-USSR venture, the nuclear test ban treaty and Kennedy's least known but most profound address on world peace. In working with the USSR, France was hardly breaking ranks with the western world, but was instead following this less well understood and historically overlooked parallel line of development. Political liberals always made the case that cooperation was good because it encouraged inter-dependence between states, making conflict less likely and giving them benefits too costly to lose [24].

General de Gaulle resigned the presidency in 1969 and died the following year. When the 50th anniversary of his death approached, the commemorations knew few bounds, starting with the film called, simply, *De Gaulle*. President Macron installed de Gaulle's wartime symbol, the cross of Lorraine, into the logo of the presidential home, the Élysée and placed de Gaulle's *Mémoires* visibly on his desk for his official photograph. Xavier Bertrand, in charge of the commemorations, spoke of how de Gaulle had asserted France's independent, unaligned voice in the world, as evidenced by the rapprochement between France and Russia [25].

References

1. Stent, Angela: *Franco-Soviet relations from de Gaulle to Mitterrand*. Harvard University, National Council for Soviet and East European Research, 1989.
2. Jung, Philippe: *Gagarin – a special relationship with France*. Presentation, International Astronautical Congress, Cape Town, 2011.
3. Tourycheva, Ekaterina: *Des palmiers en steppe – les légendes de Baïkonour*, Russia Beyond, 4 March 2014; Limonier, Kevin: *Baïkonour – le cosmodrome qui n'apparaissit sur une carte*, 6 January 2018.
4. Gruntman, Mike: *Word cosmonautics – a history*. Presentation to International Astronautical Congress, Valencia, 2006.
5. *Ariel 1: the secret history*. Space:UK, summer 2012, #35.
6. Purdy, Anthony & Burchitt, Wilfred: *Yuri Gagarin*. Panther books, 1961.
7. Nuenhist, Christian; Locher, Ann & Martin, Garret: *Globalizing de Gaulle – international perspectives on French foreign policies*. Plymouth, Lexington Books, 2010; Teague, Elizabeth & Wishevsky, Julia: *Russia at the gates – Gorbachev's European house*, in Martyn Bond, Julie Smith & William Wallace (eds): Eminent Europeans – personalities who shaped contemporary Europe. London, Greycoat Press, 1996.
8. Kragh, Helge: *The universe, the cold war and dialectical materialism*. Aarhus, Centre for Science Studies, undated.
9. Phelan, Dominic: *Jodrell Bank and Bernard Lovell's 1963 trip*. Presentation to British Interplanetary Society, 7 June 2014; *Jodrell Bank director visits Soviet space station*. Flight International, 25 July 1963.
10. Stewart, Will: *Beautiful boffin's bid to woo Newton of radio astronomy to defect to USSR*. Daily Express, 21 October 2012; *Doubt cast on Sir Bernard Lovell's brainwashing*. Physics World, 31 January 2013.
11. Azoulay, Gérard: *Correspondances Paris-Moscou*. Paris, CNES, 2010.
12. *Une satellite française autour de la lune, proposent les Russes*. Le Figaro, 25 June 1966.
13. For the early period of Soviet-French cooperation, see Rebrov, M; Kozyrev, V & Denissenko, V: *URSS-France – exploration de l'espace*. Moscow, Progress editions, 1983.
14. Palin, Michael: *Erebus – the story of a ship*. London, Arrow, 2018.
15. Lardier, Christian: *Soviet meteorological rockets history, 1946–91*. Paper presented at International Astronautical Congress, 2008.
16. Zenker, Stefan: *Swedish Space Corporation, 25 years, 1972–1997*, www.zenker.se/space, accessed 23 August 2019.

17. Cambou, F: Lavergnat, J; Migulin, VV; Morozov, AI; Paton, BE; Pellat, R; Pyatsi, AK; Rème, H; Sagdeev, R; Sheldon, WR; & Zhulin, IA: *ARAKS – controlled or puzzling experiment?* Nature, 23 February 1978; Paton, Boris & Lapchinskii, VF: *Welding in space and related technologies*. Cambridge, Cambridge International Science Publishing, 1997.
18. LePage, Andrew: *Venera 5 and 6 – diving toward the surface of Venus*. Drewexmachina. com, posted 16 May 2019.
19. Gwertzman, Bernard: *Pompidou is welcomed warmly in the Soviet Union*. New York Times, 7 October 1970.
20. *Georges Pompidou à Baikonour*, from anecdotes-spatiales.com, 5 March 2012.
21. Dupas, Alain: *Space cooperation with the Soviet Union (Russia) – a French point of view*, from US-Russian cooperation in space. US Congress, Office of Technology Assessment, Washington DC, 1995; *Trois questions à Jacques Blamont*. CNESMag #51, November 2011.
22. Senate Committee on Aeronautical and Space Science: *Soviet space programmes, 1971–5 – goals and purposes, organization, resource allocations, attitudes toward international cooperation and space law*. Washington DC, 1976.
23. Cross, Davis, Mai'a K: *International cooperation during the space race*. Paper presented to the International Astronautical Congress, Washington DC, October 2019.
24. Miller, Gregory: *The eagle, the bear and (other) dragon – US-Russian relationships in the SpaceX era*. The Space Review, 15 June 2020.
25. Marlowe, Lara: *Film portrays de Gaulle as saviour of French honour in 1940*. Irish Times, 10 March 2020.

2

Scientific cooperation

The 1966 agreement set in place a range of cooperation programmes which came to fruition long after the visit of President Pompidou. The first was a very concrete outcome: the launch of the French satellites SRET, Signe 3 and Aureole by the Soviet Union. Although France had the capacity to launch its own small satellites, the USSR provided the opportunity to launch heavier satellites into higher orbits and these missions led to extensive cooperation in the Prognoz and Interball programmes. Strains and stresses were most evident during the Venus balloon project. Space science saw the extension of cooperation beyond France and to the European Space Agency (ESA) in the areas of biology (Bion) and materials processing (Foton). Germany emerged as a significant player, concluding with the Spektr RG project.

First satellites: SRET and Signe 3

In 1970, the USSR and France signed an agreement called SRET (Satellites de Recherche et sur l'Environnement et la Technologie), whereby small French satellites would fly piggyback on larger payloads in mainstream Soviet launches. SRETs were technology rather than scientific satellites. They were small and flew piggybacked with Russian Molniya communications satellites out of the Plesetsk military launch base at the Arctic Circle. Molniyas orbited in curving, 12-hour orbits some 39,000 km high over the northern hemisphere, travelling in and out of the radiation fields twice a day. The two SRETs, which had the Russian appellation MAS, followed in their paths, measuring how radiation, vacuum and

---

The original version of this chapter was revised. The correction to this chapter is available at https://doi.org/10.1007/978-3-030-67686-5_7

© Springer Nature Switzerland AG 2021, Corrected Publication 2021
B. Harvey, *European-Russian Space Cooperation*, Springer Praxis Books,
https://doi.org/10.1007/978-3-030-67686-5_2

temperatures degraded solar cells. The lessons learnt were put to vital use in the design of Europe's forthcoming Meteosat satellites.

SRET 1 was launched on 4 April 1972, the first western probe to be launched by the USSR, with SRET 2 following in June 1975. SRET 1 was a 15.4 kg poly-hedron that followed the Molniya orbit of 480–39,260 km. The principal purpose of the mission was to test cadmium solar cells, which were 20 percent lighter than those of silicon used to date and were intended to be used later on Europe's forth-coming Meteosat weather satellite. SRET's cells worked well for over a year. SRET 2, twice the weight at 30 kg, tested systems of radiator cooling for Meteosat, as well as thermal protections and protective plastic materials.

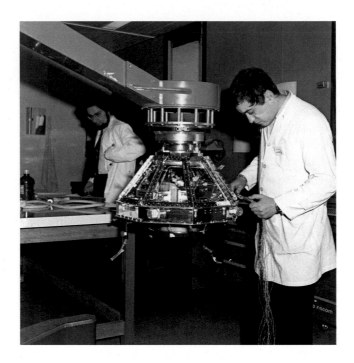

SRET 2. CNES.

The French asked to attend their launch, but this did not happen, probably because Plesetsk was off-limits. Although the French were not allowed to see the launch nor go to the launch site, the Russians would not run the risk of the SRET 1 battery not being fully charged up and recycled on the morning of the launch. Accordingly, Pierre Koutsikides guarded the satellite in its hangar the night before, making sure it was continuously charged. Faced with a long lonely night, cosmo-drome staff suddenly appeared to supply him with a bundle of provisions: pâté, sausage and the inevitable vodka, set on a tablecloth. There was also a sleeping bag.

The first French satellite to get a dedicated Soviet launch as sole payload was SIGNE 3 (Solar International Gamma Ray and Neutron Experiment, normally written lower case as Signe). 'Signe' was used to refer both to a satellite (Signe 3) and an instrument, several of which were flown in the Prognoz programme. Signe 3 was originally a satellite to detect gamma bursts called D^2B Gamma and was due to be launched on the Diamant B (hence D^2). Diamant was cancelled to free up resources for the Ariane programme, at which stage the Soviet Union stepped in to offer to launch it with the expectation of access to the data.

Signe 3 (top left). Signe 3 rocket (top right). Signe 3 team (bottom). All CNES.

Signe 3 was flown from France to Moscow in April 1977 for a launch from Kapustin Yar cosmodrome near Volgograd later in the summer. Its purpose was to study the effects of solar wind on the upper atmosphere. Signe 3 was launched on 17 June 1977 into a 459−529 km, 50.67°, 94.33-minute orbit on the R-14 Cosmos launcher. Weighing 102 kg, it was 80 cm tall and 70 cm diameter and carried both x-ray and gamma-ray equipment. Its launch was well documented, giving western observers a good look at the R-14 and its launch operations at Kapustin Yar.

The instrument package was developed by CNES Toulouse, together with the CESR and the Université Paul-Sabatier. Other participants were the Centre for Nuclear Studies in Saclay and IKI, the Lebedev Institute and the Ioffe Institute in the USSR. Signe 3 focussed on exoatmospheric astronomy, especially x-rays of galactic and extragalactic origin, trying to identify the sources of gamma bursts, high energy rays and solar ultraviolet rays. There was a 2 kg aeronomy experiment devised by the aeronomy service of CNRS at Verrières-le-Buisson to determine the influence of solar activity on the terrestrial atmosphere. Signe 3 reported on background noise in high inclination orbits, the South Atlantic Anomaly and two polar comets. The success of Signe 3, principally in detecting gamma bursts and flashes, was hailed at the 14th annual meeting in Marseilles that October. Signe 3 paved the way for instruments carried on Prognoz 5 and 6.

Aureole: first joint satellite

The first joint Soviet-French satellite was Aureole, which arose from a scientific conference in Paris in 1968 and was specifically the idea of Yuri Galperin and his teacher, the great Soviet scientist Iosef Shklovsky. Aureole was to study charged particles and auroral displays under the ARCAD programme begun in 1968. Originally, the series was to be called ARCAD (ARC Aurorale et Densité) but an hour before lift-off it was renamed and written as Aureole, for AURrora and EOLus), although ARCAD continued to be applied to the programme as a whole and the literature sometimes identifies Aureole 1, 2 and 3 as ARCAD 1, 2 and 3. The physics of Earth's environment over the Arctic was one of the key areas of expertise of early Soviet space science. Yuri Galperin was the chief scientist on the Soviet side and he later regarded Aureole 3 as having been his best satellite. His French counterpart was plasma physicist Francis Cambou (also written as François).

Yuri Galperin (left) preparing observations.

Aureole 1 and 2 were DS U2 GKA satellite models developed as part of the Cosmos scientific programme dating to 1962 and launched on the Cosmos 3M rocket from Plesetsk. Each carried an identical mixture of French instruments to study low-energy electrons and protons, where they were expert, with Soviet instruments for the high energy range, where they were knowledgeable, so they complemented one another. The satellites were a joint project between CNES, the Centre for the Study of Space Radiation in Toulouse, France and the Institute for Space Research in Moscow (IKI). Aureole 1, a 660 kg satellite, was launched from Plesetsk on 27 December 1971. Because the cosmodrome was officially secret – although well known in Britain to the schoolboys from Kettering Grammar School who had tracked signals from launches there – the French were told that Aureole was launched from Kapustin Yar, which had the benefit of not being secret. The focus of the mission was the polar lights, or the aurorae borealis. Aureole 1 orbited from 410–2,500 km, 74°, 114.6 minutes. Its observations were supplemented by ground stations in Cape Schmidt, Tiksi, Yakutsh, Dixon Island, Heiss Island, Murmansk and Nancy.

Aureole 2 was launched on 28 December 1973 with essentially the same equipment, but was focussed on the way in which the atmosphere heated the regions of the polar lights in a form of a slow-motion nuclear reaction. Aureole 2 had a new spectrometer to analyse the variations of intensity of electrons and protons and to determine the ionic composition of the ionosphere by measuring $H+$, $He+$ and $O+$. As a result, scientists were able to map the regions and identify the methods of penetration of the solar wind into the magnetosphere. The outcomes of the first two Aureole missions were published in the top article of *Annales de Géophysique*, an early example of joint Soviet-French missions publishing their scientific results under the names of the scientists of the two countries concerned [1].

Aureole 3 belonged to the next generation of DS satellites, the AUOS Z M-A-1K series, which weighed 1,000 kg and was able to carry up to 200 kg of instrumentation. Aureole 3 was proposed at the 1971 reunion and decided at the 1974 Kiev reunion, with the scientific programme confirmed at La Grande Motte in 1975. It was by far the most successful of the Aureole series. Because it was a bigger satellite with more instrument space, Areole 3 was a quantum step forward from 1 and 2, making possible more instruments, new instruments and the coverage of a greater range of natural phenomena, especially focussed on the coupling of the ionosphere and magnetosphere. The objective was to follow the mechanisms that transported solar energy to the Earth, the nature of the northern lights and the penetration of solar plasma into the polar ionosphere, with resulting radio-electric perturbation. Aureole 3 was a pressurized cylinder 2.7 m tall and 1.6 m in diameter, with eight solar panels generating an average of 50 W and a maximum of 250 W. Aureole 3 benefitted from advances in French instrumentation and computers and was the first in the cooperative programme to carry a computer. Yuri Galperin insisted on preparing special new electromagnetically clean solar panels. As a result, their electromagnetic disturbances were decreased 1,000-fold and their durability improved. The main French centres involved were those in Toulouse, Saint Maur and Orléans, with IKI on the Soviet side.

Aureole 3.

Aureole 3 carried four Soviet and seven French experiments and one joint experiment, including proton and electron spectrometers (*Kukushka, Pestchanka, Spectro*), an energetic particle detector (*Fon*), a thermal mass spectrometer (*Dyction*), an interferometric probe (*Isoprobe*), electric and magnetic field probes (ONCh-TBF), a magnetometer (TRAK) and a photometer (*Altair*). The experimental schedule was planned every Friday by IKI and the Centre d'Opérations Specialisé ARCAD 3. It was circulated to the scientific teams the following week for comment, agreed upon and then put into operation the week after. Readings were taken on both a pre-ordered basis and on command, with direct readout to French stations, recorded dumps over Soviet territory and with common scheduling of experiments. Signals and experiment readings were transmitted to Norilsk, Kirovsk, Tarusa, Toulouse, Kourou, Tromso, Kerguelen and Adelie Land in Antarctica, for analysis in IKI and Toulouse. The two countries had a system of relaying digital telephone information between Dubna and Barseneillon over 30 channels in the 11–14 GHz band, via 24-hour Soviet communications satellites.

Aureole 3 was launched on the larger Tsyklon rocket on 21 September 1981 and operated for five years to summer 1986. Daytime aurorae were found for the first time and it also found much evidence of man-made electrical activity in the form of transmitters and industrial plants. Following an industrial explosion in Alma-Ata, Kazakhstan, Aureole 3 followed the effect of the large-scale acoustic wave on the upper atmosphere and ionosphere. As a result, electrostatic VLF- and ELF-noises, as well as an intense MHD-wave, were recorded in the corresponding flux tube.

Yuri Galperin convened the many international conferences at which the mission outcomes were shared, with one in 1984 dedicated to Aureole 3 alone. Although such an approach might be considered routine now, it was not so in the early 1980s. The outcomes of the mission were published in a special issue of *Annales de Géophysique* and in a special publication by CNES in French and English [2]. The missions are summarized in Table 2.1.

Table 2.1:
Early French satellites and joint Soviet-French satellites launched by the USSR

Date	Satellite	Launch Site
27 Dec 1971	Aureole 1	Plesetsk
4 Apr 1972	SRET 1	Plesetsk
16 Dec 1973	Aureole 2	Plesetsk
5 June 1975	SRET 2	Plesetsk
17 June 1977	Signe 3	Kapustin Yar
21 Sep 1981	Aureole 3	Plesetsk

Prognoz and Prognoz M/Interball

The Prognoz programme was an important point of French-Soviet collaboration from the early 1970s into the 1990s. Prognoz was an uncertain project within the Soviet system – IKI was divided on it – but when France proposed instruments for the series it became definitely established. Prognoz, meaning 'forecast', was the principal Soviet solar observatory and there was a substantial French involvement in its missions. Prognoz orbits, which ranged from 400 km to 200,000 km, made it possible to study the interaction of the solar wind with the interplanetary medium, the shock wave and the transition zone with the magnetosphere in both quiet and violent periods. Although originally intended to study solar gamma emissions from our Sun, Prognoz opened up the study of gamma ray bursts from sources then unknown, first attributed to binary neutron stars in our galaxy and exotic sources further afield. Once again, the primary French instrument was the Signe detector, developed by CESR in Toulouse and flown on Prognoz 2, 6, 7, 9 and M. Signe detectors turned out to be a method of research spanning two decades that made fundamental scientific discoveries, with a dedicated French-Russian conference about them in Moscow in 1983. On 5 March 1979, the Signe detector discovered Soft Gamma Repeaters – repetitive gamma emissions later called magnetars.

The Signe 1 detector was flown on Prognoz 2 to detect solar neutrons and x- and gamma ray eruptions between 0.4 and 12 MeV. Its mission turned out to be anything but routine. August 1972 saw three shock waves from the Sun reach Earth, followed by dense, warm plasma. It was also called 'the great eruption', with the most violent period being 4–7 August 1972. Signe measured neutrons in the range 0.98–16 MeV and gamma rays from 272 keV–9 MeV. Prognoz 2 also carried the French experiment Calipso, an electron and proton spectrometer for the 0.2–20 keV range designed to measure low-energy particles in the outer magnetosphere. Between them, they measured the flux of electrons and gamma protons in 16 bands every 41 seconds. Calipso made it possible to estimate the speed of the solar wind during the great eruption at 1,700 km/hr, with a temperature of 10^5 to 10^7 K. Data were collected for almost six years, until Prognoz 2 lost stability in March 1978. Calipso made the first measurements of a shock wave associated with a solar eruption. In effect, these instruments on Prognoz made possible detailed measurements of the parameters of the solar wind. Calipso 2 would later be carried on Prognoz 5.

The successor to Signe 1 was Signe 2MP (Modified Prognoz) and 2MS (Modified, Sondes (probe)), the idea being to triangulate the bursts between MP in orbit around the Earth and MS in orbit around the Sun on interplanetary probes. Signe 2MP was attached to Prognoz 6, 7 and 9, while MS was attached to Venera 11–14 on their missions to Venus.

Prognoz 6 carried several Soviet-French joint instruments: Signe 2MP, Galactika, Gémeaux S2 and Helium. Prognoz 6 marked the introduction of Galactika, which had been devised by the Observatory of Crimea and the Laboratoire d'Astronomie Spaliale (LAS) of Marseille to measure the sky background in ultraviolet between 1,000 Å and 3,000 Å. It duly made 4,000 spectrograms of 26 regions of the sky between 1,200 and 1,900 Å, giving a picture of the concentration of ultraviolet across our galaxy. Signe 2MP detectors again studied solar x-ray, gamma bursts and cosmic x-rays, finding that solar flares typically released energies of between 10^{29} to 10^{30} ergs. Gémeaux S2 and Signe 2MP confirmed the existence of two phases of acceleration of particles during solar eruptions. Between them, they were able to characterize the different phases of the solar flare from pre-flares to pulses. Galactika recorded ultraviolet radiation from the galaxy, while Signe 2MP detected three confirmed gamma ray bursts over October-November 1977. Similarly, Prognoz 7 carried another Signe 2MP (x- and gamma rays from 20 keV–3 MeV), Gémeaux S2 (corpuscular radiation from the Sun and high energy particles in the high atmosphere and magnetosphere of Earth, especially gamma particles from 0.3–20 MeV, protons from 2–500 MeV and alpha particles from 30–75 MeV); and Galactika (ultraviolet rays from 120–350 nm).

Prognoz 7 (1978) and 8 (1980) marked the first cooperation with Sweden in the programme, where scientists shared their interest in terrestrial magnetism, so their aim was to study the high latitude magnetopause. Prognoz 7 marked the introduction of the ion composition PROMICS experiment, for which Rickard Lundin was Principal Investigator. Born in Lyscksele in 1944, Lundin taught physics at the University of Umeå before moving to the Kiruna Geophysical Institute where he developed hot plasma experiments on sounding rockets. He devised PROMICS for Prognoz 7 and 8 and its successors PROMICS 3 on Interball and ASPERA on Phobos, Mars 94 and Mars Express. He was able to make measurements of hot plasma in regions of the plasmasphere previously unexplored: in the boundary layer, ring current and auroral field lines **[3]**. Rickard Lundin and his Soviet colleagues subsequently wrote *Substorms with multiple intensifications – post onset plasma sheet thinnings in the morning sector observed by Prognoz 8.*

Prognoz.

Prognoz 9 carried the French Signe 2M-9 to study cosmic and solar gamma ray bursts in the 40−8,000 keV energy range. Prognoz 9 found 75 cosmic gamma ray bursts, despite an erroneous trigger which set off many false readings. Signe found an extremely energetic burst on 11 October 1983, which repeated on 13 October and thereafter (a repeating burster). Twelve repeater short bursts in Sagittarius were analysed and they were determined to have come from the same source, 10° from the galactic centre. The precision with which these bursts were marked and located was not bettered for many years. NASA's Goddard Space Flight Center (GSFC) subsequently provided website connections to the Prognoz datasets, as did IZMIRAN in Russia, a tribute to their long-lasting value.

The Russian space programme suffered from a radical reduction in funding during the transition from the period of the Soviet Union to the Russian Federation. The first contraction actually began during the Soviet period as the government responded to criticism of excessive space spending. The two big contractions were in the early 1990s (the introduction of the free market and the ensuing chaos) and 1997 (the rouble crisis). At one stage in 1997, it appeared that the programme might even collapse in its entirety. Of all the many aspects of the programme, space science suffered the most and it was more than challenging to keep scientific programmes going at all. New projects were abandoned and the best they could do was to get airborne those projects that had already reached an advanced stage during the Soviet period. Prognoz 11 and 12, also called Prognoz M1 and M2 or Interball 1 and 2, were an early example of late Soviet-period missions that flew in the Russian period, but were probably lucky to have done so at all.

Prognoz M, like its antecedents, was built by the Lavochkin design bureau in Moscow, but cooperation was extended widely. A planning conference held in Moscow in January 1987 saw Austria, Germany, Poland, Czechoslovakia and Finland signing up, as well as ESA, with the total signatories eventually reaching 20. Both Prognoz M craft carried an impressive range of instruments. Four French instruments, Ion, Elektron, Hyperboloid and Mémo, were carried on both missions, with two joint experiments, Opéra and Ikare. Elektron was to measure electrons from 10 eV−30 keV, while Opéra was to measure the electrical field from 0−50 kHz, especially the mechanisms for the acceleration of solar particles and their encounter with the magnetosphere. The French experiments were made in Toulouse (ion mass spectrometer, electron detector), Orléans (analyser of electromagnetic waves) and Saint Maur (detector of weak energy particles). Sweden provided an ion composition spectrometer for both missions, whilst Czechslovakia provided two remarkable sub-satellites, Magion 4 and 5, which moved in and out of the path of their mother ships.

Interball was the last project of Yuri Galperin. Prognoz M1/Interball 1 was launched on 3 August 1995 into an eccentric 500−193,000 km orbit, 65°, with Prognoz M2/Interball 2 following a year later on 29 August 1996 into an orbit 19,200 km out. Between them, they produced a substantial scientific return, leading to nothing less than a remodelling of the magnetosphere [4]. France published the first results of its six experiments in December 1996. Interball was interesting because of the attention given to its dissemination, facilitated by the arrival of the world wide web. IKI set up web pages devoted to the two probes, their sub-satellites, news, publications, observations, the principal participating institutes (Toulouse, Kiruna, Lindau in Germany), conferences and workshops and a data archive. The results from Interball 2 were posted by IKI on an *Auroral probe results* page under the headings of the different experiments and the Russian and European investigators (www.iki.rssi.ru/auroral), in Europe's case listing scientists from Sweden, France, Finland, Greece, the Czech Republic, Bulgaria, Poland, Slovakia, Austria, the Netherlands and Hungary. Similarly, the results of the tail probe were published (www.iki.rssi.ru/tail) with the contact details of the experiments from Russia, France, Finland, the Czech Republic, Poland, Britain, Slovakia and Greece. Results were published in such journals as *Cosmic Research* (in English, *Kosmicheskie Issledovaniya* in Russian), *Plasma 97, Année Géophysique, Advances in Space Research, Physics and Chemistry of the Earth, Journal of Geophysical Research* and various conference proceedings. Rickard Lundin, with his Swedish and Russian colleagues, published the results from the PROMICS experiment in *First results from the hot plasma instrument PROMICS 3 on Interball 2* and *First results from the plasma composition spectrometer instrument PROMICS 3 on Interball 2*, in *Annales Geophysicae* [5].

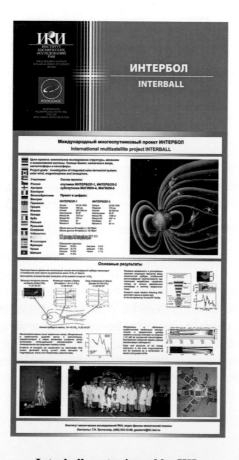

Interball poster issued by IKI.

Yuri Galperin died on 28 December 2001, outliving by only two months his great French collaborator Francis Cambou, who had died on 29 October. Cambou (1932–2001) was a nuclear physicist and founder of the Centre d'Étude Spatiale des Rayonnements (CESR) in Toulouse in 1963, which he directed until 1982. He was also the first professor of space physics in France, also in Toulouse. In 1967, Cambou threw himself into cooperation with the USSR, working especially with IKI and IZMIRAN. Later, he became an advisor to the European Commission on the education and mobility of researchers. There is a photographic record of Cambou and Galperin travelling together on trains and posing in their warmest coats and hats in front of a Mil helicopter in the far north, in snow flurries whipped up by arctic winds. Yuri Galperin opened doors for the French: he knew Georgi Babakin, director of Lavochkin and made the suggestion to him that French equipment be carried on lunar and interplanetary probes. The Prognoz programme is summarized in Table 2.2.

Table 2.2:
Cooperation in the Prognoz programme: French and Swedish instruments

Mission	Date	Nation	Instruments
Prognoz 2	29 Jun 1972	France	Signe 1, Calipso
Prognoz 5	22 Nov 1976	France	Calipso 2
Prognoz 6	22 Sep 1977	France	Signe 2MP, Galactika, Gémeaux S2, Helium
Prognoz 7	30 Oct 1978	Sweden	PROMICS
		France	Signe 2MP, Gémeaux S2, Galactika
Prognoz 8	25 Dec 1980	Sweden	PROMICS
Prognoz 9	1 Jul 1983	France	Signe 2M-9
Interball 1	3 Aug 1995	France	Ion, Elektron, Hyperboloid, Mémo Opéra, Ikare
Interball 2	29 Aug 1996	France	Ion, Elektron, Hyperboloid, Mémo Opéra, Ikare

The other project belonging to the early Russian Federation period was SCARAB (SCAnner for RAdiation Budget, though this was generally written as lower case Scarab). The Scarab programme was designed to measure the radiation balance of the Earth using three wide-band radiometers on a Meteor weather satellite. Scarab went back to an April 1985 initiative by Pierre Morel, who was responsible for climate issues in CNES, for an instrument to measure the Earth's radiation budget. It was prompted by the termination of the mission of the American Earth Radiation Budget Satellite (ERBS) launched in 1984. A CNES delegation travelled to Moscow with a concept note in May 1986, arriving just as the effects of the Chernobyl disaster were becoming known. Although inspired by the American satellite, Scarab took the concept a step further and with a wider range of imaging bands. Scarab was agreed in principle at the annual reunion in Yerevan in October 1986, the USSR having already agreed to fly it on a Meteor weather satellite. Unfortunately, the timing was poor, with financial problems beginning to affect the Soviet programme, so the mission was delayed. The Meteor was duly launched on

25 January 1994 as Meteor 3-7: the instrument failed after a year, but not before it had made an original contribution to climate research. The second Scarab was flown on the Earth resource mission Resurs 1-04 on 9 July 1998 and a third Scarab on the Franco-Indian Megha-Tropiques satellite launched on 12 October 2011.

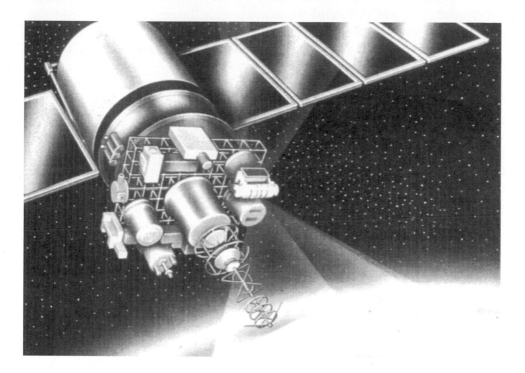

SCARAB. CNES.

Introducing Sweden: Interkosmos 16

Sweden was the second west European country to fly instruments on a Soviet spacecraft, supplying a solar ultraviolet spectrometer. This was done through the auspices of Interkosmos, which had already been assigned an important institutional role to assist in the cooperation with France. Interkosmos had been the space body set up in 1966 for cooperation between the socialist countries, but western countries were not members. They began a programme of satellites, simply called Interkosmos, in which instruments were contributed by the socialist countries in accordance with their interests and skills, the series running to Interkosmos 25 in the 1990s. Although France had been assigned to Interkosmos to facilitate its connections to the Soviet space programme, Sweden approached

Interkosmos directly, to which it responded quickly enabling the Swedes to become a small part of the Interkosmos programme.

Their project was the work of Jan-Olof Stenflo of the Institute of Astronomy and the University of Lund, who had written his thesis on solar observations from the astrophysical observatory in Crimea. He came up with the idea of a satellite-borne instrument and persuaded the new Swedish Space Corporation (SSC) to pursue the project. The purpose of the instrument was to investigate the transition zone between the chromosphere and the corona, the instrument being built by the SSC itself, Jungner instruments and Saab Scania.

The mission was set up during a meeting between Interkosmos and the SSC in Stockholm over 4−13 September 1972, after which a protocol was signed. The project manager, Stefan Zenker, also met a Soviet delegation at the COSPAR meeting in Madrid. In December 1972, a Swedish team visited the observatory in the Crimea, also meeting IKI and IZMIRAN in Moscow with Yuri Galperin being present. The instrument was duly loaded and launched on 2 June 1975, but it failed, the only failure in the Interkosmos series. Just over a year later, the backup instrument was successfully put into orbit.

The Swedish partners had not been invited to the 1975 launch, but the successful one a year later was attended by Jan-Olof Stenflo and two engineers from the SSC. They saw the integration of the satellite into the Cosmos 3M launch vehicle in the hangar, the rocket being readied for launch and the launch itself. It seems that they were the first westerners allowed into the Kaspustin Yar cosmodrome. This was Russia's original rocket site, used to test the A-4 rockets brought back from Germany after World War II. From 1962, it was used for the Cosmos programme and as a missile test site, but it did not have the same security sensitivity as some parts of Baikonour. The Swedes made a sketch of the launch site and assembly hall afterwards for the benefit of those who might be interested.

The 370 kg Interkosmos 16 was launched on 27 July 1976 into an orbit of 464−517 km, 50.6°, 94.4 minutes. Its instruments broadcast signals until 13 November 1976, but Jan-Olof's instruments broadcasted for only 12 days. This was not unusual for the Cosmos programme at this time, which tended to transmit a high volume of information but for a short period. Signals were picked up in Sweden itself from 3 August onward, but just in case they were not, the Russians sent them a box of the computer tapes. The results were published later in *Applied Optics* (1976) and *Solar Physics* (1980) and the instruments were adapted for the Prognoz programme. Swedish cooperation with the USSR did not otherwise progress much further, with Sweden using Europe's Ariane for its first satellite, Viking, in 1986. Curiously, neither Interkosmos 16 nor Prognoz were mentioned in the official history of the Swedish space programme. Interkosmos 16 was little publicized outside the two countries concerned at the time, with even the Americans noting the 'little fanfare'. The history of the SSC made the observation that Interkosmos and Prognoz gave Swedish scientists and engineers modest but important hands-on experience of developing and testing scientific equipment, such as long-life components, radiation hardening, thermal control and high cleanliness, which was much more demanding than for sounding rockets [6].

From the 1970s, there were intermittent discussions between Sweden and the Soviet Union over the potential of remote sensing for environmental monitoring in the Baltic Sea, Sweden's capacity being especially apparent in 1986 when the SPOT satellite, a French project with Swedish participation, famously photographed the burning nuclear reactor at Chernobyl. Three years later, the Soviet Union converted to the Brundtland report *Our common future* and in *Earth mission 2000*, offered to turn photoreconnaissance satellites over to environmental monitoring in cooperative programmes with Europe. That year, the SSC made an agreement with Glavkosmos whereby such imaging would be collected in Kiruna for the environmental benefit of the Baltic region and further afield. In the event, this proposal was eclipsed by the independence process and SSC ended up collaborating with the individual Baltic states for their remapping outside retreating Soviet control. In the end, Kiruna set up an Environmental Data Centre in 1993, an eventual outcome of a process launched many years earlier [7].

Cooperation in biology and materials science

The next field of cooperation to be developed was materials science and biology. This finally brought European-Russian cooperation outside the dominant French axis, introducing Germany and then later ESA on behalf of all the European spacefaring countries, large or small. Biology was an obvious area for cooperation, with the Soviet Union having run biological missions since 1973 under the Cosmos programme with Cosmos numbers. Eventually (1991), the programme acquired its own title, Bion, with Bion numbers then being applied retrospectively (Cosmos 605 became Bion 1). NASA had been involved in Bion from an early stage as it was considered a politically uncontroversial area for joint endeavour, though the use of animals later became contentious. The Bion spacecraft was essentially the original spacecraft of the Soviet space programme, which flew for military missions (Zenit), piloted missions (Vostok, Voskhod) and many other purposes. Bion offered opportunities not available elsewhere for biological missions and the series has lasted over 40 years.

An important date in French-Russian cooperation in space biology was the week-long reunion of 20–27 April 1975, held in Dubna, with Professors Nefedov and Dr Kotovskaya from Institute of Bio Medical Problems (IBMP) and René Bost (CNES) and Hubert Planel (Toulouse Faculty of Medicine) attending. This led to experiments called Bioblock (also written as Biobloc and Bioblok) and Exoblock (also written as Exobloc and Exoblok) flown on the Bion satellites, versions of which found their way onto the Salyut and Mir orbital stations. They referred both to an experiment and the apparatus involved.

The first Bioblock, installed on Cosmos 782 in November 1975, was a joint experiment to house samples to be tested for radiation, with the first experiment concerning salad seeds, tobacco seeds and brine shrimp eggs (*Artemia salina*). The results showed, for example, that cosmic rays had discernible effects in delaying the development of eggs and that there were chromosomic aberrations in lettuce seeds and abnormalities in tobacco. Bioblock 2 was flown on Cosmos 936 in August 1977, testing the effects of heavy ions on Artemia eggs and lettuce seeds, while *Ulysse,* developed by the Pasteur Institute and the IBMP, tested the immunology of rats. Cosmos 1129 carried Bioblock 4 to test the effects of heavy cosmic radiation, galactic radiation and zero-g on cultures, matched against a ground analogue. Cosmic heavy ions, called HZE particles, had the effect of reducing hatchability in cysts and stimulating the germination rate of tobacco but prompting a higher level of abnormalities. Bioblock 5 flew on Cosmos 1514 to study the effects of galactic radiation on biological objects (the missing Bioblock 3 in this series flew on the Salyut 7 orbital station).

These were largely examples of bilateral France-USSR cooperation, with ESA becoming more involved in the 1980s. This is a useful point at which to introduce the European Space Agency [8]. Its predecessor, the European Space Research Organization (ESRO) signed a cooperation accord with the Soviet Academy of Sciences in 1971, but a formal agreement to update this with ESA had to wait until 25 April 1990, just in time for the end of the Soviet Union. ESA's establishment date was as far back as 30 May 1975, with 12 founder states. A particular feature of European space activities, which can make them complex to examine, is that some projects are European and entirely under the guidance of ESA, while others are national and quite unconnected to the agency. National space programmes are also free to engage with other national programmes, either bilaterally or multilaterally, again without reference to ESA. To complicate things further, some projects share costs between ESA and national programmes and there may be projects that involve both ESA as a multinational agency and the national agencies of individual member states. Member states are all required to fund ESA's science programme, but may also opt in to other programmes (e.g. launchers). The principle of *juste retour* applies (member states should get opportunities and industrial contracts broadly proportionate to their financial contribution, known as *geo-return* in ESA). Despite its complexities, ESA opened up the prospects for highly structured forms of cooperation with a wider range of countries. ESA booked and paid for the flights, ran the mission campaigns and oversaw the instrumentation, but experiments were selected on scientific merit, not by country. The national agencies provided financial support to their scientists and helped them to develop their payloads. The European side was coordinated by the European Space Research Technology Centre (ESTEC) in Noordwijk, the Netherlands.

In 1993, ESA started an office in Moscow for the purposes of liaison, communications and cooperation (1995 is its formal opening date). The initial staff complement was nine: one French, one Austrian and seven Russian, including translators and customer help. The opening of the ESA office on Sretensky Boulevard in Moscow was clearly a milestone in cooperation. The first head was Alain Fournier-Sicre and since 2007 it has been René Pischel [9]. To this day, the ESA office is the front office of practical cooperation. Its functions range from ensuring the smooth management of joint missions to meeting people at airports and arranging visas, hotels and travel. It ensures contacts between agencies and industry on both sides and keeps up to date with the latest developments. Although policy decisions are taken by ESA meeting in Paris, its views on cooperation are likely to be influential.

Bion

CNES joined the Bion programme with Bion 3 (1975) and ESA became involved from Bion 8 (1987). The circumstances were unexpected. ESA had developed the crew-operated Biorack biological experimental apparatus for American shuttle operations from 1985, when it flew on Spacelab D1, but this multi-user system was grounded with the loss of the shuttle *Challenger* the following year and Biorack did not fly again on the shuttle until 1992–7, when it flew five times. Its successor, Biobox, flew seven times: on Bion 10; Foton 10, 11 and M-3; STS-95 and 107; and Shenzhou 8. Initially a stand-in for Biorack, ESA developed Biobox as a robotic incubator for cell culturing experiments, with no crew interventions required, that could fly on unmanned Bion and Foton satellites. From 1992, Biorack was used in parallel with Biobox on the shuttle and Bion and Foton respectively.

Possibly spotting an opportunity, the Interkosmos Council approach the grounded ESA post-*Challenger*, offering the Bion cabin for life sciences experiments. The offer was on a no-fee basis, but with the proviso that scientific results would be made available to Interkosmos and that its scientists would participate in post-flight analysis, information exchange and subsequent joint publications. The principal agency handling these missions was IBMP. As with the shuttle, Bion cabins offered the advantage of both an internal cabin and external locations, called KNA containers. ESA joined for an initial set of three missions: Bion 8, 9 and 10.

Foton M-3. Marcel van Slogteren, ESA.

A first joint planning meeting was held in January 1987, which agreed the ground rules. Experiments were discussed at the second meeting in April 1987, with ESA proposing CARAUCOS (the effects of radiation on stick insects), DOSICOS (the damaging effects of radiation on plants and bacterial spores) and SEEDS (the damaging effects of radiation on seeds), to be stacked both inside the spacecraft and outside in KNA containers. Each experiment listed a team of ESA and IBMP experimenters. With the launch of Bion 8 due on 29 September that year, preparation time was short. The experiments were prepared in Cologne, then flown to IBMP in Moscow to be handed over and transported to Plesetsk, while ESA personnel stayed at IBMP to monitor the mission. A high degree of improvisation was required, the joke being that not only did the insects age more rapidly in weightlessness, but so too did the ground team.

ESA's experiments were carried both inside the spacecraft and outside in the KNA containers. Bion 8 (Cosmos 1887) famously attracted public attention when one of the two monkeys on board freed himself from his restraints, removed his electrodes and began tampering with the controls, sparking the classic headline of 'Monkey to ground control: I've taken over!'. Worse was to come, as the capsule was misaligned at re-entry (not the monkey's fault) and came down 3,000 km off course in Siberia in temperatures of -17°C. The monkeys were kept warm at a fireside by villagers who found the cabin, but the fish experiments were lost in the cold. Army paratroopers then arrived and put the spacecraft and rescuers in a tent. Bion 8 made one *Star Trek* type experiment, throwing a 305 keV electrical field around the spacecraft for more than a full orbit, 105 minutes, to protect it from radiation.

As for the ESA experiments, Bion 8 led to ten scientific papers in English and Russian. The principal contributors came from IBMP, ESA and the German space agency. Examples of the findings included that the hatching rate of stick insects slowed in orbit, lettuce seeds hit by heavy ions developed chromosomal aberrations and the germination of plant seeds was delayed or reduced, more so with those outside the cabin. One seed, hit by a HZE particle, developed a sponge-like tumour rather than flowering normally. Abnormalities were observed in the antennae, abdomen and legs of the stick insects. ESA published a post-mission report on this and the subsequent mission **[10]**.

Biopan external container. ESA.

Bion 9 carried experiments addressing tissues, muscular skeletal growth, muscles, heart tissue, cells, metabolism and radiation in rats and monkeys. The joint ESA/IBMP experiments were CARAUCOS, DOSICOS (inside and outside) and SEEDS (inside and outside) as before, to which were added FLIES (fruit flies) and PROTODYN (rapeseed and carrot plant protoplasts). The first three experiments were prepared in Germany, the latter two in Moscow at IBMP by a joint team of ESA and IBMP investigators. The ESA team in Moscow had no further sight of the experiments until they were returned and were not permitted to travel to the launch site nor oversee final installation. ESA experimenters might initially have thought that they were distrusted as Europeans, but the same ground rules applied to Soviet scientists. This was the normal way their compartmentalized programme operated.

There was then a crisis when the launch was delayed for a week due to a battery problem, a most unusual event in a space programme that normally ran like clockwork. Again reflecting the structure of the space programme at that time, the phone call was not made directly from Plesetsk, which was a militarized area, but from the post office in the civilian town of Mirny some 40 km distant. A week's delay would invalidate some of the experiments, so 16,000 fresh stick insect eggs had to be flown in from Germany at short notice and the fruit flies replaced with a batch from the Institute of Genetics in Moscow in time for launch on 15 September. There was one problem, which was that the cabin overheated to 30°C due to the launch having been delayed for a week, with the satellite then flying in a profile of prolonged exposure to the Sun for each orbit. The experiments landed on the morning of 29 September, were unloaded within an hour, put into thermally-controlled ESA boxes powered by a car battery and were back in the agency experimenters' hands that evening. The ESA team returned home with them after five weeks in Moscow. The five Bion 9 experiments led to 21 publications and it was the last such mission under the USSR. Bion 9 findings confirmed those of Bion 8, with malformations in stick insects affected by high-energy rays and aberrations in lettuce seeds, a third of which were hit by HZE particles. There was a reduced rate of stick insect hatchings and the fruit flies laid twice as many eggs as on the ground, but plant cells would not regenerate. The rate of germination of seeds was reduced and flowering delayed, while 1,389 eggs were hit by HZE particles.

Biopan. René Demets.

A standard apparatus, the 42.5 kg Biobox, was duly developed. Similar to the earlier shuttle Biorack, it provided a controlled temperature and was equipped with a 1G reference centrifuge. Unlike Biorack, it was fully robotic, able to complete all experimental steps autonomously once a start command had been sent from the ground. Almost simultaneously, ESA developed Biopan, an external facility based on the Soviet KNAs, but equipped with thermometers and solar sensors. Biopan, developed by Kayser-Threde in Munich, was a 38 cm by 21 cm, 25 kg container mounted on the exterior of the cabin and exposed to space conditions. It was capable of hosting assorted experimental packages with a total weight of 5 kg. The lid of Biopan could be opened in space by telecommand and closed in time for landing, protecting the experiments against the heat of re-entry. Typical samples on Biopan included shrimps, tardigrades, lichens, fungal spores, yeast, algae, bacteria, plant seeds, amino acids and peptides. Biobox was developed to investigate the influence of weightlessness on cell culture growth and development, while Biopan was intended to investigate the effects of open space on dormant desiccated life and organic molecules [11].

ESA insisted on improvements, such as a modern, proper laboratory in Moscow for pre- and post-flight operations. Moslab was a prefabricated laboratory built in the Netherlands and brought to Moscow by truck in 1992. It filled an important gap, as there was no other office or location from which the ESA specialists could work, nor any direct communication links.

Bion preparations. ESA.

Planning for Bion 10 began in the old USSR, but it was eventually launched from the new Russia, the mission being organized against a background of uncertainty, even chaos, with countries changing their status and places changing their names. From Bion 10, Russia made it clear that a no-fee basis was no longer possible; there would have to be a paid arrangement and the informal arrangements that had governed previous Bions could no longer apply. After a three-year period of organizational upheaval on the Russian side, an agreement was eventually signed on 12 October 1992, only three months before lift-off. ESA documented the experience, the mission and the scientific outcomes in the 125-page *Biological experiments on the Bion 10 satellite* [12]. The practical difficulties were that 'the two parties lived typically 2500 km apart, are unable to travel freely to each other's place, speak different languages and are deprived of direct communication via telephone, fax or e-mail'. Instead, they set down a schedule of regular meetings, one of which took place in the middle of the coup in the Soviet Union. It was a joint project, in the sense that all experiments had members from the two sides and there were two end-of-mission reviews, the final one being held aboard a barge.

Although Bion 10 took place at the most politically unstable time of the whole cooperative experience, it was an object lesson in how the practicalities of cooperation could be overcome, revealing many of the gaps that had to be closed if cooperation were to work. Plesetsk still had a combination of 'secretive atmosphere' that added a 'flavour of adventure' to the project. Organizations in the early Russian Federation were in some disarray and ESA found itself dealing with a multiplicity: TsSKB in Samara, the spacecraft makers and then the Russian space agency which went through three forms: RKA, then Rosaviakosmos, now Roskosmos. ESA's partners on the Russian side, for example, IBMP, seemed to have little control over how to ship the European instruments to Plesetsk, leaving ESA officials negotiating shipping with the railway company at the station. IBMP did not have the management skills or authority to organize all the Russian side of the operation and could not control either the spacecraft builder (TsSKB), nor the flight control centre. This was not a criticism because it had never been designed for such a role. In the end, in 1993 the new Russian space agency stepped in and appointed a new organization called 'the Cosmos Group' to handle all the arrangements. A formal agreement called *Framework agreement for the launch, operation and retrieval of ESA microgravity multi-user facilities and experiments* was worked out. Order emerged from chaos.

The Russians remained insistent that only people with specific tasks to do should be at the launch or landing sites, with no tourists allowed. The Bion 8 and 9 experiments were passive, without electrical interfaces with the spacecraft, but with Biobox there was a greater level of complexity, so an ESA presence during the pre-launch preparations became important. ESA now received a grandstand view of how Russia operated its space programme in harsh weather, with the plane landing in Plesetsk in a violent snowstorm on its third attempt. They also got to attend the landings and ESA personnel were on hand for the landing of Foton 11 in 1997, although the landing site had moved by then. In 1995, Russia tired of parachute bags full of cash being demanded as a landing fee, so touchdown sites were moved from Kazakhstan, Kazakh territory, to the very southern part of Russia, just north of the Kazakh border.

Bion 10 carried ten different ESA and IBMP experiments (BONES, MARROW, OBLAST, FIBRO, ALGAE, FLIES, CLOUD, WOLFFIA, SEEDS and DISICOS). The first four, accommodated together in ESA's Biobox, were dedicated to understanding bone weakening in microgravity at cellular level. ALGAE, FLIES and CLOUD were stand-alone experiments studying the growth and development of insects and small plants under weightlessness. In WOLFFIA, the effect of space radiation on tiny duckweed plants was investigated, while SEEDS and DOSICOS were radiobiological studies, both re-flights of similarly-named experiments on Bion 8 and 9. ESA's payload was 56 kg, including the Biobox. Bion 10 saw the start of CNES experiments with pleurodeles, flown subsequently on Foton 10 (1995) and Foton 11 (1997) and as Fertile on Mir (1996, Cassiopée; 1998, Pégase), Génésis (1998, Perseus) and Aquarius (Andromède). One set of salamanders made a five month stay on orbit. They were considered especially suitable for assessing the effects of weightlessness on the immune system and anti-bodies.

There was an emergency nine days into the Bion 10 mission when the cabin began to overheat past the 28°C critical point. This was due to it entering an orbit that exposed it to almost continual sunlight, caused by a combination of timetabling issues. When it reached 31°C, the mission was terminated and the Bion brought down two days early. The results from Bion 10 were mixed, as the temperature rise had compromised some experiments (e.g. ALGAE). Furthermore, the airtight seals of three of the four KNA containers had broken, making the results unusable. Despite this, seven experiments yielded significant results, with 24 scientific papers from the mission published later and some experiments subsequently re-flown on Foton missions. There were important results for ESA from the experiments with flies, duck weed and bone mineralization.

After Bion 10, ESA left the Bion programme for Foton. Bion 11 followed in 1996 as a Russian mission without ESA. It carried two monkeys, as had previous Bions (6−10), but when one of them died, the programme concluded. Under pressure from animal rights defenders, the Americans made it clear that they could not cooperate in such a programme.

There was a long gap before the next Bion mission, Bion M-1, which was launched on the newer Soyuz 2.1a rocket on 19 April 2013. The cabin had a number of improvements and offered both a higher altitude (575 km) and, with solar panels, double the flight time. Although the majority were Russian experiments, Bion M-1 also carried the European experiments Omegahab (water biology, developed by Germany's Kayser-Threde) and Biokon (growth of biological samples, Italy). For the first time, there was round-the-clock monitoring of the heart rate of mice in a CNES/IBMP experiment, Mouse Telemetry on board Bion (MTB). Bion M-1 also carried IBMP's first experiment in meteoritics, a follow-on from the STONE experiment initiated by ESA on Foton cabins.

Bion M-1 down. ESA.

Bion M-1 recovery tent. ESA.

Bion M-1 also launched six small hitchhiker satellites, including the Student Oxygen Measurement Project (SOMP), a 1 kg cubesat from the Technical University of Dresden with an electrolyte sensor for atomic oxygen; and BEESat 2 and 3, a pair of 1 kg Berlin Experimental and Educational Satellites cubesats, each with a small camera for Earth observation. BEESat 2 would test attitude control and BEESat 3 would test communications. The mission also had its problems, because when the cabin returned to Orenburg on 19 May it became evident that failures in the life support systems had led to the deaths of all eight Mongolian gerbils, 39 of the 45 mice and the fish, with the only survivors being the geckos, slugs and snails.

An agreement on Bion M-2 was drawn up between CNES and IBMP on 12 December 2018. This was an international mission also involving the United States, Germany, Japan, Hungary and Bulgaria, but France was the first to sign an agreement. The purpose of the mission, scheduled for 2023, is to examine the effects of space radiation on the living organism. The cargo comprises 65 mice (to be filmed by camera) and a batch of fruit flies. The spacecraft is expected to fly in an orbit of 800−1,000 km for 30 days, reaching the radiation belts and picking up in that time a radiation dose equivalent to three years on the space station. The mission director for both was Vladimir Sychev. CNES is supplying a second mouse telemetry experiment, MTB-2. A summary of the series can be found in Table 2.3.

Table 2.3:
Bion cooperative missions

Mission	Date	Duration	Partners
Cosmos 605/Bion 1	31 Oct 1973	22 days	
Cosmos 690/Bion 2	23 Oct 1974	21 days	
Cosmos 782/Bion 3	21 Nov 1975	20 days	NASA/CNES
Cosmos 936/Bion 4	3 Aug 1977	19 days	NASA/CNES
Cosmos 1129/Bion 5	25 Sep 1979	19 days	NASA/CNES
Cosmos 1514/Bion 6	14 Dec 1983	5 days	NASA/CNES
Cosmos 1667/Bion 7	11 Jul 1985	7 days	NASA/CNES
Cosmos 1887/Bion 8	29 Sep 1987	13 days	NASA/CNES/ESA
Cosmos 2044/Bion 9	15 Sep 1989	14 days	NASA/CNES/ESA
Cosmos 2229/Bion 10	29 Dec 1992	12 days	NASA/CNES/ESA
Bion 11	24 Dec 1996	14 days	NASA/CNES
Bion M-1	19 Apr 2013	30 days	NASA/CNES
Bion M-2	Scheduled		NASA/CNES

Foton

The Foton spacecraft was the same design as Bion (i.e. Zenit, Vostok), but was intended for materials processing missions and did not carry the life support system developed for the monkeys on Bion. It was built in the same location, so working with Foton was not a dramatic change. Like Bion, Foton was originally part of the Cosmos programme, but became civilianized with its own identity. Its big selling point was that it was a free flier, meaning no attitude control was applied from the moment it entered orbit to when it left, providing the purest possible zero-gravity experience.

Predictably, France was first on board the Foton programme, beginning with Foton 5 launched on 26 April 1989 and recovered near Orenburg on 11 May. This was followed by an agreement on 16 June 1989 between CNES and Licensintorg, the foreign arm of Glavkosmos, to fly CROCODILE (CRoissance de Cristaux Organiques par DIffusion Liquide dans L'Espace) on Foton 6 in spring 1990. CROCODILE was designed to produce high quality organic crystals for non-linear optics to pave the way for new optical equipment and amplifiers. Lack of thermal convection, which could be provided in zero-g, was a significant help for production. Also carried was a FFR185m electrophoresis experiment in cooperation with Spain and Belgium under the European Union's *Eureka* programme, for the production of pancreatic cells to produce insulin, hormones and renal cells.

Although the emphasis was on material processing, from Foton 7 the series began to fly biological cargoes that did not require life support systems. Because of pressure from animal rights defenders in Germany, ESA was encouraged to distance itself from mammal experiments, so Foton suited this well. For Foton

7, Russia offered access to the Kashtan system for protein crystallization, but on a pay-by-flight basis. This was different to Bion, where ESA planned experiments together with IBMP. Kashtan, meaning 'chestnut' in Russian, was developed by the Splav centre in Moscow and was an incubator operating at 4°C with a Perspex chamber and 52 protein sample positions. ESA was offered 20 samples at a cost of €4,500 each, but that included the all-in costs associated with the mission, a modest sum compared to shuttle operations. A first meeting was held between ESA and Glavkosmos in Essen in June 1988. Detailed discussions on the first mission took place in April 1991, with a view to a flight only five months later.

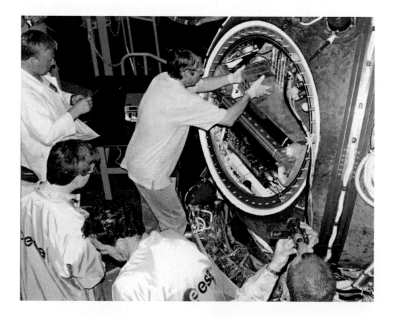

Foton being loaded. ESA.

European experimenters were expected to deliver their cargoes three months ahead of launch to the TsSKB in Samara, where they would be tested in a reduced mission simulation before the Foton was sent by train to Plesetsk. Samples could be loaded up to 72 hours before launch. There was almost no infrastructure available for the Europeans, who had to book phone calls back to the Netherlands 24 hours in advance from the *Intourist* hotel. The experimenters had no first-hand experience of the mission, simply handing over the experiments beforehand and collecting the results afterwards. Entering the Splav building and the Centre of Medical Biotechnology in Moscow was as close as they got. Data were transmitted to mission control (TsUP) in Moscow and TsSKB in Samara.

From 1999, with FluidPac on Foton 12, ESA developed a tele-support system in the cabin for transmissions of telemetry to ESA at Esrange in Sweden as well. Foton became such an important collaborative programme that ESA provided the main part of the volume, number and mass of experiments, overseeing the whole mission in cooperation with TsSKB. ESA subsequently issued a dedicated publication, *Focus on Foton – a unique platform for experiments in space* (2001).

ESA experiments were carried on nine Fotons from 1991–2007. The first ESA experiment, on Foton 7, was developed by Wim Jansen, project manager and veteran of Bion 9, with Naomi Chayen of Imperial College London as lead scientist [13]. The outcomes were disappointing, with only half the samples crystallizing and some damaged by leaks, but there was an unexpected finding that the crystals were affected by salt concentrations. Three scientific papers were published. On the same flight, CNES flew the 15 kg SEDEX (Synthèse Enzymatique de DEXtrane). This was an entirely new experiment proposed by Bioeurope in Toulouse, its purpose being to study the behaviour of enzymes in microgravity (dextrane is a form of glucose used as a substitute for blood plasma). Also on that flight was the German *Flugdatenrekorder* (flight data recorder), provided by Kayser-Threde.

Foton landed. ESA.

By the mid-1990s, cooperation missions in the Foton programme had begun to settle down. Foton 8 in 1992 was the qualification flight of ESA's Biopan (Biopan 0). As planned, the lid of Biopan opened by telecommand, but it would not close. After several attempts and repeated telecommands over several orbits, the lid was finally closed and locked. The problem was investigated and resolved, so Biopan was declared operationally ready.

The first operational Biopan mission – Biopan 1 – was Foton 9, a 17-day mission launched on 14 June 1994 from Plesetsk and recovered on 30 June. Six ESA experiments were carried to test the effects of solar ultraviolet and cosmic radiation on organic molecules (including DNA) and desiccated organisms. These were DUST (effects of solar ultraviolet radiation on peptides), SURVIVAL (endurance of living organisms in the space environment), BASE (damage to cellular DNA from cosmic radiation), SHRIMP (effects of space radiation on shrimps), VITAMIN (efficiency of radio-protective substances) and MAPPING (radiation measurements behind varying levels of shielding). Also on Foton 9 was Gézon (GErmanium ZONe), a materials processing experiment by CNES using the Splav centre's Zone 4M furnace for the study of capillary convection of germanium coated with gallium.

Foton 10 was a €1 m mission for which ESA and CNES provided payloads independently of one another. It ended in disaster. Launched on a 15-day mission out of Plesetsk on 16 February 1995, the Foton carried Biobox 2 with three cell culture experiments (MARROW, OBLAST and FIBRO) that were re-flights of Biobox 1, plus three biological mini-facilities from ESA (FLIES, ALGAE, BEETLE). FLIES investigated insect ageing in space, ALGAE measured cell proliferation of unicellular algae in space and BEETLE was intended to see how biological clocks in desert beetles were affected by zero gravity. A fresh contribution to the programme was Instrument de BIologie Spatiale (IBIS, written as Ibis), a robotic incubator from CNES. Equipped with a 1G reference centrifuge, Ibis was designed to investigate the effects of weightlessness on biological samples. The cabin returned to Earth perfectly on 3 March, landing on a river bank near Orenburg to the south-west of the Urals, but when it was lifted out of the recovery area the next day the helicopter flew into a blizzard. Gusts of wind seized the capsule, causing it to sway alarmingly and the Mil crew dropped the payload from an altitude of more than 100 metres, with an impact of 400–500G. The experiments were smashed.

A Foton 11 replacement mission was launched on 9 October 1997, largely as a 14-day re-flight of Foton 10 (this was part of the arrangement if something went wrong). Launched from Plesetsk, it carried a small, 154 kg cabin measuring 1 m across called MIRKA (*Mikro-Rückkehrkapsel*, also written as Mirka), which was developed by Kayser-Threde and provided by the former German space agency DARA. This was flown at the front of the standard, main re-entry cabin and ejected separately before retrofire, making its own re-entry to be recovered on the steppe of southern Russia. MIRKA carried experiments regarding heat shield instrumentation, heat flow, pyrometers and a new type of ablative material. The main cabin carried Biobox 3 (three experiments: MARROW, OBLAST and FIBRO, a reflight of Biobox 2 on Foton 10); three stand-alone experiments (FLIES, ALGAE, BEETLE, a re-flight of Foton 10); Biopan (six experiments, a re-flight of Foton 9); and three CNES packages, CROCODILE 2 (crystal growth in microgravity), Ibis 2 (cellular biology) and Beta (micro-vibrations). Both the main Foton 11 cabin and the MIRKA touched down safely on brown steppe grass near Orsk on 23 October, only 2,000 m apart from one another. There were no Ibis results, due to its power supply having failed shortly before launch. Rather than face a lengthy delay, it was decided to proceed with a dead payload **[14]**.

Foton cargo. ESA.

Foton 12 followed on 9 September 1999, with CNES providing Ibis 3 (four experiments) and ESA a 240 kg payload comprising Biopan 3 (four experiments), FluidPac 1 (a massive new apparatus for fluid physics, three experiments), STONE 1 (ESA's first experiment in meteoritics, with three rocks screwed into Foton's heat shield) and two stand-alone biological experiments (ALGAE and SYMBIO). Foton 12 included a furnace, a machine for measuring microgravity, a biological container and a device to measure electromagnetic emissions.

The French team was given a real treat, being permitted to watch the launch from a protected shelter only 800 m from the rocket itself, much closer than previous ESA teams. The Soyuz, as usual, performed as advertised, taking off only 1/100th of a second late. Biopan 3 was opened from 20 hours into the mission to the 303-hour point, exposing the YEAST, DOSIMAP, VITAMIN and SURVIVAL experiments to vacuum, solar ultraviolet and cosmic radiation. Foton 12 came back a day early, because the Ibis shut down when the centrifuge stopped and the temperature fell. It landed in green-brown grasslands 133 km north west of Orenburg 15 days later, a mere 2,000 m from the welcoming team. A detector package, Quasi Steady Acceleration Measurements or QSAM), first flown on Foton 11, measured the shock of landing at 38G, just within the design limit of 40G. Within minutes, rescue teams were alongside unloading the experiments.

Foton 12 included experiments on cells, fluid physics, micro-organisms, particle radiation and meteorites. The last was probably the most unusual, as Foton 12 carried the first meteorite experiment, designed to test the extent to which rocks could survive re-entry. Basalt, dolomite and sedimentary Martian-type rock were carried, with the results suggesting that they could survive but would be almost unrecognizable on the ground. Various types of rocks were exposed to re-entry to test why some types of meteorites survived entry at great speed into the Earth's atmosphere and others did not. Foton 12 was the first flight for ESA's FluidPac, a new 182 kg facility to provide real-time monitoring of fluid physics experiments, relaying the progress of the experiment directly to the ground including Kiruna in Sweden. It contained three experiments: BAMBI, MAGIA and TRAMP. New results were obtained from processing tin and aluminium and thermo-capillary convection. Overall, ESA was very pleased with the outcome [15].

Russia now offered an improved version of Foton and ESA signed up with Rosaviakosmos on 11 April 2001. The cabin was improved and called the Foton M series, with an increased payload (660 kg), higher orbit (400 km) and better performance lithium batteries. The first was named Foton M-1 to mark the modifications ('M').

Foton M-1 was loaded with 44 biological and materials processing experiments from Europe and Russia, covering everything from fluid dynamics to growing crystals in microgravity, on what was planned to be a 15-day mission. This was the first time that ESA had booked the entire 355 kg payload capacity space aboard, with 75 ESA engineers attending a conference in Samara to plan the improvements. The scientific programme comprised six French, one German and one Dutch-led experiment. The manifest included FluidPac, carrying four fluid physics experiments, BAMBI, ARIEL, DAGOBERT and SIMBA; the Agat materials science furnace with six materials science experiments (Germany); Ibis 4 with six biological experiments (CNES); and CRAMINIO and STONE, to test the effects of re-entry of meteorites (ESA). Biopan 4 (ESA) carried a record nine space exposure experiments, developed by scientists from Italy, Russia, the Netherlands,

Bulgaria, Spain, Germany and Austria. These experiments involved sophisticated equipment: Polizon, for example, was an electro-vacuum furnace working at 1,200°C while generating a rotating magnetic field of 5mT **[16]**.

The Foton M-1 mission began on 15 October 2002. No sooner had it cleared the pad than observers 800 m distant noted smoke streaming from one of the strap-on rocket engines before it went into cloud. At 29 seconds the sound stopped, with the thundering roar of the rocket giving way to silence and the rocket dropping back through the clouds in reverse. The engine had apparently failed, so the computer shut down the whole rocket. The Soyuz U launcher fell back in a curve that brought it back toward the spectators and it exploded in a fireball only 600 m away. The huge explosion killed one of the rocket troops in his cabin and injured many others, with the fires taking some time to be extinguished. It was a rare failure of the Soyuz rocket. The wreckage was recovered from the snow at winter's end, with parts of what was left of the experiments returned to ESTEC in the Netherlands. The Ibis experiment, though, seemed to be jinxed. Ibis 1 had been destroyed on Foton 10; the power supply on Ibis 2 failed before launch; Ibis 3 was shut down prematurely; and now Ibis 4 had been destroyed on launch. No Ibis 5 had been built, so the doomed Ibis team went home.

Foton M-1 wreckage. TsSKB via René Demets.

Foton Biopan 4 recovered. René Demets.

TsSKB's mission manager, Valeri Abrashkin, quickly assembled the still-shocked Europeans about an hour later and told them to begin preparations for a re-flight. His strength and confidence soon had a calming, reassuring effect. The new flight would be Foton M-2, which launched from Baikonour on 31 May 2005, the first Foton from that site. It carried a similar 385 kg payload of experiments in biology, fluid physics, material processing, meteorites, radiation dosimetry and exobiology. Foton M-2 made a smooth landing 16 days later and ESA specialists were on hand to retrieve the cargo.

The two experiments that attracted the most external interest were STONE 5 and LICHENS. STONE 5 arose from strange events in Antarctica several years earlier. In 1981, American investigators found meteorites that apparently came from Mars, challenging previous views that meteorites could only come from the asteroid belt. The idea that they could have been shot into space and reach Earth as the result of a high-speed impact on another planet had been seen as outrageous – but it was now being seen as a possibility and, moreover, a big boost to panspermia theorists. By 2008, NASA had collected 17,000 meteorites from Antarctica, preserved in the cold, proportions of which came from the Moon and the asteroid belt but with 39 identified as Martian. In a dramatic development, one of them appeared to have fossilized worm-like life forms that had survived crashing through the Earth's atmosphere and had been preserved in the rock ever since.

Many people were, naturally, sceptical about the possibility of Martian life. Now, in the STONE 5 experiment, rock samples resembling Martian meteorites were embedded with microbiological life to see if and how they could survive re-entry. Three types of rock were tried: igneous basalt (lost), sedimentary dolomite (burned up) and simulated Martian regolith (basalt, carbonate and sulphate, which survived). The rock slices in STONE were just 1 cm thick and were placed on the hottest point of the Foton heat shield, the stagnation point. Thicker samples risked destabilizing re-entry. Panspermists pointed out that survival might have occurred if fatter rocks had been placed further away from the stagnation point.

The other experiment, appropriately called LICHENS, involved lichens (*Rhizocarpon geographicum*) taken from 2,000 m up in the high Sierra de Gredos mountains in Spain and installed on Biopan 5 to test their survivability. Despite their extreme dehydration in vacuum, the lichens recovered metabolic activity within 24 hours of their hydration in the laboratory after the flight. Although they had spent 15 days in orbit, the lichens made a recovery of more than 90 percent on their return to Earth, showing how they could survive in vacuum, extreme temperatures and solar and cosmic radiation. This presumably gave panspermists new heart. A summary of the results was published in the November 2005 *National Geographic Magazine* (Spanish edition, November 2005). Even expressed in the cautious language of scientists, the results were dramatic:

'After flight, the survival rate and the photosynthetic activity of the samples were found to be identical to the controls that had been kept on ground. As such, the LICHENS experiment demonstrated for the first time that complex multi-cellular macroscopic terrestrial organisms are able to cope with full exposure to the open space conditions. Most of the lichenized fungal and algal cells were able to survive after full exposure to solar UV, vacuum and cosmic radiation; the lichen upper cortex seemed to provide adequate protection against solar irradiation. After post-flight re-hydration, the metabolic activity of the lichens was restored within 24 h. These results would suggest that complex life forms, adapted to tolerate extreme conditions on a certain planet, could resist interplanetary transfer through space' [17].

The outcomes were published further afield in specialized magazines and journals, such as *Astrobiology* and in the popular press in Spain, the home of LICHENS lead investigators Rosa de la Torre and Leopoldo Sancho [18]. A press conference on the outcomes was held later in the Space Expo museum next to ESTEC, where even a theologian offered a commentary. LICHENS results were confirmed by related experiments. MARSTOX likewise tested the ability of extremophile spores and cells to survive in space in Martian soil analogues and some of them did indeed survive. ORGANICS 2 found that organic molecules of the type found in meteorites, comets and cosmic dust (polycyclic aromatic hydrocarbons), were quite stable in space [19].

Foton: Albert Marenny and four KNAs. ESA.

This research was further developed on Foton M-3 in 2007. Foton continued to take cargoes that might otherwise have flown on the Bion programme, such as cockroaches, snails, lizards, butterflies, gerbils and fish. The idea of cockroaches was put forward by students in Voronezh Medical Academy. The roaches had enough time to conceive during their 12 days in orbit and subsequently gave birth to healthy baby cockroaches on their return. There was an exotic experiment called Fotino developed by young European scientists to lower a 5 kg cabin on a 30 km tether to test a new re-entry technique (this seems to have worked, but the cabin was not found to prove it). The organization had become much smoother this time. For Foton M-3, in sharp contrast to Foton 7, ESA hired a planeload of scientists and engineers, with accompanying payloads, cool boxes and chemicals and flew directly from Rotterdam to Samara and then on to Baikonour.

Returning to the survival of life forms, the wanderings of NASA's rovers *Spirit* and *Opportunity* across Mars since 2004 had confirmed that Mars had previously been a warmer, wetter, watery world, with remnants of life likely to be found in sedimentary rocks. Thus far, the focus of tests had been on the black, basaltic type Martian meteorite rocks found in Antarctica, so had they been testing the wrong kind of rock? The STONE 6 experiment aboard Foton M-3 was an attempt to fly two samples of sedimentary rocks containing known bacteria within them through the 1,700°C heat of re-entry. One was a 370 million-year-old sample of mudstone from the Orkney Islands in Scotland and the other a 3.5 billion-year-old mud sample from Pilbara, Australia.

Foton M-3. ESA.

Both survived, although severely ablated into a whitish colour. The bacteria did not survive, but their 'Pompeified' remains could be discerned. The first lesson was that Antarctic scientists should now not only look for black basaltic rocks but also white-coloured sedimentary meteorites – more difficult to find in the snowy landscape. The second was that to survive entry through Earth's atmosphere, the wall of protection must be at least 2 cm thick. Foton M-3 also carried water bears (also known as tardigrades), which on Earth had developed the ability to survive desiccation. This was the first instance of animals, as distinct from lichens and bacteria, surviving in open space. Survival rates, including reproduction, were high, but fell dramatically for those also exposed to solar ultraviolet radiation: that was the killer, not the vacuum nor galactic radiation.

A feature of Foton was the broad range of partners who joined for individual experiments. For example, the LETVAR 2 experiment on radiation exposure on Foton M-2 brought in scientists from Austria, Germany and IBMP in Moscow, while PERMAFROST 2 was undertaken by a research team drawn from DLR in Germany and the Institute for Physical Chemical and Biological Problems in Soil Science of the Russian Academy of Sciences. For RADO 2 (radiation dosimetry on Foton M-2), the participants were IBMP in Moscow, the Atomic Institute in Vienna and the Atomic Energy Research Institute in Budapest. Bulgaria, world experts in radiation measurement, contributed an RD3-B3 spectrometer, enabling their sources to be characterized (e.g. outer belt electrons, inner belt protons, galactic cosmic rays) **[20]**. RD3-B3 (Radiation Risks Dosimeter Radiometer Dosimeter) was a small, light automatic device which measured not only cosmic particles but also solar irradiance over four ranges of wavelengths. The solar sensing part was built with the University of Erlangen in Germany, making it a Bulgarian-German experiment. It was introduced on Foton M-1 and also flew on Foton M-2, M-3 and Bion M-1, being subsequently developed as R3D-E on EXPOSE for the International Space Station (ISS).

ESA left the Foton programme after Foton M-3 and the series concluded with a Russian-only mission, Foton M-4. ESA left the Bion programme at the same time, largely because its science funding was concentrated on the ISS, once Europe's *Columbus* laboratory there became available from 2008. The movement of space science onto orbital stations was a feature of both the Russian and American space stations. In this case, the decision may have been mistaken, because the ISS did not offer the same opportunities or environments as Bion or Foton. Given the significant outcomes of both Bion and Foton, the scientists involved had good reason to be aggrieved. In the view of one critic, the ISS became 'a sponge that absorbed everything'.

Europe built up considerable experience of experiments aboard Bion and Foton and the mission descriptions given show the range of experiments conducted. They were especially important when American opportunities were limited after the two shuttle groundings and before *Columbus*, but Bion and Foton mission durations, facilities and profiles offered advantages in any case. Although the extremophile experiments were the most eye-catching, the others were very much the bread-and-butter of advanced biological and materials research, providing a unique opportunity for Europe and Russia to work together. The LICHENS and STONE experiments attracted worldwide interest, being published all over the world. ESA's ESTEC centre had never seen such a level of external interest in a biological experiment. As one put it, 'the person in the street nowadays knows that tardigrades can stay alive in open space'.

These missions provided opportunities for Russian scientists – like David Gilichinsky and Albert Marenny – to work with western colleagues. Albert Marenny was a radiation physicist who worked in the Research Centre for Spacecraft Radiation Safety. He wrote *The effects of heavy ions on biological objects*, among other things the outcome of the Bioblock experiments. David Gilichinsky (1948–2012), head of the Laboratory of Soil Cryology in Pushchino, was one of the world's great scientists. He was a pioneer in studying microorganisms in Siberia and Antarctica, with all that that meant for the survival of life forms in the space environment, who was published in *Astrobiology*. He most famously grew *Silene stenophylla* from a seed that had been stored away by a squirrel in Siberian permafrost some 30,000 years ago. For the Foton programme, he was a guest experimenter on Foton M-3. The two men were hardly known in the west, but should have been, given that Russia had so long been a pioneer in a field that dates back to Vladimir Vernadsky and Nikolai Vavilov. The Foton missions are summarized in Table 2.4.

Foton: David Gilichinsky. René Demets.

Table 2.4:
Foton cooperative missions

Date	Mission	Partners
16 Apr 1985	Cosmos 1645/Foton 1	
21 May 1986	Cosmos 1744/Foton 2	
24 Apr 1987	Cosmos 1841/Foton 3	
14 Apr 1988	Foton 4	
26 Apr 1989	Foton 5	CNES
11 Apr 1990	Foton 6	CNES
4 Oct 1991	Foton 7	CNES/ESA
8 Oct 1992	Foton 8	CNES/ESA
14 Jun 1994	Foton 9	CNES/ESA
15 Feb 1995	Foton 10 (lost after landing)	CNES/ESA
9 Oct 1997	Foton 11	CNES/ESA
7 Sep 1999	Foton 12	CNES/ESA
15 Oct 2002	Foton M-1 (lost at launch)	CNES/ESA
31 May 2005	Foton M-2	CNES/ESA
14 Sep 2007	Foton M-3	CNES/ESA
19 July 2014	Foton M-4	

Mars and Venus

As we have seen, space science cooperation in the 1980s extended into the European countries belonging to ESA, with Germany making its first significant entry to the field. Despite this, the Franco-Soviet axis remained the dominant one and the key focus was the long saga of the Venus balloon.

As noted earlier, France was expert in balloons, while their great promoter, Jacques Blamont, combined directorship of the CNRS Aeronomy Service (deputy director, 1958–61; director, 1961–85) with his senior role at CNES dating back to the original agreement. The balloon campaigns from Kiruna meant that there was now a shared Soviet-French knowledge base in ballooning. Blamont's ambition was to use balloons to explore the thick atmosphere of Venus and he had first suggested the idea to Soviet colleagues in late 1967 after the success of their Venera 4 mission. In the course of the early 1970s planning for cooperation on Venus missions, Mstislav Keldysh told Blamont that he was not interested in a fleet of little balloons, but – to use the Soviet terminology – one big Floating Aerostatic Station. Keldysh made Mikhail Marov, one of the great Soviet planetary scientists, his *pourparler* and Blamont believes that Marov was the inventor of the idea – and the name – of the Floating Aerostatic Station. The Soviet space programme tended to think large, rather than small.

Mikhail Marov.

The first proper discussions on the balloon took place in spring 1972 and at COSPAR held in Madrid that May, involving Blamont, Marov and Georgi Petrov, the director of IKI. The annual Franco-Soviet reunion took place in Tbilisi, Georgia from 18–27 September 1972. Marov startled the meeting by announcing at the outset that the USSR had accepted the French proposal for a French-built balloon with a gondola of 100 kg, of which 40 kg would be instrumentation, to float at 55 km at 100 m/sec for four to five days, with signals relayed through an orbiter. Mission studies would start immediately with a view to launch in 1981, with Roald Kremnev put in charge of the Russian-French technical working groups and liaison with the unidentified Chief Designer of Lunar and Planetary Probes – in practice, the head of the Lavochkin Design Bureau.

For Blamont, all his birthdays had come together. This would be a big project that would impress ordinary people in both countries and signify that the hatchet over Roseau was now buried. The balloon would float in the middle of the cloud layer, which Venera 8 reported was between 49 km and 65 km. The design was challenging: the French were used to balloons assembled on the ground which then ascended, but this one would descend from above. The Soviet atmospheric entry probe would brake and open a large parachute, then taps would open to inflate the 9.5 m diameter, 275 m^3 balloon with 25 kg of helium and it would climb to its operating height. Blamont insisted that the balloon had to be isentropic, able to respond to vertical up-draughts and down-draughts and adjust its inflation. The balloon division at CNES now set about constructing an envelope that could withstand the temperature of Venus (<90°C) and its sulphuric acid composition, choosing a 100 micron sandwich of reflective aluminium and Teflon respectively. The Soviet orbiter, the relay, would circle the planet at 1,000–60,000 km, inclination 140°, period 24 hours and would carry a payload of 100 kg of instruments. The cost to France of the mission was FFR750m a year but once set down as an international agreement, Blamont observed that the Department of Finance could not touch it. The French calculated that the cost on the Soviet side was around €120 m. The project acquired a name, Éos.

The French convened a colloquium of 90 scientists from 25 institutes in the castle of Gif-sur-Yvette in January 1974, with the instruments agreed at a second meeting of 30 French and 53 Russian scientists in Moscow that April. After the sudden death of President Pompidou, in the following month Valéry Giscard d'Estaing was elected President. Previously a minister for finance, he had ordered an end to what he called the 'silliness' of the Ariane rocket and had refused to sign the agreement to set up ESA, which would have meant spending more money. Spaceflight would be left to 'Les Grands' (the two superpowers), CNES would have a science-only role and France would supply small instruments to the big powers. Michael Debré, prime minister when CNES was founded and a spaceflight supporter, was now out of government, but one of his associates, Jacques Darmon, became *chef de cabinet* to the new minister for finance, Michel d'Ornano, who got his post because he had been the new president's campaign director but was otherwise at sea with his many dossiers. Darmon organized the pushback and in October gave the new president a weekend reading dossier, after which d'Estaing emerged to announce that Europe had to do space (ESA was saved) and France had to do Ariane, but Diamant was cancelled (the USSR flew its satellite, Signe 3, instead) and the budget line for the Venus balloon, Éos was reduced to almost nothing. The French space budget would flatline for the rest of his presidency, but most of it would go to Ariane and the ESA subscription.

Éos took the full force of the hurricane. When CNES President Maurice Lévy went to the Kiev annual reunion in October 1974, he announced that France could no longer fund the project. He did get approval from the new president to float the idea of a French spaceman, for which the cost would be modest. For the Russians, this must have been another bruising experience for them, the second in six years. At the next reunion, at La Grande-Motte in September 1975, the French offered to assist the Russians in any low-cost ways that they could on their Venus missions – including balloons – through studying, developing and providing the envelope and some instruments.

Many years later, Blamont attributed the loss of Éos to the biases of the ministry of finance. Its minister once told previous CNES president Jean-François Denisse over dinner that CNES had 'interesting' projects… but not the one to send a balloon to Venus'. Blamont had a letter hand-delivered to the new president asking him to reconsider, but with no joy. Maurice Lévy found his budgets clipped by the cabinet every week and when CNES staff protested, both he and the director general of CNES were dismissed. French space efforts in the 1970s were focussed on building Ariane and the applications satellites it would launch, which absorbed the flatlined budget and left nothing over for scientific missions. The situation was comparable to the United States, where the shuttle absorbed so much of the space programme that space science there also suffered, with the Americans unable even to afford a mission to intercept comet Halley, then approaching Earth. Despite the demise of Éos, the Soviet Union had continued to fly French instruments and even satellites, doing so free of charge.

The international picture was difficult at this point. In the United States, the Space Science Board made proposals – called the Purple Book – for a renewed American interest in Venus, a project that became Pioneer Venus. Blamont unsuccessfully re-proposed his balloon idea there, but instead became principal investigator for its nephelometer experiment. NASA invited Europe to participate in Pioneer Venus, only to then revoke the invitation, though European scientists could still participate as individuals (three did so, including Blamont). All this took place against the background of a long European-American dance about funding and building the Space Shuttle. Despite promises of an important role, Europe instead ended up with an uncertain one in constructing Spacelab, a relationship sufficiently unequal for Blamont to call it a 'policy of crumbs'.

Éos was dead, to the point that even talking about it was not a good idea. The challenge for its protagonists was to try resurrect something whilst ensuring that it did not look like Éos. For the formal conclusions to the Leningrad reunion in 1976, the French inserted a paragraph somewhere in the middle referring to future participation in Venera (the Russian name for the Venus programme). In spring 1977, France officially notified the Russians that CNES would be a participant in a Soviet-directed mission. This was not a bilateral programme, which required government approval, but an agency contribution to another agency's projects, which did not. As a result, the politicians were bypassed and CNES Toulouse set up a project group for what became known as Venera 84 but was really the new Éos. Expected to launch in 1983, it could be promoted as a commemoration of the first ever balloon flight, from Paris, 200 years earlier. The CNES budget allocated was modest, at FFR160m out of its total budget of FFR2.662bn. Thirty engineers were assigned to the project full-time in Toulouse, rising to 50 from 1982. Blamont kept pitching his project to anyone who would listen, with it even being picked up by the anglophone press [21].

Venera 84 was led by a French working group under Gérard Rivière, with the Russian one of 100 engineers under Roald Kremnev. There was a scientific committee presided over by Jacques Blamont at the French site and Mikahil Marov in Russia. At this stage the new director of IKI, Roald Sagdeev, entered the picture. He brought the disadvantage that he was not interested in Venus – he was a plasma man, not a planetologist – but the advantage that he favoured international cooperation and obtained permits for his leading scientists such as Vasili Moroz and Vladimir Kurt to travel abroad.

IKI.

Most of these Venera technical meetings took place at the wood-panelled IKI. The French had to argue for any technical information, which was then given by someone whose role was unknown to them. The French sat in deep armchairs with low tables in the meeting room, often with people who said nothing and whose identity was unknown. They never dealt directly with the place where the spacecraft was actually made, the *zavod* (factory), which was forbidden to foreigners and they had to deal with intermediaries. Vladlen Vereschetin once told Blamont that even if there were a revolution, they would still never, ever see the inside of a *zavod*. The French never saw designs, nor written documentation and they cheekily asked if the Russians had no paper on which to print them. Although they thought that the way in which they were kept at arm's length was bizarre, they realized that the same rules applied to their Russian scientific colleagues. Despite there being 2,000 scientists, engineers and technicians working in IKI, there was only one photocopier – a French one – with the key kept in the office of the director. The French counter-proposed the integration of the instruments in IKI. Although the French had their own difficulty in getting projects past their government, they were aware that IKI also had that problem and were required to deal with an interministerial space commission which had to approve mission outlines, projects, budgets (e.g. the launchers allocated) and quality control.

One scientist who spent years on collaborative programmes commented on how little technical documentation they received, which was strictly limited to the experiments involved. All the information was there, 'but stored in the brains of the technicians involved'. The assembly of the rocket, cabin and experiments was done 'without manuals and checklists – or at least, they were not used on the shop floor'.

By the time of the 1979 reunion at Ajaccio on Corsica, Venera 84 had evolved to the point of agreeing the instruments for the balloon (28 kg) and orbiter (80 kg). Everything was going fine. CNES had made the sandwich and tested the inflation of the balloon in Venus-like conditions. News reached the reunion that Brezhnev had died and many dinnertime toasts were made to honour his passing, before they found out that this was merely a rumour that was untrue, something we would now call fake news. Then Blamont made what he later called the greatest blunder of his life. With the United States having abandoned its mission to comet Halley, he suggested over pre-dinner drinks that Venera 84 could observe the comet from its vantage point of orbiting Venus, even if 40 million km distant. Not particularly interested in Venus itself, Sagdeev thought about this on the way back to Moscow. His colleague Vladimir Kurt suggested that, rather than orbiting Venus, the mother craft delivering a balloon could instead fly on to intercept the comet. Unfortunately, because fuel and payloads were limited, it was hardly possible to do both, so it quickly became an unfortunate choice of balloon *or* comet.

Unknown to the Europeans, Sagdeev had to get the change of course past the Military Industrial Commission. His winning argument was that the Venus Halley mission would be a demonstration of the Soviet Union's ability to deliver multiple independently-targeted warheads across the solar system, something so spectacular as to deter the Americans on Earth. Afterwards, one of the military experts voiced his scepticism, showing that he had not really believed Sagdeev, but had kept quiet during the meeting when it mattered.

Having succeeded, during a routine progress review meeting in Moscow on 24 January 1980, Sagdeev and Kremnev made a theatrical proposal to revise the mission with a comet interception. To avoid having to modify the existing mission agreement and upsetting the French, the Soviet side presented the comet interception as a Soviet-only venture, but this only annoyed the French even more. At COSPAR in Budapest in June 1980, the internationally-minded Sagdeev invited not only France to contribute instruments but also Germany and the socialist block, turning it into a bigger international enterprise than anything NASA had ever contemplated. There was an inconclusive meeting in Moscow straight after COSPAR. Valeri Barzukov, director of the Vernadsky Institute of Geochemistry and one who had enviable political access (he was vice-president of the scientific and technical department of the central committee of the Communist Party of the Soviet Union (CPSU)), declared the balloon to be the enemy of Soviet science. Mikhail Marov set up a peace-making dinner for Sagdeev, Barzukov and Blamont, but it ended in deadlock. Marov held out for the balloon, but he had lost his protector, Keldysh, who had died in 1978. This phase coincided with preparations for the next five-year plan, with a warning from government to keep costs down. In July, meteorologist Yuri Marchuk of the Academy of Sciences was brought in to mediate and told Sagdeev and Petrov to come up with a compromise in two weeks. On 10 August, Sagdeev tracked down Blamont to his holiday location in Saint-Martin-d'Uscladelles, where he told him that the outcome was that the two mother ships would drop landers and continue on to Comet Halley. Instead of a big aerostat, there would be three small balloons each of 3 m diameter. That was it. The telex to CNES President Hubert Curien that followed invited the signing of a replacement agreement at the next reunion in Baku in October. On 13 August, the Russians were shocked when CNES Director General Yves Sillard sent a telex that simply said, '*Non!*'

It is possible that CNES thought it was in a negotiating position here and did not realize either the finality of the compromise on the Soviet side, nor the dangers of opening up a chasm between the two parties. The attention of CNES at the time was very much focussed on preparing for and funding the first piloted French flight to Salyut, so it may not have given the balloon project the consideration it deserved. The French Venera 84 scientists travelled to the next reunion at a hot, waterless Baku that autumn, overseen by over-watchful local police. Boris Petrov had died at the end of August, another ally gone, but Sagdeev and Kremnev tried to salvage something for the French by suggesting at least their involvement in tracking the balloons. Blamont persuaded CNES to delay a formal response – the earlier telex notwithstanding – until February. In January, Blamont put together a counter-proposal for France to supply the three smaller 3 m balloons, as well as a

5 kg gondola with instruments to measure temperature, pressure and light, a wind meter and a battery and transmitter. The cost would be FFR120m, a negligible adjustment to the budget allocated.

Blamont and his colleagues made their presentation to CNES management in Toulouse on 27 January 1981, but Yves Sillard simply responded by saying 'We could do it, but we won't do it'. He never gave a reason, but Blamont explained this as Sillard's lack of interest in space science and his exclusive focus on Ariane and application satellites (Earth observations, telecommunications). None of the CNES managers stood up to object. The budget line for Venera 84 was taken out and applied to the funding for human spaceflight − although a little was left for instruments to be supplied for the Venera mother ship − and the team was dispersed. Its only legacy was the sandwich design developed by a company in Lyon, one which it subsequently applied to sails for yachts. Sillard appeared to think that without the French, the USSR would abandon the mission, but that merely showed how little he knew them. Hubert Curien explained the French decision to Roald Sagdeev in person in Moscow on 2 February. Sagdeev never really forgave him and later referred to the project having been a success *despite* CNES. He invited Blamont to advise him on the now-Soviet balloons under a contract between IKI and the aeronomy service, which bypassed CNES. Of course Curien found out, but gave his blessing as long as it did not incur any costs to CNES. Blamont came up with the FFR1.5m from his own laboratory funding.

VEGA mission. CNES.

Sagdeev feared that the French withdrawal would sink the project, but his masterstroke was to appeal over their heads to the international community for participation. To his astonishment, no less than nine countries volunteered, which made the whole project both more viable but also more complicated. This included some American scientists who appeared to be unaware that there was a cold war on. The University of Chicago's John Simpson proposed a dust detection meter which he had hoped to fly on the American comet probe (cancelled) and then Europe's Giotto (too late). Simpson was subjected to menacing interrogation about giving secrets to the Russians, which he countered by promising to use components from the local Radio Shack store that, if the Russians copied them, 'would set them back years' (in fact, the instrument may not have been a significant scientific addition). Once news got into the American press, there was a furious governmental reaction, with President Ronald Reagan's advisor, Richard Perle, also called 'the prince of darkness', accusing NASA of undermining foreign policy.

Jacques Blamont soon found himself working at IKI with meteorologist Vyacheslav Linkin, one of the foremost scientist engineers of his day and Viktor Kerzhanovich, the man who discovered the super-rotation of the Venus atmosphere and who was responsible for radio-electric systems and tracking. Their first decision was to reduce the balloons from three to one for each spacecraft, which was simpler. The Lavochkin design bureau went ahead with developing the balloons but found the task more difficult than expected, largely due to their inexperience. Instead of a sandwich, the bureau used Teflon hand-painted with liquid Teflon, which actually worked well. Blamont had no direct contact with the bureau and never even found out who the people were. The Soviet balloons had a mass of 25 kg floating, the gondola or nacelle being 6.5 kg. The tiny payload mass meant a small transmitter capable of only 4 bytes/second for every 210 seconds of each 330-second transmission, with interrogation of the instruments every 75 seconds. It would transmit on 1,667 GHz, the bandwidth for radio-astronomers already used by 20 radio observatories.

Russia still hoped that France could help with tracking the balloons, because NASA had a worldwide network. The Russians had Yevpatoria (70m dish), Ussurisk, Baikal, Simeis (26m), Puchino (22m) and Bears' Lake (64m), but they were no good when their side of Earth was turned away from Venus. NASA had no formal arrangement to work with the USSR – even discussing its possibility during the Reagan administration could be considered treasonable – but there were agreements in place for NASA to work with CNES and CNES had agreements to work with Interkosmos, so that is how it operated. Blamont helped to draw in the stations that cooperated with NASA: Canberra; South Africa (officially, the USSR did not acknowledge its participation); Arecibo (Puerto Rico, 213 m), even little São Paolo (14 m). Such participation was very much offline, was nearly halted by Richard Perle's intervention and the protocols were only signed after the mission was over. NASA officials met with their Soviet counterparts in Moscow in February 1985 for the first time, their awkward unfamiliarity with their surroundings plainly evident to their European colleagues (by contrast, CNES already had a permanent hotline with IKI). NASA even put in a financial commitment of $4 m, 30 times more than CNES.

On his side, to oversee the mission science, Sagdeev created the MNTK or, in English, the International Scientific and Technical Committee. This was rather like the NASA Science Steering Groups, but with bureaucrats, representatives of each participant country and the unidentified figures of the design bureaux meeting every three months. This was a real brick out of the wall for the old Soviet way of doing things and although it actually worked well, Interkosmos found it hard work and there was no MNTK for the later Phobos mission, likely a mistake.

Overshadowed by the balloon episode, France remained active in the rest of the VEGA mission. France would fly eight experiments (four on the lander, four on the bus) and two TV cameras (narrow and wide angle), totalling 50 kg in all. The lander experiments were a mass spectrometer, an aerosol collector, an ultraviolet spectrometer and temperature/pressure gauges. On the orbiter, the experiments were the imaging system, an infrared spectrometer, a three-channel spectrometer and a plasma wave analyser. For the comet interception, IKI set down the specifications for the telescope that France would make to track the comet, whilst IKI would provide the CCD camera and detector and the Academy of Sciences in Budapest the calculator – what we would now call the computer. The Americans were astonished to learn that the USSR already had CCD technology, provoking a find-the-spy hunt, with the French ambassador in Washington DC called in to explain how his government could have supplied it to the Soviet Union. In reality, the USSR had developed its own CCD technology years earlier and used it on military reconnaissance satellites, but did not want that known. Likewise, the Americans did not believe that the USSR could have synthetic imaging radar until they saw the radar maps of Venus made by Venera 15 and 16.

Once the French had supplied the telescope, a company in Leningrad – possibly the Arsenal design bureau – made its own version, which IKI announced would replace the French model. Following French protests, a compromise of one French and one Russian telescope was agreed and both worked equally well. The Czechs were due to supply the all-important scan platform but were behind schedule, despite additional funding provided by their government on the orders of its general secretary. Again, the Russians announced that the platform would be replaced by a parallel version which they had prepared. As with the French, the Czechs took great offence, but the MNTK later learned that the Soviet version had failed its vibration test, so the Czechs got to fly after all. Czechoslovakia and Hungary were especially important, as the only countries to supply subsystems, with the others supplying instruments. The original payload was 125 kg, but ended up at 240 kg. The USSR contributed three instruments for the study of the gases, cometary particles and the solar wind, while other instruments came from Germany (Heidelberg), Hungary, France, Austria, Poland, Czechoslovakia, ESA and the US. According to Blamont, the IKI side had a hard job keeping a project with so many international partners on schedule. The main axis on the European side was a young engineer, Josette Runavot, who dealt on the Russian side with Anatoli Perminov, someone they later learned was *pourparler* for Vyacheslav Kovtunenko, a missile and satellite designer who worked for OKB Yangel. The Europeans found the Soviet labyrinth difficult to understand because the power relationships were never clear, with key figures whose role was rarely explained (nor sometimes even their name divulged) moving in and out of meetings. In the end, the Europeans depended on the goodwill of individual IKI officials to sort things out.

VEGA.

Integration continued to be a problem area. Even the representatives of the socialist countries had yet to darken the doors of a *zavod*. For the Mars 1973 mission, the Russians came up with a solution and set aside a separate integration area. To their horror, the Europeans now saw the instruments they had meticulously prepared in clean room conditions being hand-carried and manually bolted on in an ordinary room. The Russians then came up with a new solution: integration not in Moscow but at Baikonour. The precedent had already been established by the army of French journalists let into the preparation halls for the first piloted flight in 1982. Now, the *zavod* sent the spacecraft to Baikonour where the integration teams worked round the clock to get them ready, hoping that there would be no show-stoppers or problems that could not be fixed at such a remote location. At least they could go there now.

The greatest European-Soviet venture was now about to take place. Despite the difficult French-Russian history over VEGA, top French officials were invited to watch the December 1984 launch from mission control in Moscow, including the new president of CNES, Jacques-Louis Lyons and his predecessor, now minister for research, Herbert Curien. They were put up in the pre-revolutionary National Hotel on the edge of Red Square and brought to Kaliningrad, now Korolev, in a convoy of six black cars, with the route blocked off by police. In mission control were the members of the MNTK, unknown generals in military and civilian dress, Anatoli Alexandrov (President of the Academy of Sciences), the state commission for science and technology, the military-industrial commission and nine government ministers including the head of the Ministry for Machine Building. The image of the Proton on the pad came on an hour before launch, televised by a camera 1,500 m from the rocket which almost gave them the impression of being there. They could follow its nitric acid-fuelled launch for 43 seconds before the Proton disappeared into cloud. Signals poured in from downrange at Tomsk, Tobolsk, Ulan Ude, Ussurisk and a tracking ship in the Sea of Japan and, at 26 minutes, its orbital parameters appeared on screen. They were not perfect, so fresh commands were issued by tracking ships off Argentina and Dakar to re-set the time for the trans-Venus burn. At mission control, all the signals were reported in real time: ignition, blok D separation, deployment of solar panels and then the instruments. In less than two hours, it was on its way. The launch marked the first public release of photographs of a Proton launch. The second VEGA followed.

VEGA balloon, but a Soviet one

VEGA balloon instruments.

Arrival at Venus took place in June 1985. The first word of signals from the VEGA balloons came from NASA's Jet Propulsion Laboratory (JPL) in California, whose dish in Australia was the first to pick up the transmission – a function of the rotation of the two planets. Applause erupted every time signals arrived at a new tracking station, as they followed the little balloons around the planet. They received 46 hours of data before the batteries ran out, all from a tiny 6.5 kg payload 100 million km distant. The signal from the second VEGA arrived at the exact second hoped for. For the French balloonists, it was sweet justice to see almost 20 years of work rewarded with such an effortlessly successful outcome. France's friends at NASA had helped, but so fragile were American-Russian relations that American participation was unpublicized and even a celebration over a Coca-Cola for each member of the team was prohibited. The Russians handed over all the magnetic tapes from the balloons and the French carried them back to Toulouse.

In Moscow in July, President Mikhail Gorbachev had held a Kremlin conference on Soviet science, with the balloon success at Venus setting such an upbeat tone for the event that four IKI scientists were at once awarded state prizes. The American, Russian and French scientists convened in Toulouse in August to study the data, the flags of each country flying at the entrance of the Centre Spatial. They found that the balloons had each travelled more than 12,000 km, passing through day and night, the highlight being the discovery of violent vertical wind draughts, especially over Aphrodite. The French had fitted a tiny lightning detector to the instrument package, although in the event none was detected.

Overlooked in all this was the success of the landers. The VEGA landers carried ISAV, developed by Jean-Loup Bertaux with Vasili Moroz, which measured SO_2 from 60 km right down to the surface, important in assessments of the level of past or present vulcanism on the planet. A later version of this instrument flew to Titan on Huygens.

Within days of the June success in Moscow, ESA decamped to Kourou, Guyana, for Europe's first deep space mission. The early years of ESA had been dominated by building the Ariane launcher, Spacelab and applications programmes, with space science a lesser priority. This changed with the arrival of Ernst Trendelenburg, from a great Berlin family whose grandfather invented the hospital bed on wheels and whose father was a director of Siemens. Young Ernst served the Wehrmacht in Paris during the occupation, going on to Westinghouse, ESRO and then ESA. He came from ESRO to be head of scientific and meteorological programmes when ESA was formed and was then Director of Science until 1983. Like the French, who voiced the concept, but unlike the Germans who did not, Trendelenburg believed that Europe should be an assertive political power, one able to stand up to the USSR and USA. He championed the idea of a European mission to comet Halley, which subsequently became the basis of a structured European scientific programme. Or, rather, the idea came from his ageing mother who had actually seen the comet in 1910. If she could remember it, he figured, then a mission there would surely have broad appeal.

Trendelenburg spoke to Padua professor Giuseppe Colombo, who had made calculations showing that a small spacecraft could take a direct route to intercept Halley, proposing that the mission be named after Giotto in honour of his drawing in the Scrovegni chapel in Padua of the comet's visit in 1301. After a detour of unsuccessful negotiations with the Americans, ESA approved the project in 1980, allocating an Ariane for Europe's first interplanetary mission. The Americans, displeased with Europe's wanton display of independence, pulled out of the negotiations for the Jupiter-Sun joint mission (it later flew as Ulysses). Germany got the five most important experiments on the Giotto mission. Sagdeev invited Trendelenburg to Moscow at a time when there was no formal agreement between ESA and Interkosmos. He arranged an evening for Trendelenburg with a Russian general where they reminisced, over vodka, about World War II panzer battles. Trendelenburg moved on in 1982, but Germany's involvement in cooperation with Russia was to grow from his time onward. CNES now hired a Concorde to fly VIPs out to Guyana to witness the Giotto launch on Ariane. While there, Curien apologized to Blamont for having pulled France out of the balloon project earlier in Baku.

For the interception of the comet itself in March 1986, Roald Sagdeev created a media event contrasting glaringly with anything that had gone before. Hundreds of foreign journalists were invited and were let loose — without chaperones — in the control centres, institutions and laboratories that many Soviet scientists had themselves never seen. By contrast, JPL kept to strict rules, giving the initial data to scientists to study in private for a few hours before the first press conferences were organized. In Moscow, on the other hand, raw data were released as they came in, on screen, to scientists, media and anyone else who happened to be there. Journalists were publishing the results immediately, while some unscrupulous scientists turned up on the spot to claim authorship of scientific papers for a mission in which they had no previous involvement. Sagdeev presided over this free-for-all, believing that it would be a glorious moment for Soviet space science, which it was. In the afterglow, John Simpson even got a medal. The photographs of the comet interception were indeed fabulous. The American Broadcasting Corporation got live broadcasting rights, with commentary by an ecstatic Carl Sagan. Sagdeev made a blatant pitch for cooperation with the Americans, leaving the Europeans in the shade, but this came to nothing, at least during the Soviet period. At this point, there was much talk of future European-Russian collaboration, but this also came to nothing, not least because Europe had so little money for its science programme. Despite official reservations about the dangers of live broadcasts, the success of the mission and the resulting worldwide admiration was the final rebuke to the old secrecy [22].

The press now decamped to Darmstadt for the Giotto interception, the final act of the drama. In Paris, a huge FFR12m celebration was organized for 7,000 in the Cité de Sciences et de l'Industrie erected over the old abattoirs of La Vilette, relaying the signals from Darmstadt. Not to be outdone by the media spectacular the previous summer in Moscow, orchestras played on several floors and drink and food were in abundance, with the guests guided by young women adorned with feathers to symbolize comet tails. In Darmstadt were Sagdeev, Kremnev, Barzukov, Moroz, the head of NASA James Beggs and the top European officials such as Reimar Lüst, head of ESA and Ernst Trendelenburg, ESA's former director of planetary science. Unfortunately for ESA, a dust particle destroyed the camera 10 seconds before the point of closest interception, deflating its hopes for a visual spectacular that would exceed VEGA.

For European-Russian cooperation, the interception of Comet Halley was the second triumph in a year. France was responsible for the camera system which took the spectacular images of the comet, published in false-colour reds, yellows and greens to show its shape. The infrared spectrometer, IKS, was a French responsibility and it was able to detect carbon dioxide and abundant organic compounds in the nucleus, as well as measure its temperature (350 K) and albedo (3%). The French APV-V instrument measured the cometary plasma and electromagnetic instabilities, while the TKS ultraviolet instrument detected radicals ($C2$, $C3$, OH and CN). PUMA measured the impact of grains from the comet and their speed (78 km/sec), composition (mono-atomic), chemistry (C, H, O, N) and carbon level (3%). The comet was determined to be a dark, rotating, irregular object, 8 km by 15 km and 70°C, emitting dust, carbon, HCN and water.

A week later, President Gorbachev invited six prominent scientists, led by Sagdeev, to discuss the mission over tea, agreeing the importance of future international cooperation. Sagdeev was awarded Hero of the Soviet Union and Linkin the Lenin prize, while medals and honours descended on their colleagues. The French delegation returned with their luggage stuffed with nine-channel magnetic tapes. These were then run on perforated card on the computers in Toulouse, each taking a half hour. The first results were shown on slides to a working group meeting only two weeks later, such a turnaround being considered exceptional at that time.

The success of VEGA created a great sense of optimism for the future of planetary exploration, reflected in the annual reunion at Trouville in October 1987 which discussed not only the upcoming Phobos mission, but also future plans for the exploration of Mars and the asteroids. A feature of the Venus missions was that they were accomplished over a period of about five years, from inception to publication of the scientific results, with no delays and clockwork efficiency. This proved to be much shorter than subsequent missions, which took many times longer. There was a sea change at ESA, with space science having been overlooked because of the heavy financial commitments to Ariane, Spacelab and applications. Now there was the prospect of a coherent scientific programme, even if, as Blamont put it, they had to survive a 'Russian roulette cascade of committees and working groups'. In France, too, prospects rose. From 1979, the CNES budget began to recover after many arid years, rapidly so with the arrival of François Mitterrand as President in 1981. Modernization was a motif of the new regime, while Hubert Curien became Minister for Research in July 1984 in the government of Prime Minister Laurent Fabius. The USSR remained the leading country in space science, with the American interplanetary programme experiencing lean times until the shuttle could be paid for. Moreover, the Americans seemed uninterested in structured international cooperation.

Phobos

The Soviet Union now felt it could return to Mars, which it had left alone since 1974. The USSR announced its intention of preparing an even more ambitious project to send two probes to land on the little Martian moon Phobos at the Grande Commission in 1984. Despite the difficulties with VEGA, IKI had invited France on board as early as December 1984, even before its official approval. In fact, Sagdeev enlisted the French in trying to persuade his government of the merits of the mission. The lengthy meeting of the state commission to approve the mission actually took place in the hours that followed the VEGA launch, but finished in time for a doubly happy post-launch celebratory dinner. Twelve countries joined, making the foreign contribution worth R60m of the R272m project, the value of the French contribution being about €5.3 m. France was the leading European contributor, with the ISM (infrared spectrometer), IAS (mineral composition of the surface), AUGUSTE (structure of the atmosphere), DION (Phobos surface composition), LIMA D (ions), FREGAT (mapping Phobos), STENOPEE (libration of Phobos), IPHIR (solar pulses), LILAS/APEX (gamma bursts), PWS (ions) and a micro-camera. LILAS/APEX were successors to Signe 2MS, developed by EP Mazets of the Institute of Physics and Technology in Leningrad and built by CESR in France. Phobos was a high point of European-Russian cooperation, involving 20 countries including the socialist block states (e.g. GDR, Bulgaria, Czechoslovakia). There were multi-national teams for individual instruments, such as the LIMA D laser (USSR, Austria, Bulgaria, Czechoslovakia, Finland, GDR, Germany), the DION ion spectrometer (Austria, Finland, France, USSR), the magnetometer (Austria, GDR), the wave plasma analyser (ESA, Czechoslovakia, France, USSR) and the ASPERA spectrometer (Finland, Switzerland, USSR).

Phobos.

Both probes were launched in July 1988. Sadly, Phobos 1 was lost the following month (a ground controller accidentally sent up a command to close the probe down) and in March 1989 Phobos 2 failed in the final stages of closing in on the Martian moon, though not before sending back photographs as it neared. CNES and Interkosmos reviewed the results of the Phobos missions in Paris on 23–27 October 1989 and whilst this might have appeared to be an attempt to salvage something from a mission that was perceived not to have ended well, in fact there was a strong scientific outcome which it was important to capture. Starting with cruise studies, APEX captured neutrons at 2.23 MeV while IPHIR got 160 days of study of solar oscillations every five minutes and LILAS and APEX obtained 200 candidates for cosmic gamma bursts. AUGUSTE obtained numerous spectra and profiles taken during occultations by the Sun and found both aerosols in the Martian atmosphere and an ozone layer between 40 km and 50 km. As for the surface of Mars, ISM obtained 4,000 images in near infrared, covering 15 percent of the surface between 30°N and 30°S and showing variations in hydration and water molecules in the Martian rocks. Vallis Marineris was rich in hydrated minerals, while older and darker terrains were rich in calcium. ISM images of Phobos suggested that its dark surface was little hydrated, implying that it was a captured carbon asteroid but also that it was quite heterogeneous. FREGAT got 37 images of the moon Phobos, the most detailed taken to that point. Ondes and Plasma measured the shock wave when the solar wind met the Martian electromagnetic environment.

Mars 8 and Phobos SR

Despite the disappointing outcome to the Phobos mission, the Soviet Union and later Russia invited international cooperation on its next Mars probes. French participation was formally invited at the annual cooperation meeting at Yerevan in October 1986 and settled at the reunion in Saint Jean de Luz in October 1988. The Mars 94, Mars 96 and Mars 98 missions that were subsequently designed went through numerous iterations for an orbiter, rover and balloon (a French balloon seemed to be an idea that would not go away). In the event, Mars 94 was delayed and became Mars 96, was then renamed Mars 8 and was the only one launched. The Yerevan meeting ultimately led to an agreement in September 1991 for France to provide a two-camera navigation system and steering algorithms for a six-wheel 75 kg Russian rover, Marsokhod, to be run by Groupement pour les Essais en robotique Mobile Spatiale (GEROMS) which set up its own test site. From 5–16 December 1994, a Marsokhod built by the Lavochkin design bureau in Moscow was tested there. In the event, Marsokhod was a victim of some of the many re-designs.

Marsokhod.

France was still not willing to give up on the balloon idea and formed a study team in spring 1988. At the 1988 cooperation meeting, Jacques Blamont proposed a balloon of 30 kg with a gondola payload of 8 kg. The balloon would rise at 9am local time each day to as high as 6,000 m and descend as the air cooled in the evening and the balloon deflated, to be anchored for the night. It would drift hundreds of kilometres each day, or about 2,000 km over ten days, taking high-definition photographs with a resolution of 10 cm, so it was an ambitious experiment. The balloon would be 21 m tall, 5,500 m^3 in volume and the envelope 6 microns thick, with a snake-like cable to anchor the balloon by night. The scientific package outlined consisted of three cameras, an infrared spectrometer, a gamma spectrometer, an x-ray fluorescent spectrometer, a weather station, a thermal analyser, a magnetometer and a radar sounder radar. The Russians also talked to the American Planetary Society about a balloon, which may not have pleased the French [23].

The first warning of money problems came at a meeting in Moscow in January 1991. The dire financial trouble that subsequently engulfed the Russian space programme was well in evidence at the end of the Soviet period. In April 1991, it was decided not to launch two spacecraft in 1994, but to divide them between October

1994 (two small stations and two penetrators) and November 1996 (balloon and rover). In effect, the Russians pushed back the complex balloon and rover parts of the project so that they could at least get something airborne, which would now be a single spacecraft for the next launch window, Mars 94. For the moment, the overall atmosphere was still optimistic, but during the April 1994 reunion in Toulouse, both missions were put back a further two years to 1996 and 1998 respectively. The first, though, was almost ready and IKI asked for delivery of the French instruments so that they could now be fitted. These were EVRIS (stellar oscillations), LILAS 2 (solar and cosmic gamma bursts) and RADIUS MD (solar x-rays and then x-ray radiation around Mars) for the coast phase, plus SPICAM (atmosphere), OMEGA (atmospheric composition) and ELISMA, DYMIO and FONEMA (ionosphere and plasma) for after Mars arrival. ELISMA combined institutes in France, the Netherlands, Moscow, Sofia, Warsaw, Lvov and Britain to study natural and low-frequency waves of thermal plasma. The camera system was an interesting combination of the two Germanies: a narrow-lens camera originally designed in the GDR with a high-resolution one from the federal republic.

Even in its simplified form, the 1996 Mars 8 mission was the most ambitious ever sent there: a six-tonne probe carrying two landers, two penetrators, an orbiter and equipment from 20 countries. In fact, it was the largest, heaviest Mars probe ever built. On the 40 kg small landers, the magnetometer was built by France, the meteorology sensors by Finland and the alpha proton x-ray spectrometer by Germany. Étude de la Variabilité et de la Rotation des Intériers Stellaires (EVRIS) was a pioneering instrument to measure vibrations, periodic movements and variability in stars. Its 9 cm-diameter telescope with a high-precision pointing star sensor could point toward 15 brilliant stars for 20 days at a time. The penetrator's data processing and communications control system was developed by the University of Kent in Britain. The key western investments were by France in balloon development (€91 m) and by Germany in the imaging system, the High Resolution Stereo Camera and the Wide Angle Opto-electronic Stereo Scanner, able to take in 500 km at a time at perigee and two thirds of Mars at apogee (€29 m). At one stage, the Russians tried to save money by scrapping the ARGUS scan platform in favour of a body-mounted camera system, but the Germans stepped in and insisted that the cameras would only be of real value if they could be pointed.

As preparations for launch got under way, it became increasingly apparent that this might be the last Russian Mars mission for some time. French patience with its balloon idea being repeatedly bumped ran out in October 1995, so it withdrew from the Mars 98 project. The official reason was that 1999 was forecast to be a stormy one in the Martian atmosphere and not propitious for the French cameras and radars. Even though the French thought they could adapt the experiments to harden them, there was not enough time to do so, but the writing may already have been on the wall for the project by then in any case.

Right up to the end, there was doubt whether it would even be possible to get Mars 8 airborne. Baikonour was hit by electricity outages because the wages of the power workers were so far in arrears. The integration of Mars 8 with its launcher was completed in the hangar by candlelight, with the workers keeping warm using gas cylinder stoves.

Many of the international collaborators were in Baikonour for the launch on 16 November 1996. The Proton rocket lit up the cold night sky, pushing its cargo into Earth parking orbit for the trans-Mars injection. On the ground, the vodka corks popped open to celebrate the successful launch. They should have waited. Over the south Atlantic, the upper stage of the Proton, the blok D, should have fired the probe out of Earth orbit on a curving trajectory to Mars. In what we can only presume to be a random failure, the blok D failed to fire. Mars 8 thought it had done so, separated from the blok D and fired its engine to give it what would have been a final trans-Mars kick. Instead, the manoeuvre pushed it into an elliptical orbit of Earth and the probe crashed into the Andes two days later. Blok D ended up in the Pacific off Easter Island. This was the crushing end of the Russian Mars programme for the time being.

The only consolation from the mission was that some of the instruments found their way onto other missions later, like Europe's Mars Express launched by Russia. EVRIS was the basis of COROT (COnvection, ROtation et Transits planétaires), the planet-searching mission launched by Soyuz 2.1.b from Baikonour in December 2006.

Mars 8 left a final historical footnote, Vesta. Originally intended to follow these Mars missions, this project was defined at the France-USSR reunion of September 1984 in Samarkand as two spacecraft in 1991 to pass Venus and use gravity assist to fly on to four asteroids, of which Vesta was to be the largest. When detailed mission studies were made in 1987, it appeared that it would be easier to re-route the mission via Mars, dropping a probe there. When CNES found that the commitment to Vesta would exceed its budget, it drew in ESA, whereupon the mission was further redefined with two flybys of Mars and then an excursion to the main asteroid belt to visit six asteroids and two comets over seven years, with landings and penetrators. With the delays to Mars missions, Vesta withered and died. Asteroid and comet missions were eventually accomplished by ESA (Rosetta), the United States (Dawn to Ceres and Vesta) and Japan (Hayabusa 1 and 2). The Vesta profile was taken up by China for a mission projected for the late 2020s, Zheng He.

American misadventure: Netlander

After the failure of Mars 8 and the unhappy experience of recent Russian cooperation, France quickly turned to the Americans as partners. There was a new minister for education, research and technology, Claude Allègre, who convened a seminar at Arachon, the outcome of which was what CNES called 'orientations majeurs'. He announced a €400 m project with the United States for a three-part mission: a Mars return vehicle to be landed on Mars in late 2003 by Delta III to collect samples and fire them into Martian orbit; and two Ariane 5 launches in 2005 to send a spacecraft to collect the samples, make a second sample collection and land four Netlander surface probes to form a monitoring system. The samples would be fired out of Mars orbit in July 2007 and reach Earth in May 2008 [24]. It could be said that this was the most complex set of Mars missions ever approved, quite a *grand projet*. Its short but unhappy history, its circumstances verging on Greek tragedy, led to its swift disappearance from the historical record and a subsequent collective amnesia, but as a story of cooperation it deserves a mention.

Claude Allègre drafted a letter of intent and Memorandum Of Understanding (MOU), making the US partnership a done deal. Allègre served under Lionel Jospin's socialist government from 1997 to 2000 and was a man of idiosyncratic views on vulcanism, global warming and Greek philosophy. He was a controversial politician, a geochemist by background (*Introduction to geochemistry,* 1974). Arguing against vulcanologists who took the view that the volcano there was not dangerous, he persuaded the authorities to evacuate Guadeloupe in 1974 (the vulcanologists were right). His main philosophical text was *La défaite de Plato* (*The defeat of Plato,* 1996), an attack on conceptualism. Not only did he favour doing business with the Americans, but he also wanted to pull France out of piloted spaceflight projects with the Russians and ordered their suspension. He was scathing of human spaceflight as 'pleasure rides for the rich'.

Netlander. CNES/Ducros.

The mission was defined in more detail at an international symposium in Paris on 5 February 1999 and at a NASA/CNES workshop in Toulouse over 7−9 July that year. The mission was given a blaze of publicity by CNES, with a special dossier in its magazine, including an interview with the minister. Although it was a French-American project, France sought to draw in either ESA as a whole or its member states individually (Finland offered some surface packages). In summer 1999, the national aerospace research agency, the Office National d'Études et de Recherches Aérospatiales (ONERA) got to work testing the heat shield. NASA and CNES made a joint presentation of the mission at the International Astronautical Congress that autumn in Amsterdam. Mission planning had made substantial progress and a first launch date had even been set (May 2003). The lander, ascent stage, soil canister, instruments and potential landing sites were outlined. Design had reached a detailed stage on the Netlanders, how they would land and their instruments.

Despite a big fanfare of publicity at the start, reports on its progress dried up in 2000 and the project fizzled. The House of Representatives cut NASA's budget by 10 percent ($1.4bn from $13.5bn) in a partisan stand-off between a Republican congress and the Democrat Clinton White House, including a pointed cut of $75 m to future Mars missions, though that was hardly fatal. Like most cancelled projects, Netlander had no distinct end point, nor was its obituary ever written. Its grave was so deep that a subsequent history of sample return missions did not even mention the project. Following the failure of two missions to Mars in 1999, NASA redefined its Mars strategy on the basis that collecting samples would be premature if one did not know the best locations to look for them in the first place, hence the two rover missions, *Spirit* and *Opportunity*, in 2004 and the Mars Reconnaissance Orbiter in 2006 **[25]**.

More importantly, there was a new minister responsible for the space pro-gramme for the next two years, Roger Gérard Schwartzenberg (2000−02). As for Claude Allègre, he supported Nicholas Sarkozy against socialist Ségolène Royal in the later presidential election, possibly hoping for an appointment by Sarkozy, but he probably ruined his chances by publishing a badly-timed book *against* climate change (*L'Imposture climatique*). In 2003, CNES acquired a new direc-tor general, Yannick d'Escatha and a new Minister for Research and New Technologies, Claudie Haigneré, who presented a fresh space plan to the govern-ment on 15 April 2003, which was adopted. She signalled a reorientation around France's traditional partner, marked by the appointment by CNES of an 'Advisor for Bilateral and Multilateral Affairs (Russia)' and a CNES representative at the French embassy in Moscow (Catherine Ivanov).

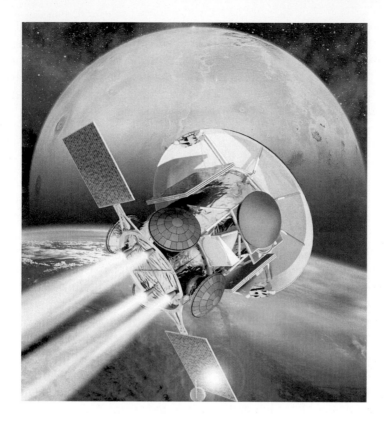

Mars sample return, 2005. CNES.

It took Russia a long time to marshal its strength for a fresh attempt at Mars, not trying again until 2011 with the Phobos Sample Return mission, also called Phobos Grunt. European participation seems to have been limited, possibly allowing memories of Mars 8 to heal, but ESA agreed to make Mars Express available as a signals relay. There was a contribution by the Centre for Space Research in Poland which built a 2 kg recoilless jackhammer mini-soil sampler called *Chomik* (hamster), requiring only 1.5w of electricity, though it is not clear if it actually flew. The project suffered the financial delays experienced by all Russian scientific projects at that time, had already been put back from 2009 and moved from a Proton rocket to a less expensive Zenit 2SB. The Chinese were decisive in getting it airborne, providing funds for a sub-satellite called Yinghuo to be included to orbit the planet.

The Europeans must have been glad that they did not get involved in Phobos Sample Return. It was eventually launched at night on 8 November 2011 but in an exasperating repeat of Mars 8, the upper stage did not fire and it remained stubbornly stuck in Earth orbit. At one stage, ESA managed to contact Phobos Sample Return through its tracking station in Australia, but Russian ground controllers were never able to do so to override the problem. The spacecraft eventually crashed into the Pacific off the coast of Chile in January 2012. An inquiry blamed a badly designed computer control system with poor components, compounded by a communications system that could only work in deep space (and not in low Earth orbit), exacerbated by the lack of marine tracking systems at the critical point of the Mars injection burn over South America. At the heart of the problem was a space programme still struggling with the legacy of the financial and organizational chaos into which the country had been plunged in the 1990s. There was one silver lining in this dark cloud: Russia took the view that going-it-alone was not a good idea and that mission success might be more likely in partnership with others.

Observatories

The final area of European-Russian cooperation in space science – observatories – proved be one of the most successful. In 1976, the USSR began the development of a series of 'great observatories': Astron, Kvant, Gamma, Granat and Spektr. The Americans had a 'great observatories' programme too, but this was a conscious effort not just to close a gap where Soviet performance had lagged behind, but to move ahead. An unusual combination of circumstances led to the Integral observatory joining this set.

The first of these, Astron, was an ultraviolet observatory. It used an OST telescope from the Salyut manned space laboratory applied to a Venera Venus probe design and was launched by the Proton rocket on 23 March 1983. Astron had its roots in a proposal by Andrei Severny, director of the Crimean Observatory, who presented the idea of an ultraviolet telescope in orbit in 1975. It was put under joint study by institutes in Marseilles and Nauchny and eventually involved Lavochkin, the observatories of Geneva, Marseilles and Crimea and the French companies Matra and Crouzet. Its 80 cm telescope was twice that of the preceding International Ultraviolet Explorer and was the largest space telescope before Hubble.

Astron telescope.

The 200,000 km orbit of the 3.5-tonne Astron carried it outside the Earth's radiation belt and it was over the USSR for 20 out of 24 hours. The main instrument was a 400 kg, 80 cm main mirror ultraviolet telescope, *Spika*, which was 3.5 m long, built by the Crimean Astrophysical Laboratory and equipped with a French spectrometer called UFT, used to observe stars in the 1,200 to 3,000 Å range. The French role in the project does not seem to have been widely publicized, but undoubtedly contributed to its success. The four principal scientists associated with the ultra violet telescope were Andrei Severny (Crimean Astrophysical Observatory), Georges Courtes (CNRS Laboratoire d'Astrononie Spatiale) and Alexander Boyarchuk and Viktor Ambartsumian (Crimean Astrophysical Observatory). Ambartsumian was one of the great veteran Soviet astronomers who had started at Pulkovo observatory in Leningrad and had somehow survived the purges there.

Although intended for a year-long mission, about 200 communications sessions were held with Astron every year until its fuel ran out in 1991 and there was a high scientific return from the mission. Observations centred on Hercules X-1, Orion, Taurus and Leo and the accretion of material from red giants to neutron stars. Astron made possible a star catalogue published by the Observatory of Crimea. Two years into its mission, Astron observed comet Halley and another two years later, the 1987A supernova.

Kvant: Britain's moment

Next up was Kvant. This was a departure in two respects. First, Kvant was attached to the crewed orbiting space station, Mir, where it arrived in 1987, operating both automatically and under the command of the cosmonauts on board. Second, the participation drew in countries not hitherto much involved in cooperation with the USSR: Britain, Switzerland and the Netherlands. In one of its first cooperative projects, Germany had been introduced to Kvant in 1981.

Kvant carried the British-Dutch COMIS (COded Masked Imaging Spectrometer) or TTM in Russian (*Teleskop s Tenevoi Maskoi*), developed by the University of Birmingham in Britain and the Space Research Laboratory in Utrecht, the Netherlands. It was 2.5 m long and 40 cm diameter, with a 540 cm^2 detector to follow x-rays in the 2−30 keV range with a precision of 1 arc second. The British were joining a European scientific package already on board Kvant which the Russians called Roentgen. This comprised TTM, Sirene 2, HEXE and Pulsar X-1 from IKI. HEXE was developed by the Max Planck Institute of Extraterrestrial Physics in Munich, with the University of Tübingen, for the 15−20 keV range, using four identical detectors with 200 cm^2 area each. The

French-built Sirene 2 (in Russian, GSPS for *Gasovyi Stsintillyacionnyi Proportsionalnyi Spectrometer*) had a detector area of 300 cm^2 and worked in the 2–100 keV range. Kvant also had a 40 cm ultraviolet telescope called *Glazar*, a cooperative project between observatories at Byurakan in Armenia, Marseilles and Geneva. This was capable of 8-minute exposures down to 17th magnitude, intended to search for galaxies and quasars from 1640 Å, as well as for studying ultraviolet fluxes. *Glazar*'s creator was Academician Viktor Ambartsumian, a world-reknowned expert in explosions in the nuclei of galaxies. *Glazar* had first been tested out on a CNES balloon from Aire-sur-l'Adour in Landes, but an orbiting location offered a long-term observing opportunity. *Glazar* took more than 270 images of celestial objects, including 115 of Supernova 1987A in the Large Magellanic Cloud, but suffered from limited pointing opportunities. The instruments promised to be six times more powerful than anything flown to date.

Despite Britain's important role in Kvant, it received little or no attention in the general domestic media, the principal coverage being in *Spaceflight* and *Spaceflight News*. British participation arose from a 1981 invitation by the head of IKI, Roald Sagdeev, to Kees de Jager, the head of SRON (Stichting Ruimte Onderzoek Nederland, or the Netherlands Institute for Space Research), in effect the Dutch space agency. SRON in turn felt that this was a project best undertaken with British colleagues and contacted Peter Wilmore, whose Department of Electron Physics in the University of Birmingham had built up considerable expertise in x-ray astronomy on Skylark rockets from Woomera, Australia, as well as Ariel 5 and Spacelab 2.

The leaders on the British side were Gerry Skinner and Domar El Anam. They agreed a division of responsibilities, with SRON providing a detector of the type later used on the successful Italian-Dutch BeppoSAX satellite and Birmingham providing the structure, mask, electronics, tracker and test equipment based on the Skylark experiments. In Britain, precautionary enquiries were made about the government's view, the outcome being that it would not object, provided that it was kept low profile and there could be no question of Science Research Council funding. To minimize the visibility of British participation, the documents governing the experiment were Dutch-Soviet only. Meetings were held in Moscow and Utrecht, with one in Birmingham and it was agreed that the host country would pay the costs of its visitors. These visits marked the first time that Rashid Sunyaev, Russia's leading scientist in this field, visited the west. He had even had an effect named after him, the Sunyaev-Zeldovich effect (Yakov Zeldovich was his teacher).

Kvant and Mir.

The next step was to see how the equipment could be installed on what the Russians vaguely called 'a future planned space mission'. An early task was to see how the TTM could be attached. The British representative was invited to meet a Soviet engineer whom he had never met before, nor did again, over breakfast in a cheap Paris hotel. The British provided the dimensions of their instrument and the engineer provided details of their standard electrical interfaces. The Russians would not say exactly where the instrument would be attached, commenting, 'just tell us what you need and we will provide attachment points'. The British provided already-flown parts from the Skylark programme which by then was winding down, but the Russians were not told this. This was nearly the project's undoing, because having been in space they had absorbed detectable levels of radioactivity (trace thorium and uranium). This was later evident in tests before they were shipped to Moscow, but the Russians were not told this either. Low-cost commercial electronics were used to avoid repercussions if the Russians inadvertently got anything of potential military use.

It was not until the Mir space station was aloft (February 1986) that the British and Dutch scientists were informed that their instruments would fly on a module called Kvant that would be attached it, though they received no advance notice of Kvant's launch in March 1987. When the time came to turn on the TTM, they were invited to the Institute of Space Research in Moscow. The British had developed what they called a micro-computer to take the data, but 'micro' in those days was the size of a central heating radiator. In order to bring it with them, the scientists had to get an export licence, which was granted only on the condition that it had to be locked if out of their sight in daytime and left in the British consulate overnight, to prevent its secrets being stolen and its boards duplicated. On the Russian side, IKI already had an IBM 360 to take the data, but the British micro-computer was faster.

Kvant become operational in June 1987 and when it was turned on, TTM first made x-ray observations of the centre of the Milky Way. It was the largest complex of its kind in orbit at the time and some astronomers came to regard TTM as the best x-ray telescope of the decade. The instruments were pointed from mission control in Korolev at the behest of the science teams, normally when the cosmonauts aboard Mir were asleep. The instruments had their highest level of use from May to September 1989 when the station was temporarily unoccupied. The TTM had a FOV of 8 × 8°, operated in the range 2−30 KeV, had an angular resolution of 2 arc-minutes and an effective collecting area of 625 cm² **[26]**.

Mir and Kvant normally flew locked to the local vertical, so for these observations the station had to be re-orientated toward its target objects. The entire station had to be pointed at the target using gyrodynes spinning at 10,000 rpm, requiring 90w power each and able to achieve accuracies of 20 arc minutes, one arc minute and 5 arc seconds as required. This was a time and energy-consuming activity that became increasingly challenging as additional modules arrived at Mir. One western scientist observed how the systems fought each other and control could be lost entirely. As a result, observing periods were less in number and duration than they would have liked.

It was an eventful, crisis-punctuated mission. First, Kvant failed to hard dock when it arrived at the station in April 1987. Two cosmonauts, Yuri Romanenko and Alexander Laveikin, spacewalked and found a plastic bag stuck in the docking system. Its removal made hard docking possible. Then, in autumn 1987, the TTM which had been working since May stopped functioning, apparently due to an electrical failure in the detector. Its high voltage generator discharged. The Dutch feared that this was the end of the experiment because it was on the exterior of the station, connected by manacle rings, connectors and screws from the Skylark parts, was not designed for repair and was on a part of the station without hand-holds. The Russians were very impressed with TTM and did not give up easily, however. Sagdeev brought his western colleagues out to a concert in Moscow where, suitably relaxed, he popped the question as to whether this could be done. The Dutch built a 40 kg replacement detector and the British provided two tools to release the manacles.

The Birmingham team was in Moscow mission control, TsUP, in June 1988 when Vladimir Titov and Musa Manarov, on a year-long mission, spacewalked to TTM with a repair kit. They had to cut through the skin of Kvant and then encountered a cable which been installed on the fight model but not the ground version, which caused consternation but was ultimately unimportant. They then found that they could not unscrew the attachment over the TTM, which was stuck due to a resinous substance. All this took place with intermittent radio contact and none of the television cameras now typical of ISS spacewalks. They managed to saw their way past the screws, but it was physically hard work. Then the screw to undo the manacle ring connector belt – a well-rehearsed part of the operation – jammed and snapped. The cosmonauts had to abandon the repair and return, quite demoralized.

Kvant.

The British team had a grandstand view. The wall in TsUP showed the ground stations and tracking ships in the southern ocean, with rings around them that indicated when communications could be received from the cosmonauts, though not in between. A television crew from Birmingham was there too. A visit was also arranged for Star Town, where they saw a full-scale working model of Mir and Kvant for the first time, including the section where the TTM was installed.

A new toolkit was sent up on the unmanned Progress 38 freighter and a second spacewalk took place on 20 October, this time with a television camera. This second attempt went smoothly, with the old detector removed and completely replaced. Within 24 hours, the Birmingham scientists had new and better images, starting with the galactic centre. Although subsequent repairs to the American

Hubble Space Telescope attracted much media attention, Kvant was actually the first orbiting telescope to be so repaired. In a strange sequel, the Birmingham scientists were asked to help out by providing a hand drill for Mir based on one advertised in a British engineering catalogue, the page being faxed to them from Moscow. They duly supplied the drill, added a requested battery from a radio spares company and despatched it. They still do not know how it was used.

By sheer good luck, Kvant coincided with the greatest supernova in anyone's lifetime, 1987a. Astron first spotted the supernova on 23 February 1987, but Kvant had instruments more suited to supernova observations. Mission control was persuaded to give priority to supernova observations from Kvant, even though it meant that other experiments on the space station were delayed. They were able to follow 1987a rise to a peak in August and then decline. Supernova 1987a was expected to release large amounts of x-rays and gamma rays, which would become evident from the debris of the explosion. The German HEXE detected hard x-rays from August 1987 onward, while the British-Dutch TTM was able to show that low energy x-rays did not escape the explosion. TTM found a number of new x-ray sources, listed as 'KS' in catalogues ('KS' for 'Kvant Source'). For the record, these were KS J1716-389, KS 1731-260, KS 1732-477, KS 1741-291 and KS 1741-293, the figures indicating their coordinates. All were x-ray binaries close to the galactic centre. The lack of low-energy x-ray sources from the supernova was an important finding. The outcomes were published in Astronomer telegrams, International Astronomical Union (IAU) circulars and Soviet journals. The Roentgen complex as a whole made 300 observation sessions from June to September 1987, with 115 dedicated to the supernova. There were 2,200 sessions by 1990, or 23 days total.

Data were sent down from Kvant to tracking stations, forwarded to TsUP, then IKI and then put on tape for the Birmingham scientists to collect. The tapes were returned through visiting scientists and some 2,400 found their way to the Netherlands and were later passed on to Birmingham. Western impressions were that Soviet space science was very strong on theory, if technically not as good as western countries, but ever easier to work with as *glasnost* (the policy of 'openness') spread throughout the system.

TTM might have achieved more had more observing time been available. Mir's northerly 51° orbit was not the most suitable and the most useful observations were likely to be made near the equator. It was actually possible to get more x-ray science over a shorter period on the week-long missions of Spacelab on the American space shuttle, which could carry larger and more sensitive instruments and offered more dedicated observing time from its more southerly (28°) orbit. Most of Kvant's x-ray observations were in 1987, thankfully an astronomically most rewarding period. Mir's pointing ability declined and observations eventually became possible only when a known source came into the field of view and could act as a reference source. Subsequent modules arriving at Mir (Krystall, Priroda, Spektr) had their own pointing demands and the cosmonauts were left to resolve competing priorities, so it was more difficult to get time for TTM observations.

From the British point of view, Kvant was an unusual combination of a high-risk and low-cost contribution. By way of sequel, the Dutch space agency went on to use this experience for the highly successful wide-field camera for the Beppo SAX satellite. For the Germans, the Roentgen suite led to Spektr RG, 30 years later. The British team joined the ESA-NASA INTEGRAL project, but when the Americans withdrew, the British role was so limited that they pulled out and it ended up as an ESA-Russia mission. Kvant opened the door for Britain to lead the Spektr X project (below).

Kvant was an example of highly successful European-Russian cooperation, one in which Britain played a leading role. The politics, though, fascinate. The most astonishing aspect of the British participation was that although it was permitted by the authorities, it was hardly encouraged, went unfunded officially and had to be almost hidden from view, with no record of its coverage on state television or radio. It goes unmentioned in most British space histories. Although the American astronaut repairs to the Hubble Space Telescope are the stuff of legend, few have heard of the heroic repairs of Kvant by Manarov and Titov.

Gamma

Gamma was defined as a project ahead of its formal adoption in the great observatories programme. It went back to 1965, when Gamma was one of a two-part programme of free-flying modules that could be docked to a large orbiting space complex, the other being an infrared telescope called Aelita intended to fly in 1981−4. Aelita was Franco-Soviet from the start, devised by Rashid Sunyaev and Yakov Zeldovich on the Russian side and by Geneviève Debouzy, Michel Rogeron and Patrice Brudieu on the French as a 1 m infrared telescope, 3.5 m long and 1.6 m in diameter, to fly in a 15-hour orbit. The French were in eminent company, as Zeldovich was a nuclear physicist who worked with Andrei Sakharov and Igor Kurchatov. He was one of the great Soviet astronomers and cosmologists and co-author of *Relativistic Astrophysics* (1967). His student, Sunyaev, went on to lead the Spektr RG project (below). Following the success of Europe's Infra Red Astronomy Satellite (IRAS), however, Aelita was called off.

The intention for Gamma was that it would dock with an orbiting space station in between periods of free flight, to allow cosmonauts to retrieve film and carry out servicing tasks. In 1972, Gamma was redefined as an ambitious high-energy gamma ray observatory that would better the performance of the American SAS-2 and the European COS-B. International cooperation was invited and France joined the project in 1974. Gamma was officially approved by the government on 17 February 1976, but it went through further refinements, for example being visited by Soyuz spacecraft for servicing. In the end, it turned out to be simplest to fly Gamma as an independent mission constructed from a Progress space freighter. In 1982, the design was finalized as a Progress with a French telescope and solar panels for power. At a time when there were frequent changes of plan in the Soviet space programme, it is possible that the French connection kept the project alive.

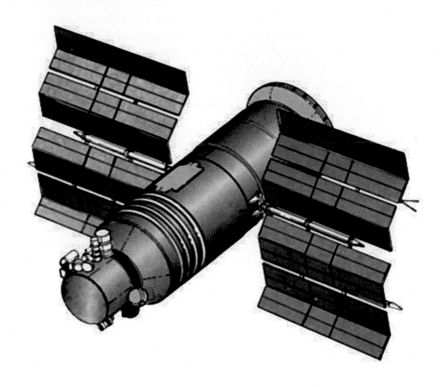

Gamma.

A team of 100 worked on the project, drawn from the Centre for Nuclear Studies in Saclay and CNES Toulouse. The Russian side involved IKI, the Physics Institute of the Academy of Sciences (FIAN) and the Energiya design bureau. Meetings occurred every two months, alternating between France and Russia. There were many delays on the Soviet side. At one stage, Italy joined the project to take responsibility for the solar sensor, but then dropped out to be replaced by Poland. Integration coincided with *glasnost*, so scientists and engineers were able to participate in the final integration at Baikonour.

Weighing 7,320 kg – typical for Progress – Gamma was launched on 11 July 1990, climbing to an orbit of 416 by 434 km. The French contribution was a substantial one: the gamma ray telescope *Gamma 1*, 50–5,000 MeV for highly energetic particles and with a resolution of 20 arc minutes; the Vidicon camera system of the telescope; the Spectre 2/Pulsar X-2 instrument to detect cosmic x-rays from 2–25 keV; and the Disque instrument to measure weak gamma rays from 20 keV–5 MeV. Gamma produced substantial scientific results over its 12 to 18 month lifetime. Gamma started by pointing at the centre of our galaxy, a candidate black hole location where there were high energy sources, in particular Sagittarius A, a powerful radio source. The instrument for imaging bursts did not deploy, despite months of trying, which was a disappointment. High energy spectra were obtained and some of the most valuable were of 5GeV protons coming from a solar eruption. First results were presented on 29 June 1991 at CNES in Paris in the presence of the minister for research and technology, Hubert Curien.

Granat

Following the election of President Mitterrand in spring 1981, who was committed to an increase in funding for scientific research, CNES convened a summit in the Alps to take advantage of any new opportunities that might arise or be created. This settled on a priority project, SIGMA (Satellite d'Imagerie Gamma Monté sur Ariane, also written as Sigma), to photograph and map the gamma ray sky.

Granat.

This was very much an instrument in search of a home and was originally due to fly on the Ariane 4 in 1985, hence the name. However, the spending party was over and funding dried up a year later when the franc was devalued in July 1982 and Sigma had to find a new home. That autumn, the annual Franco-Soviet reunion took place in Kishinev, Moldova and the idea of making Sigma a joint project was proposed sometime between a visit to the local wine collective farm and the opera. The meeting agreed a team of Rashid Sunyaev, Roald Sagdeev and Yakov Zeldovich on one side, with Jacques Chêne *chef du projet* on the other, to supervise a team to construct a telescope, the core of the satellite, in Toulouse and Saclay. It was originally called Astron 2 and was based on a Venera spacecraft bus. With a new home as a joint mission on the big Proton rocket, the cost for the French fell from FFR400m to FFR80m.

The principal Russian contributor was the Frunze Design Bureau, which built four ART-P imaging x-ray telescopes and an x-ray telescope mounted on what was called the *Podsolnuk* installation, designed to swivel around to gamma bursts. In addition to Sigma, there was a gamma ray burst detector called PHOEBUS (Payload for High Energy BUrst Spectroscopy, also written as Phoebus). Denmark was also on board with WATCH, an all-sky monitor intended to detect transient sources. Different instruments observed different parts of the sky: Sigma at 30−1300 keV; IKI's ART instruments at 3−100 keV; PHOEBUS at 100 keV−100 MeV; KONUS at 10 keV−8 MeV; TOURNESOL at 2 keV−20 MeV; and WATCH at 6−180 keV. Granat was 3.5 m long, 1.2 m diameter and had a big memory for its day, 128 Mb.

Granat Sigma the telescope.

Sigma was the most complex telescope of its time, weighing over a tonne and with 60,000 parts. It broke the old rule whereby western equipment was simply handed over for integration by Soviet-only personnel. The French pointed out that there was nothing simple about Sigma and that they really needed to be there throughout the integration to ensure that nothing went wrong. Soviet resistance to having scientists involved in integration was so well embedded that, as one scientist put it, 'you almost needed the permission of the Central Committee [of the Communist Party] to make an exception'.

Generally, though, the old secrecy was melting away across the science missions. One scientists recalled how he 'was so close to the launcher on the pad that I could knock on its metal skin', but that the closest he got to a rocket at Cape Canaveral was 5 km. At landing sites, the secrecy was almost gone. The Soyuz rocket plant in Samara, the *zavod*, was different, with no photography and security observing. It was also unpredictable in the same place: on one occasion, it was possible to walk freely around Plesetsk, on another, only escorted visits to the park were permitted.

Granat shortly before launch. Roskosmos.

Thus, the Sigma team found itself inside the Lavochkin design bureau in Moscow in May 1989 and inside the Proton hangar in Baikonour that December. The French were struck by the extreme precision of the programme on the one hand — the launch took place to the second as set down a month before — and by the crudeness of the test to ensure that the solar panels would deploy on the other. Without gloves, a technician simply pulled the panels out and pushed them back in by hand. He explained that he had done this 25 times before and the panels had always worked. Granat was launched on 1 December 1989 into a highly elliptical orbit of 2,000 by 200,000 km.

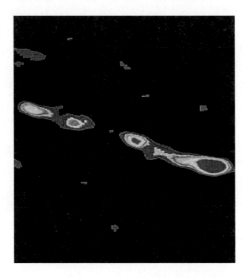

Granat source in the centre of the galaxy.

The CNES team then flew on to Yevpatoria, Crimea and its 70 m tracking station. This had done no tracking since the Phobos mission that March, followed by new teams working monthly shifts thereafter. Granat became one of the most successful of the great observatories. Its first results were presented at Saclay over 10–13 December 1990 to the country's best astronomers, reporting on intense interruptions, positron annihilations, intense gamma-ray sources and black holes. Sigma focussed initially on the central regions of the galaxy, where it identified more than a dozen gamma sources starting with the anodyne IE 1740, which was originally thought to be a black hole but was found to have bipolar jets and was characterized as the first micro-quasar. In October 1994, Granat was put in Tournesol mode, rotating every 13 minutes but scanning the whole galaxy for emission sources every six months.

France came to play an important role in keeping the mission going in the worsening financial environment of the mid-1990s. First, the Yevpatoria dish ran out of funding, so France stepped in to pay mission control to continue its vital work. Then the Russian space agency ran out of money again in spring 1997 when Granat was making exceptionally useful observations of the galactic centre. Once again, France came to the rescue to provide funding for tracking until September 1998. By this stage, Granat was overheating, its fuel tanks were dry and only one transmitter remained. Granat eventually gave out on 27 November 1999.

Granat and Astron commemorated.

More than 400 scientific papers were published from Granat, encompassing the galactic centre, radio sources, galaxies, gamma rays, black holes, quasars and great annihilation events. Its findings were advertised by IKI, the University of Liverpool, *Advances in Space Research*, the Astrophysics Data System (ADS) service and by NASA's Goddard Space Flight Center (GSFC).

INTEGRAL

Most European-Russian collaboration arose from the Soviet Union inviting western cooperation, or European countries finding their way into Soviet projects. INTEGRAL (also written lower case as 'Integral') was different, as it began as a European project. Approved in 1993 under the *Horizon 2000* programme, its name stands for INTErnational Gamma Ray Astrophysical Laboratory. It was seen as the direct successor to Granat, whose outcomes were persuasive at this stage. INTEGRAL was intended as a gamma ray observatory successor to the American Compton gamma ray observatory. The objective of the mission was to take detailed, high-resolution spectral images, a hundred times better than before, of previously identified objects, with a focus on violent cosmic events in the centre of our own galaxy, neutron stars, binaries, bursts and black holes. INTEGRAL then became an ESA/NASA project, but first the Americans and then the British withdrew. As a large observatory, 5 m tall and weighing four tonnes, INTEGRAL required an Ariane 5 rocket, the cost of which became ever more problematical and which would not be available until the late 1990s in any case.

Spotting an opportunity, Rashid Sunyaev was authorized to offer ESA the Proton rocket and, by way of encouragement, brought the INTEGRAL team on a tour of the Khrunichev factory floor where it was made – another brick in the *zavod* wall gone. Proton was so powerful that it could offer a high orbit that would bring the delicate observatory outside the radiation belts. One of Russia's problems was that with its own great observatories nearing the end of their working life and with no money to fund new missions, its scientists were beginning to lose their own supply of fresh, home-made scientific information. Rather than charge a launch fee for INTEGRAL, Russia offered a free launch in exchange for 25 percent observing time.

Numerous ESA member states became involved: France, Germany (spectrometers); Denmark (x-ray monitor); Switzerland (data centre); the Netherlands (testing); and Italy (construction and prime contractor). The INTEGRAL users group, chaired by Ireland, included representatives from France, Italy, Spain, the Netherlands, Germany, Denmark, Switzerland and Russia, but not Britain, save for the University of Southampton which was a collaborating institute. The principal investigators came mainly from Germany, Spain, France and Italy.

INTEGRAL was duly launched on 17 October 2002, a clear day, with the Proton billowing out the tell-tale brown nitric smoke against the autumn steppe. As it went through maximum dynamic pressure, the Proton shook off three successive steam clouds, one in the shape of an onion ring. INTEGRAL was soon circling the Earth at 400,000 km out every 64 hours.

The first results from INTEGRAL were presented at a press conference in Paris in December, where it was reported that the instruments had been calibrated and that it had detected its first gamma ray bursts and observed a transient pulsar. INTEGRAL became one of ESA's greatest success stories – and remains so – and was subsequently re-funded into the 2020s. INTEGRAL has provided enormous volumes of invaluable original data on x-ray sources, the Milky Way, galaxies, black holes, pulsars, gas jets and gamma rays. It found gamma ray emission from the Crab Nebula, titanium emissions from the 1987 supernova, gamma ray bursts 400 million km distant – the brightest source ever found – in Cygnus and the first gravitational wave event. In a more open information age, its data are downloaded to and made available by Versoix (Switzerland) and IKI's Russian Data Centre, with a mirror archive at Goddard in Maryland.

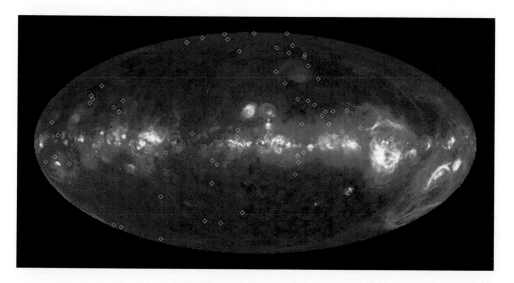

INTEGRAL sky map. ESA.

Spektr

Spektr was the name of the programme to follow the 'great observatories' and went through numerous evolutions from its formal approval by the Soviet Council of Ministers in 1981. Emblematic of the funding problems affecting the Russian space programme, the first did not fly until 2011 and the second in 2019. Had they started sooner, they might have given Russia the lead in deep space observations. This time, in a new development, the USSR invited in western countries to share in the project definition stage. A concept conference was held in October 1987, attended by astrophysicists from Britain, Denmark, Finland, the GDR and Italy. The original idea was for at least three observatories for radio, x- and gamma ray observations and ultraviolet.

The collapse of space science funding in the 1990s meant that no progress was made for ten years. In 1997, the Russian Space Agency finally announced R200m in funding and, in a transparently obvious move to spread costs, specifically invited participation by ESA. Costs would also be kept down through the use of a new, standardized bus called Navigator developed by the Lavochkin design bureau and by the use of cheaper rockets than Proton, such as Zenit and even the smaller Soyuz. The Spektr project was then re-defined as four spacecraft:

- Spektr R (Radioastron)
- Spektr X, later renamed Spektr RG (Röntgen Gamma),
- Spektr UV (Ultra Violet)
- Spektr M (Millimetron).

ESA provided some start-up funding for Spektr RG to match the R200m. When nothing happened, Russia asked that the debt be considered written off as part of the free launch of INTEGRAL, which it was.

Spektr R was first off the ground, launching on 11 July 2011, exactly 30 years to the day since the programme was first approved, something of an unwanted record. Its creator was one of the great Soviet astronomers, Nikolai Kardashev, known from the 1960s for his writings on extraterrestrial civilizations. The idea of Spektr was to put a radio observatory into a high orbit and use Very Long Baseline Interferometry (VLBI) with other observatories, some as far away as Australia, to achieve high angular resolution of deep space objects. Spektr R used the Zenit rocket, the Fregat upper stage and the new Navigator bus to reach a distant orbit of 332,000 km. Its orbit was perturbed by the Moon, so multiple engine firings had to be made to ensure that it reached its five-year lifetime (2016). The probe actually lasted longer, with contact eventually lost in 2019.

The European equipment contribution comprised a c-band receiver (the Netherlands and Germany) and k-band receiver (Finland), with further participation by India and Australia. For observation time, Spektr was thrown open to global participation with the worldwide practice of annual Announcements of Opportunity (AO). There were significant results from Spektr R, the first comprehensive set being presented at the biennial COSPAR congress in Moscow in 2014, focussed on active galactic nuclei, the interstellar medium, pulsars and water megamasers [27]. The scientists involved in Spektr R went to considerable lengths to publicize the results of the mission, with a dedicated website on which the scientific papers were posted in English and Russian.

Although primarily a radio observatory, Spektr R included a plasma package, Plazma F, a descendant of experiments carried out on Interball and before that, Prognoz. Plasma F similarly took advantage of the observatory's high-apogee orbit to monitor plasma and energetic particles in the interplanetary medium. Plasma F comprised the Bright Monitor of the Solar Wind (BMSW) plasma spectrometer (Czech Republic, Greece, Russia), the MEP 2 energetic particle monitor (Slovakia, Greece, Russia) and the separately installed MDDZ detector of meteorite particles. They combined a mixture of institutes, universities and companies: the Institute of Experimental Physics in Košice, Slovakia; Charles University and the Institute of Atmospheric Physics in Prague, Czech Republic; the Democritus University of Thrace, Xanthi in Greece; GCGC Instruments in Chemnitz, Germany; IKI; the Institute of Nuclear Physics at Moscow Lomonosov State University; and Baikal State University, Irkutsk. The plasma package was able to take 32 samples per second. The principal results concerned the jet character of the solar wind, the measurement of interplanetary shock waves and the helium content of the solar wind (BMSW) and a new type of strong oscillations of energetic ions flux in the bow shock, the results being published in *Solar System Research* and *Kosmicheskie Issledovaniya* [28].

The level of European cooperation in the next project, Spektr RG, was higher and focussed on Germany. This was not how it started, as the original partner for this observatory, when it was called Spektr X, was Britain, one of the great false starts of this story. The British National Space Centre (BNSC) visited the Institute of Space Research in Moscow from 29 September to 1 October 1986. Led by its dynamic first director, Roy Gibson, the delegation met Roald Sagdeev and the Interkosmos council. IKI proposed joint activities in the areas of submillimetre astronomy, life sciences, radio telescopes and material sciences, with an initial agreement or protocol signed between Gibson and Sagdeev [29]. A ten-year agreement was signed the following March between UK foreign secretary Geoffrey Howe and Soviet foreign minister Eduard Shevardnadze. Britain was offered two tonnes of experimentation aboard, described by the director of the Mullard Space Laboratory, Len Culhane, as 'spectacularly large' compared to the hundreds of kilos maximum on European launches. Several British institutes at once expressed interest, including Mullard, the Rutherford Appleton Laboratory, Birmingham University and the University of Leicester [30]. In effect, Spektr X was offered as a flagship British-Soviet project in an area where Britain had internationally recognized expertise and was already demonstrating its success with the TTM telescope on Kvant. Before they could even get started, however, Spektr X came to an abrupt, shattering end. Three months later, in the House of Commons on 24 July 1987, Prime Minister Margaret Thatcher formally rejected the development plan of the BNSC, voicing the view that money to finance space research should come from the private sector. Roy Gibson resigned in protest and Britain effectively withdrew from the front line of space exploration for the next 20 years.

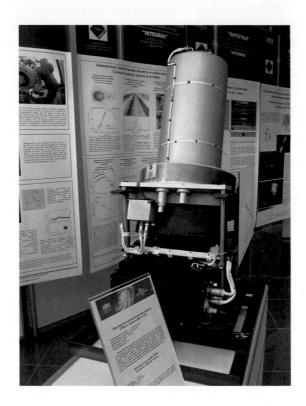

Spektr X

Redesigned and renamed as Spektr Röntgen Gamma (RG) by *Kvant* astrophysicist Academician Rashid Sunyaev, this would be the first-ever Russian mission to the Lagrange 2 (L2) point some 1.5 million km from Earth. Its original element was a Russian telescope, the ART-XC (5–30 keV band). Rashid Sunyaev's ambition was to confront the core challenge of the understanding of the universe, namely dark matter, with a whole-sky survey of up to 100,000 galaxy clusters over four years, followed by pointed observations for three years. He expressed the hope that the mission would make it possible to draw up a map of the universe, with millions of sources, including three million black holes and 100,000 galaxy clusters. Spektr RG would weigh 2,647 kg.

Rashid Sunyaev.

Due to the near-collapse of funding for new astronomy missions in the 1990s, Spektr RG made very little progress. The key to breaking the deadlock lay in a small German telescope, ABRIXAS (A BRoadband Imaging X-ray All-sky Survey), launched from Kapustin Yar on 28 April 1999. This was a 470 kg, €20 m satellite using seven nested telescopes with a single CCD detector to make an all-sky x-ray, surveying the medium energy x-ray range at $1-10$ keV with 30 arc second resolution. It was built by OHB in Bremen for the Max Planck Institute for Extraterrestrial Physics in Garching and the Astrophysical Institute in Potsdam, with data processing at Leicester University in Britain. It was expected to find 10,000 new sources, including diffuse x-ray sources and time-variable sources, many hitherto obscured by dust and gas clouds. Sadly, although placed in the right orbit, the spacecraft's battery overcharged after three days and then failed. The mission was lost and the unfortunate enterprise forgotten.

Except in Russia, that is. Six years later in 2005, Roskosmos spotted an opportunity to involve Germany, inviting a larger, 800 kg German telescope called the extended Röntgen Survey with Imaging Telescope Array (written as eROSITA) from the Department of Extraterrestrial Physics of the Max Planck Institute ($0.2-12$ keV band). The value of the eROSITA contribution from the Max Planck Institute was €100 m, half of which came in turn from the German space agency, DLR. A formal agreement was signed in 2007. As with Spektr R, a Zenit launcher would be used rather than the more expensive Proton.

Spektr RG eROSITA. DLR.

Spektr RG proved to be a slow mission to put together, with most delays being financial on the Russian side. There was an unexpected European delay in 2013, however, when the electronics of eROSITA were adjudged to be insufficiently protected from radiation at L2 and had to be replaced at a cost of a further €5 m. Eventually, the key components came together. The ART telescope arrived at the Lavochkin design bureau in December 2016 and the eROSITA in January 2017. There was then a further complication. Because of the conflict in the Ukraine, the supply of Zenit rockets had halted and the last one had been so long in Baikonour as to run out of warranty. At this late stage, Roskosmos moved the spacecraft to a Proton with the Blok DM-3, despite reservations on the part of both the German and Russian scientists and the memory that it was this block that had failed on Mars 8. Now, though, there were no other options.

Spektr RG team. Roskosmos.

Further delays with the BRK radio system pushed the launch back to September 2018 and more delays meant that Spektr RG did not reach the pad until June 2019. Nerves became ever more frayed when it was discovered that in the course of attaching Spektr RG to the Blok DM, the electrician had inadvertently drained all the power supply out of the spacecraft, so that all that would be delivered to L2 would be a totally dead spacecraft. The whole stack had to be rolled back to the shop for repair. On-the-pad delays like this were virtually unknown in the Russian space programme but may have been the outcome of having more rigorous quality control systems in place. Exasperatingly, a second attempt to launch on 12 July was called off eight hours before lift-off, when the State Commission wanted the insulation checked and the battery re-checked. Finally, on live TV, the Proton took off into a glorious blue summer sky on 13 July. The relief for the Institute of Space Research and the Max Planck Institute must have been glorious too. Days later, ground telescopes spotted Spektr RG heading for L2 and few days after that, the magic moment of 'first light' (the first telescopic observations) was received.

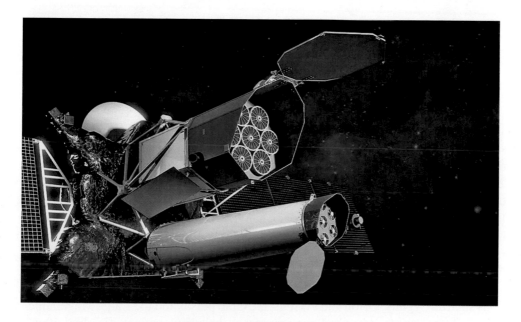

Spektr RG drawing. Roskosmos.

Spektr RG nearing completion. Roskosmos.

Over the following months, there was a growing set of reports on Spektr RG's scientific mission and its progress in scanning the sky. In spring 2020, Spektr RG's 1.5 million km distant L2 orbit passed out of range of the Russian tracking system, so ESA's three 35 m dishes in Australia, Spain and Argentina came into play, downloading 6.5GB of scientific data over 16 communication passes. In June 2020, IKI and the Max Planck Institute published a new all-sky x-ray colour-coded map of the energetic universe in the 0.3−2.3 keV band, with the centre of the Milky Way in the middle. The image held a million x-ray sources and showed hot gases in the galactic centre, active galactic nuclei and clusters of galaxies. Further inspection showed stars with hot coronae, super-bubbles, flares, merging neutron stars, binary neutron stars, white dwarfs, stars swallowed by black holes and supernova remnants. An accompanying map identified well-known constellations such as Cygnus, Orion and Perseus, the Crab pulsar and the Large Magellanic Cloud. Astronomers described the images as beautiful and exotic. Individual images showed neighbouring galaxies mingling with one another, while a blue, red and yellow map made by ART-C showed the Coma Cluster **[31]**. After such a long wait, it was a Russian-European astronomical triumph.

Spektr RG — interacting galaxies. ESA.

Two remain in the series: Spektr UV, due 2022 and Spektr Millimetron (2030). Funding of Spektr UV began in 2004 and was intermittent. There was off-on discussion of European participation, but it has been plagued by the problem of sanctions against Russia. The detectors would have been supplied by the e2v company in Britain, but they in turn relied on some American components. When the Americans realized this, they cancelled their participation, so the Russians had to look elsewhere for detectors. Germany was invited to contribute a spectrograph, but did not have the right specifications. No further European participation has yet been agreed but Russia is determined to push ahead and a Proton Blok D was ordered for the mission.

Spektr Millimetron, as its name suggests, is intended to explore the submillimetre universe (4–20 mm) from the vantage point of L2. Its design concluded in 2010 and its construction began in 2014 at the Reshetnev design bureau in Krasnoyarsk, supervised by the Lavochkin design bureau. The first indicator of European participation came when Roberto Battiston, head of the Italian space agency ASI, announced in 2017 that Italy would build the detectors. Echoing the previous experience, carbon fibre for the antenna's panels was to have been supplied by a British company, but in 2019 the company was bought out by Americans who promptly cancelled the contract, obliging Russia to go to the Sumitomo company in Japan instead. In November, Larissa Likhachova of the Lebedev Physical Institute announced hopes of drawing the Republic of Korea, China and France into the project, with an agreement already in draft. Detailed technical drawings were already being made and mock-ups built so that foreign participation could be finalized. In 2020, Roskosmos announced that Italy, China, the Republic of Korea, France and Sweden had now joined the programme. With its 10 m mirror operating close to absolute zero, Spektr Millimetron will use the millimetre and infrared range to research the early days of the universe, possibly uncovering clues to the presence of life. It will investigate the first stars and galaxies, AGNs and supermassive black holes **[32]**.

Spektr M in construction. Roskosmos.

JUICE/Laplas

No Russian probe had ever ventured beyond the orbit of Mars, whilst the Americans had sent a probe to Jupiter as far back as 1972 and Europe's Huygens had even landed on Saturn's moon Titan. Ideas for such a mission appeared in drawings at the end of the Soviet period, so it was only a question of time before they would be revived. The idea of a European-Russian probe to Jupiter dated to an ESA study in 2007 of a mission called Laplace (also written in Russian as Laplas) as a flagship goal for its long-term plan, *Cosmic Vision 2015–2025*. A European mission to Jupiter offered a collaborative opportunity to reach further into the solar system and Russia stated an early interest in contributing, offering a lander to touch down on the moon Europa.

ESA chose otherwise and decided the following year to partner with the United States, agreeing a two-spacecraft project using the Ariane 5 and the Atlas V. Given the recent French experience with Netlander, ESA should perhaps have known better, because budgetary restrictions forced NASA out of the project in 2011.

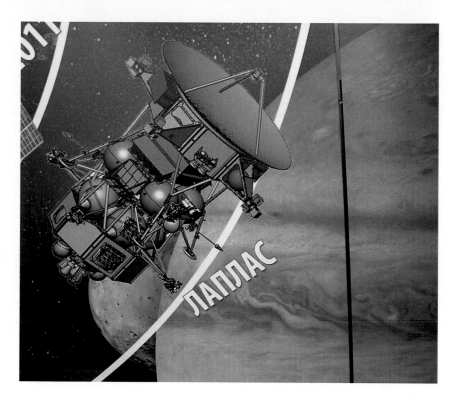

Laplas.

Russia seized the opportunity to pitch its original proposal again in 2009. ESA's new mission outline included a Russian annex for a 400 kg lander on Europa, with 20 kg scientific instrumentation, launched separately. The lander would be modelled on the Phobos Grunt lander then in preparation and the mission would be coordinated with the European orbiter which would provide data relay [33]. Lev Zelenyi and Oleg Korablev at IKI upped the Russian game with a revised title, *Europa lander mission – a challenge to find traces of alien life*, with a package of 20 instruments outlined, a larger lander at 1,210 kg and a small orbiter of 395 kg, a Proton M Briz M rocket and an electrical propulsion module.

This was then modified by both parties. ESA renamed the project JUICE (JUpiter ICy moons Explorer), dropping the 'Laplace' title. Meanwhile, Russian scientists calculated that difficulties with radiation at Europa would be too severe and that it would be better to go to Ganymede with a lander, which acquired the name Laplas-P (presumably P for *posadka*, lander). Russia also volunteered a nuclear power plant for JUICE, which would be helpful in avoiding European environmental protests. Russia held a day workshop on the Ganymede lander in Moscow in March 2013 and the government allocated R50m (€1 m) for a technical development study for November 2015.

Progress was made with its instruments. In 2014, ESA agreed that one of the eleven instruments would be a tetrahertz radiation detector contributed by the Moscow Institute of Physics and Technology, developed by Alexander Rodin. The aim of the detector was to characterize wind patterns on Jupiter and analyse volatile compound gases leaking out of icy cracks on the surface of Europa.

JUICE then disappeared from both the Russian and European literature as a joint project somewhere around 2014–16. In the best tradition of cancelled projects, there was no defined termination. There was no Russian participation in its 'red book', the mission design. From what is known, the mission proved too much for the Russian lunar and interplanetary space budget, which had a growing list of increasingly postponed projects. The Institute of Space Research was told to choose between a Laplas and a much-delayed long-duration Venus lander, Venera D. The latter had been on the drawing board since the early 1980s and even if prioritized would still not fly until 2026. The mission never received the level of integrated planning of ExoMars then in development (see chapter 5): it would be a European orbiter and Russian lander, compatible with one other but developed separately. Although it was an opportunity lost, the two sides separated amicably before having committed significant resources, time or effort. Europe's JUICE mission proceeded separately.

Luna 26, 27, 28

In the meantime, some modest opportunities opened up in the lunar programme. The Soviet lunar programme ended in 1976 but numerous attempts were made to restore the Moon to the programme in the Russian period, with new missions proposed in the 1990s and a schedule adopted from the early 2000s. Two decades later they were still waiting to fly, although the sequence of the missions had been well known for some time: a lander (Luna 25, 2021), an orbiter (Luna 26, 2024, originally called Luna Glob), a south pole sampler (Luna 27, also called Luna Resurs, 2025) and a polar soil sample return (Luna 28, also called Luna Grunt, 2028).

However, one instrument did make it to the Moon, Igor Mitrofanov's Lunar Exploration Neutron Detector (LEND), which entered lunar orbit in June 2009 aboard the American Lunar Reconnaissance Orbiter (LRO). Although LRO became best known for the astonishing quality of its photographs of already-landed spaceships and Apollo landing sites, this was a mission with a high science return. Its objective was to search for ice in the permanently shadowed lunar polar regions. Within ten years it had compiled a map of lunar ice and was still transmitting. LEND found shallow but significant levels of water ice in craters (Cabeus) or underneath them (Shoemaker) and identified a key crater to target in the future: Boguslavsky. Another of Igor Mitrofanov's instruments, Dynamic Albedo of Neutrons (DAN), was fitted to the *Curiosity* rover which landed on Mars in August 2012. It, too, found water ice in the subsurface, up to 5 percent, a higher level than in the Atacama desert on Earth.

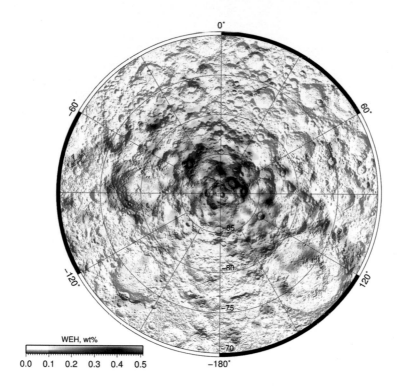

LEND map of lunar ice. Roskosmos.

In 2017, ESA announced its participation in the Luna 27 mission, though the formal agreement had to wait until a ceremony held on 30 January 2020 at the Leonardo offices in Milan, Italy. Luna 27 formally became part of ESA's *Vision for lunar exploration*, which included cooperation with the United States on the Lunar Gateway project, for which Europe would build the service module for the piloted Orion spacecraft, as well as cooperation with Russia on the Luna programme. Their principal agreed area of cooperation was a drill to extract ice volatiles and the allied tasks of sample handling and analysis. The task was to drill to 2 m, deliver the sample to a mini-laboratory and analyse it to identify its mineral composition and any water therein. It acquired the name PROSPecting for Exploration, Commercial exploitation and Transportation (PROSPECT), with the drill named ProSEED and the analyser PROspA. It was budgeted at €31.5 m. The samples would be placed in one of 25 miniature ovens for imaging, heating, chemical analysis and the extraction of oxygen, with a view to knowing how they might be harvested for water, oxygen and fuel.

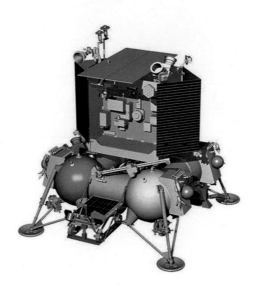

Luna 27. ESA.

Those most involved were Leonardo SpA of Nerviano in Italy and the Open University at Milton Keynes in England, where a sensor was developed to identify lunar volatiles. Switzerland joined with a Lunar Ablation Mass Spectrometer (LASMA) to make elemental and isotopic analysis of 12 soil samples. There was also a Neutral Gas Mass Spectrometer (NGMS) to analyse their volatile content. Richard Fisackerly was appointed project manager. The first drill tests took place in 2020 in Nerviano (home of Leonardo) and also at the University of Padua, drilling at -150°C into anorthosites, rocks, gravel and fine dust samples, even real meteorites. In March 2020, Hannah Sargeant of the Open University was named on the *Forbes 30* Innovation list for under 30s for her work on PROSPECT, ProSEED and ProSPA in developing systems to extract lunar water [34].

ESA then came onboard with the landing system, called Precise Intelligent Landing using Onboard Technologies (PILOT). This was a descent and landing navigation and hazard avoidance system, a sensor package to ensure a precise landing (Airbus, Bremen). For this, Eventec, a company in Estonia, with Neptec, a British company, developed a lidar-based lunar clock with an accuracy of one trillionth of a second to bounce laser pulses as Luna 27 came in to land, time being a proxy for distance above the surface. The lidar would sweep the surface to find even terrain amidst rocks and slopes and thereby enable Luna 27 to come in to the best landing site near the south pole, at least 85°S. PILOT would first be tested as a demonstrator (PILOT D) on the Luna 25 lander and was shipped to Moscow in

2016. Russian and French scientists (Toulouse and Reims) contributed to the Gas Analytical Package to measure the chemical composition of the sample, while the Royal Observatory of Belgium, Kazan University and the Institute of Applied Astronomy in St Petersburg joined to contribute X- and Ka-band radio beacons to improve knowledge of the gravity field and determine the position and velocity of the Moon to millimetre level.

Luna 27 explained by David Parker. ESA.

Although cooperation between Europe and Russia was most structured on Luna 27, it appeared to open the door to other opportunities. In 2020, it became known that Switzerland was contributing a mass spectrometer to Luna 25 and four spherical reflectors to Luna 26. Slovakia also joined Luna 26, with the Institute of Experiment Physics in the Academy of Sciences in Košice contributing ASPECT L, delivering an engineering model in October 2019. By 2020, Luna 26 had 13 instruments from four countries. The Czech Republic joined Luna 27, with Charles University Faculty of Mathematics and Physics contributing a solar wind monitor (BMSW-LG) and the Institute of Atmospheric Physics of the Academy of Sciences a wave experiment (LEMI). Both instruments arrived in Moscow at the end of 2019. The German space agency, DLR, then let it be known that discussions had opened for German participation in the Luna 28 sample return mission scheduled for 2028.

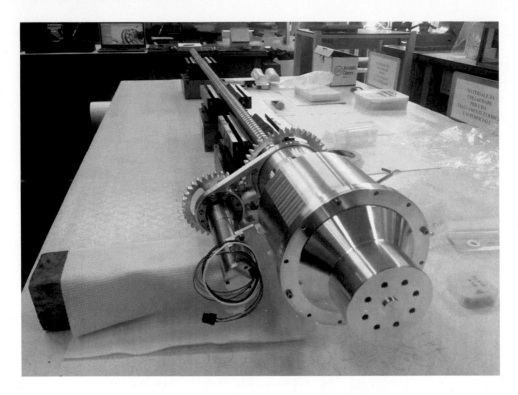

Luna 27 drill. ESA.

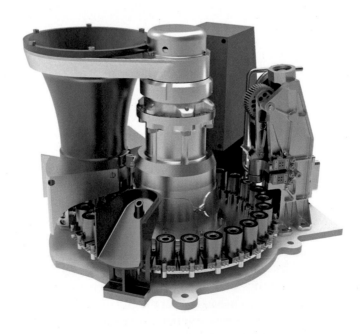

Luna 27 PROspA. ESA.

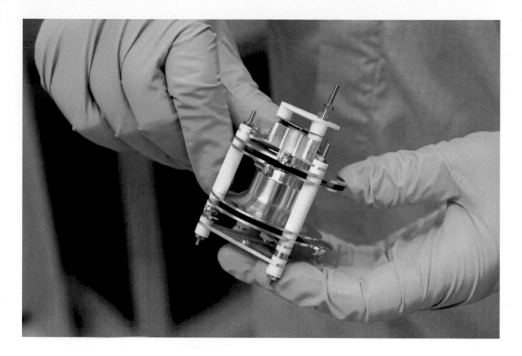

Luna 27 sensor. ESA.

Mercury was one of the least-probed planets in the solar system, having been visited only twice before, by America's Mariner 10 (1973) and MESSENGER (2004). On 28 October 2018, Europe launched the third visitor, Bepi Colombo, although it is not due to arrive until 2025. The Bepi Colombo mission study specified the need for a magnetometer and Russia supplied a Mercury Gamma Ray and Neutron Spectrometer (MGNS) and, with France, the Probing of Hermean Exosphere By Ultraviolet Spectroscopy (PHEBUS) instrument to probe the exosphere. This became Russia's first opportunity to fly instruments to Mercury. For MGNS, which would operate through the cruise as well, the principal investigator was Igor Mitrofanov, a veteran of Soviet-period planetary missions. The co-principal was Leonid Gurtvits of the Joint Institute for VLBI in Europe in Dwingeloo, the Netherlands. The aim of MGNS is to study the elemental composition of the sub-surface whilst searching for volatile elements, including any water or ice that might exist in shadowed craters, as well as triangulating gamma bursts from Mercury with other spacecraft. PHEBUS had a French principal investigator, Eric Quémerais, with co-principals from Japan and Russia (in this case, Oleg Korablev of IKI). PHEBUS will also look for water ice but its main purpose is to examine the composition, dynamics and characteristics of the planet's exosphere.

Although it was an application rather than space science, one unique cooperative venture was COSPAS SARSAT. The success of navigation satellites from the 1960s opened up possible applications for the use of satellites to receive distress calls from ships at sea. The Eole programme had shown how satellites could pick up signals worldwide and transmit them to the ground. Canada, France and the United States signed an agreement for the SARSAT system (Search And Rescue SATellite) in 1978. Coincidentally, the USSR was developing the COSPAS system and, in the spirit of détente following the Apollo Soyuz Test Project, there was sufficient goodwill to approve a marriage of the two, in effect their interoperability, which was signed in Leningrad on 23 November 1979. The first satellite under the system was launched on 30 June 1982 and the first rescue assisted − of three survivors of a light plane crash in British Columbia − took place only days after the service opened for business in September 1982. The number of people saved from ships, yachts and light planes is now estimated in the thousands. The beacons have been fitted to the European MetOp and Galileo satellites, Russia's Nadezhda and GLONASS satellites and the American NOAA satellites.

COSPAS SARSAT. Roskosmos.

Interkosmos after Interkosmos

Russian space science was one of the many casualties of transition in the 1990s. Russia was able to run some Soviet-period projects to conclusion (e.g. Interball, Koronas), keep others alive despite implausibly long delays (e.g. Spektr) and find other ways to stay in space science (e.g. INTEGRAL). Koronas was a programme of solar observatories conceived as far back as 1964, but brought to conclusion in the Russian period. It comprised Koronas I (1994), Koronas Fizika (more commonly called simply Koronas F, 2001) and Koronas Foton (2009). Koronas F was extremely successful, building up an archive of a million images of the Sun and its corona, obtaining x-ray, gamma and ultraviolet data on solar flares and taking fresh spectroscopic measurements of the corona, leading to a 483-page summary of the results. Led by IZMIRAN, the mission involved Ukraine and the former Interkosmos countries of Poland, Czech Republic and Slovakia, as well as drawing in Britain, France and Germany [35].

Unfortunately, there was little room for new projects, but one small way was through micro-satellites, one of the unexpected additions to the worldwide space industry in the 1990s. These were made possible by enterprising scientists and engineers taking advantage of miniaturization to design and build small spacecraft and fly them as passengers on existing missions. At a time when rockets and satellites were becoming ever larger, Surrey Satellite Technologies (SSTL) in Britain started an inspired counter-trend of ever-smaller satellites, which offered reduced costs both for the satellite itself and by hitching a lift on rockets with some spare payload capacity (piggyback).

Russia's first micro-satellite was the 80 kg KOMPAS 1 (Complex Orbital Magneto Plasma Autonomous Small Satellite, or COMPASS, in English) in December 2001, which was launched piggyback on a Zenit 2. This was followed by the 21 kg Kolibri 2000, launched from a Progress freighter, a different launch method. From an early stage, international cooperation was invited. KOMPAS 2 was launched on a Shtil rocket from the submarine *Ekaterinberg* on 26 May 2006 – another unconventional launch method – its purpose being to search for electromagnetic pulses that warned of impending Earthquakes, a research pioneered by Sergei Pulinets. The experiments on KOMPAS 2 came from IZMIRAN in Russia and the Eötvös Lorand University, Budapest, Hungary (VLF wave analyser), the Lviv Centre for Space Research in Ukraine, the Space Research Centre of the Polish Academy of Sciences in Warsaw and the Institute of Space Physics in Uppsala, Sweden (High Frequency Analyser).

Early Russian micro-satellite.

These small missions became opportunities to re-involve the Interkosmos countries in the post-Soviet Russian programme. The role of the socialist block countries in space research appears to have declined sharply after the period 1989–91. Up to that point, there had been frequent space science flight opportunities with the USSR on the Interkosmos programme, which concluded in the 1990s with Interkosmos 24 and 25. The leading countries in that programme were the GDR, Czechoslovakia and Bulgaria, with Hungary and Poland middle-ranking and Romania the least visible. Most of the socialist block Interkosmos countries re-orientated their relationship around ESA, generally after the period of accession to the European Union (2004, 2007). The Czech Republic became an ESA member in 2008, Romania in 2011 and Poland in 2012. Estonia and Hungary joined in 2015, while Bulgaria, Latvia, Lithuania and Slovakia set up cooperation agreements. Several set up space agencies, such the Romanian Space Agency (ROSA 1991), the Hungarian Space Office (1992), the Czech Space Office (2003) and the Polish Space Agency (2014). Bulgaria already had a Space Research and

Technology Institute, set up in 1987 but with institutional origins going back to 1969. The institutional re-orientation around ESA was most evident in the case of the Czech Republic. Although the past connection to Interkosmos was mentioned, neither its lengthy 2010 space plan nor its iteration of space projects that year listed cooperative projects with Russia [36]. It is quite possible that the older scientists, who grew up during the socialist period, kept up their contacts with their Russian colleagues and they would have shared a knowledge of the Russian language, while English-speaking younger scientists might have been more orientated toward the western, English-speaking scientific world.

The next micro-satellite was the Chibis M project ('lapwing' in Russian), a 40 kg micro-satellite developed by IKI as a successor to Kolibri, sent to the ISS flying at 400 km altitude by the Progress M-13M freighter in October 2011. Chibis M was built in-house by IKI in its 'Special Engineering Department'. The mission worked out well, with Chibis M exceeding its one year warranty period threefold. Its overall objectives were to investigate transient plasma wave processes in the atmosphere called Transient Luminous Events (TLEs), generally associated with lightning, some of which have popular names such as elves and sprites. There were five instruments on Chibis M: a Roentgen gamma detector; an ultraviolet and infrared detector; a radio frequency analyser; a digital optical camera; and a magnetic wave complex. The project not only involved Russian institutes (Skolbeltsyn Institute of Nuclear Physics and Lomonosov State University) but also the Institute for Space Research in Lviv, Ukraine, Eötvös Lorand University in Budapest, Hungary and the Institute for Atmospheric Physics in the Czech Republic. The Hungarian instrument was the magnetic wave complex (MWK) to study electromagnetism in the 0.1–40 kHz range and the inter-relationship of plasma wave processes in the ionosphere. This was the instrument principally responsible for investigations of space weather and electromagnetic waves and was a descendant of equipment originally flown on Interkosmos 12 (1975) and 24 *Aktivny* (1989). Csaba Ferenz was the Hungarian scientist involved. The solar panels had an area of 53 cm^2, whilst orientation and navigation used GLONASS. It was an attempt to achieve significant results using less expensive small spacecraft.

Instead of being launched directly from the station, as had been the case with many flocks of satellites, Chibis M was kept attached to the front of Progress when it departed on 25 January 2012. Before undocking, ISS cosmonauts Anton Shkaplerov and Oleg Kononenko connected its battery charger and the electrical command line for it to be detached. Progress then lifted its orbit to 500 km to deploy Chibis M for its mission. Chibis M de-orbited on 15 October 2014, with telemetry received right up to the previous day.

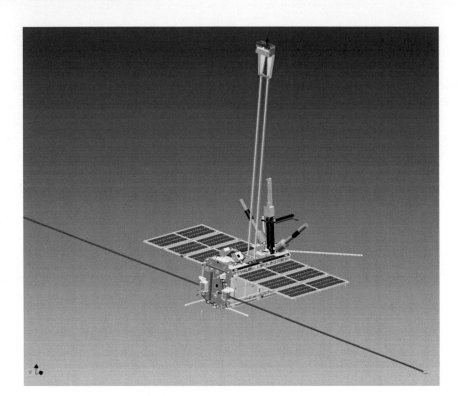

Chibis M. Stanislav Klimov.

No less than 30MB of data were downloaded using a 2.2 GHz transmitter, to data receiving locations in Russia (Kaluga and Tarusa), the Czech Republic (Panska Ves) and Budapest. Results of the mission were presented at the COSPAR conference in Moscow two years later and further results at the 10th small satellite symposium in Berlin the following year. Important outcomes were obtained in the study of lightning in the atmosphere, thunderstorm formation, electrical fields, VHF pulses and upper ionosphere low-frequency electromagnetic structures. Chibis M found that radiation from high-energy power lines leaked into the ionosphere above the F2 layer, causing extremely low-frequency electrical fluctuations. Chibis M recorded more than a hundred short, powerful, lightning discharges and also discovered a multi-band spectral resonant structure at middle and low latitudes in the ULF band, caused by lightning-generated pulses travelling up and down the ionosphere and reflecting back. It also looked at Trans Ionospheric Pulse Pairs (TIPPs), which were normally pairs of bursts separated typically by 150 microseconds. One percent were solitary TIPPs, but Chibis data, using ground reflection theory, suggested these were really the coalescence of two separate VHF bursts. Although Chibis M detected 400 TIPPs. it did not find their point of origin [37]. The chief Russian scientist involved was Stanislav Klimov.

Chibis M committee video conference. Stanislav Klimov.

The next small spacecraft was the 250 kg *Sergei Vernov*, launched with Meteor M-2 on 8 July 2014 to study space weather and the radiation belts. Csaba Ferencz and Peter Szegedi of Eötvös Lorand University provided the SAS3 data system and they subsequently co-authored a paper on swallowtail whistlers and thunderstorms possibly caused by seismic activity in the west Pacific [38]. The mission also included the Space Research Centre of the Polish Academy of Sciences in Warsaw. The *Sergei Vernov* provided good data for six months, especially on intense radiation at sub-auroral latitudes, before suffering an unexplained failure. It found what were called 'haystack' events of very low frequency waves around both poles and detected several TGFs, possibly connected with electron precipitation, prompting lightning but not thunder. The Polish Space Research Centre published its images of high latitude structures, the equator structure, the auroral region and even heating over Tromsø, Norway [39].

The accumulated results of these micro-satellite missions were presented to the COSPAR symposium on small satellites in 2019 by Stanislav Klimov, in association with his colleagues in Ukraine, Poland and Hungary. He traced these studies back to Aureole 3 [40].

Hungary was, in effect, the leading country in 'Interkosmos after Interkosmos', its two main contributions being the Plasma Wave Complex (PWC) and the SAS data system. The Space Research Group at Eötvös Lorand University in Budapest dated back to lightning studies from satellites in 1974 (Interkosmos 10) and its involvement culminated with Interkosmos 24 *Aktivny* at the end of the Soviet period (1989) before resuming in the Russian period with KOMPAS 2 (2006) and *Sergei Vernov* (2014). The purpose of the PWC was to identify the specific characteristics of ionospheric and magnetospheric radiation, characterize space weather and global lightning activity and obtain the electromagnetic parameters of space weather in the ionosphere from quiet to disturbed conditions.

The second big Hungarian contribution was SAS, or the Signal Analyser and Sampler, which worked alongside what was called the Data Acquisition and Control Unit ('Block of Storage of Telemetry' is also used). Essentially, there is no point in building a set of sophisticated instruments unless there is a system for interrogating them and then handling, transmitting and storing the subsequent data for the benefit of scientists. Here, *Aktivny* was critical, as it saw the group develop its first SAS unit (SAS1). They then provided SAS2 for KOMPAS 2 and SAS3 on both Chibis M and *Sergei Vernov*.

For ISS, the Plasma Wave Complex was expanded into what was called the Obstanovka complex, as a set of eleven instruments to test the electromagnetic environment around the space station – especially in low frequencies – and to take plasma wave measurements of Sun-Earth relationships in the F2 ionospheric layer. Obstanovka means 'environment' in Russian and the complex was installed on the Russian segment of the station (see Table 2.5). The leader of the project was Stanislav Klimov.

Hungary-Russia meeting. Roskosmos.

Table 2.5:
Obstanovka I complex

Instrument	Participants
Combined wave sensor	IKI and Lviv Institute for Space Research
Fluxgate magnetometer analogue	IKI
Fluxgate magnetometer digital	IKI and Lviv Institute for Space Research
Langmuir probe	Solar-terrestrial Influences Laboratory (STIL), Sofia, Bulgaria
Spacecraft potential monitor	Space Research Institute (SRI), Sofia
CORelating Electron Spectrograph	Space Science Centre University of Sussex, England
Radio frequency analyser	Space Research Centre, Warsaw
	Swedish Institute of Space Physics
Signal analyser and sampler	Eötvös Lorand Space Research Group, Budapest
Booms with sensors	Energiya
Thermoregulation system	IKI
Data Acquisition and Control Unit	Institute for Particle and Nuclear Physics, Hungary
Block for Storage of Telemetry	Institute for Particle and Nuclear Physics, Hungary

Source: IKI

In effect, the Hungarians were able to operate a wave complex in three different orbits: 51.6° at 400 km in the case of ISS; 52° at 500 km in the case of Chibis M; and in polar, 98.4° orbit from 640 to 830 km for the *Sergei Vernov*. Obstanovka I found significant changes in low-frequency noise between the shadowed and illuminated parts of the ISS orbit. The level of noise was lower for Chibis M, but it was able to pick up sub-auroral low-frequency electrostatic noise in both hemispheres and sub-auroral hiss in disturbed conditions on the northern hemisphere. Of the Bulgarian-made Langmuir probes, one was located at the front of the station where plasma would be undisturbed, the other at the rear. Between them, they could measure the effect of the large station disturbing the plasma through which it travelled by up to two magnitudes. The measurements showed a jump when the station went into eclipse, likely caused by the effect of darkness on electrons collected on the solar panels. It also changed when transiting areas of high plasma concentration around the equator. Electrical potential jumped during the solar storm of 23 March 2013 [41].

Bulgaria was probably the next most prominent contributor, its principal field of expertise being radiation science where it was a world leader. Bulgaria contributed the DP experiment to the Obstanovka 1, part of the Plasma Wave Complex. DP-1 and 2 were installed on the outside of the station on the Zvezda module in April 2013 by spacewalking cosmonauts Pavel Vinogradov and Roman Romanenko. The L- and T-shaped arms extending from the station were sometimes visible in photographs of the ISS, its hull and Earth in the background taken by the cosmonauts on board. Liulin was the name of Bulgaria's principal radiation-measurement device, with a total of 13 such devices installed from 1988 on Mir, Mars 96, the ISS, Foton M2 and M3, Phobos SR and Bion M1. The most recent is Liulin ISS2, developed between the Space Research and Technology Unit of the Bulgarian Academy of Sciences, IBMP in Moscow and the Energiya design bureau. Such measurements are essential in ensuring that the radiation exposure of cosmonauts in Earth orbit and further afield is within predictable and safe limits. Liulin seems to be an example of an instrument developed during the Interkosmos period that evolved and made a successful transition into the modern period **[42]**.

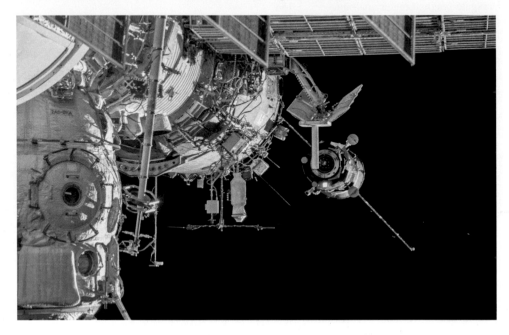

ISS external instrumentation. Roskosmos.

Whilst it did not contribute to these missions, Poland was another example of a country in transition. Its Space Research Centre (SRC), part of the Polish Academy of Sciences, was established in 1977 and organized Poland's participation in the Interkosmos programme, Vertikal sounding rockets and Prognoz, especially in solar studies. From 1991, Poland cooperated with ESA, including some of the latter's most prominent missions such as Rosetta. Cooperation with Russia continued in the form of providing instruments for Interball 1 (plasma wave analyser), Mars 96 (low frequency analyser), Interball 2 (plasma wave analyser and radio spectrum polarimeter), KOMPAS 1 and 2 (radio spectrometer) and the *Lomonosov* small satellite mission. Poland's contribution to the Koronas programme of solar observatories built on its experience in the Interkosmos programme, providing a high-resolution spectrometer for Koronas I, the RESIK x-ray spectrometer for Koronas Fizika and the SphinX x-ray spectrophotometer for Koronas Foton. The chief scientist involved was the SRC's Janusz Sylwester, the distinguished professor of solar physics.

Space Research Centre, Warsaw. CC Chmee2.

Interkosmos after Interkosmos: next

In 2020, Alexander Bloshenko, science director in Roskosmos, announced new collaborative plans involving the ISS (Obstanovka II), a Chibis successor called Chibis AI (2023) and two micro-satellites based on Chibis called Trabant (2024). In effect, they will resume the earlier experiments on Obstanovka I, Chibis M and *Sergei Vernov*. The 40 kg Chibis AI will fly at 460–520 km, the two 60 kg Trabants at 480–600 km and Obstanovka at the same altitude as before. Although it will primarily involve the old Interkosmos countries, an experiment with Italy will map the night atmosphere in ultraviolet, especially when it is lit up by high-energy electromagnetic flashes. Improved information would be applied to ensure greater safety for aircraft from lightning strikes.

Chibis AI was approved in 2012 and passed its design review in 2016, with Hungary joining in 2017 (IKI held a press conference on 21 April 2017). The mission is devoted to the study of the energetic mechanisms that generate intra-cloud discharges and gamma ray flashes in the atmosphere above thunderstorms, in particular their global mapping and occurrence and how their energies are dissipated. Chibis AI will carry a Radio Frequency Analyser and Neutron and Gas Spectrometer, one of whose aims will be to find the origin of TIPPs. To deal with the problem of identifying their point of origin, another measuring instrument called 'Kite' will be installed on the ISS itself to locate the point to within 10 km. Chibis AI will look at greenhouse gases and chemical compounds generated by high-altitude lightning and further explore the consequences for ionospheric plasma. It is intended that Chibis AI will be more accurate, reliable in event triggering and have a longer lifetime. Kolibri and Chibis M operated during the solar maximum, but Chibis AI will be at the solar minimum.

Hungary was unsurprisingly charged with developing the magnetic wave complex, comprising a spectral analysis processor and fluxgate magnetometer. For Trabant, it will develop two electromagnetically clean micro-satellites to study irregularities in the ionosphere, while Obstanovka II will study how the station inter-reacts with the ionosphere. The Trabants are to study plasma dynamics and space weather in the equatorial low latitude ionosphere at 510 km.

The Trabants will mark the largest Hungarian footprint ever. They will have a magnetic wave complex (MWC-T, Hungary) and Russian instruments for plasma density (DPP-T), electron density (FIEC-T and BAES T) and positively charged particles (Aries T). The aim is to open a new page in gradient studies of geophysical fields and low-frequency electromagnetic radiation, studying multi-scale turbulence in moving dynamic plasma. Trabant is predominantly a Russian-Hungarian project, with Hungary providing the memory block, transmitter and control systems.

The preliminary design phase for these projects was concluded in 2019. Trabant's purpose was defined as 'the experimental investigation of the fundamental problem of multi-scale turbulence in a moving dynamic plasma'. Television will show the departure of the two spacecraft from the Progress and Trabant will have its own gas jet propulsion system using nitrogen with two nozzles, with the two spacecraft flying between 100 m and 100 km apart, an approach that was verified by Czech Magion satellites in the 1980s.

Further in the future of magnetospheric science, the Strannik mission will include the Czech BMSW-S (2025) and Resonans, the Slovak DOK-M contributed by the Department of Experimental Physics in the Academy of Sciences in Košice.

Space science: conclusions

This chapter traced the trajectory of European-Russian cooperation in space science from the 1970s to 2020. It can be seen how it broadened out from being an exclusively French field of activity to one engaging other European countries such as Germany, Britain and Sweden and then ESA as the institutional actor for all its member states. Starting with Earth satellites to study the magnetosphere (Aureole), it expanded into solar observatories (Prognoz), biology (Bion), materials processing (Foton), interplanetary missions (VEGA, Phobos) and astronomy (Astron, Kvant, Gamma, Granat, Spektr). It is set to continue with the Luna programme and future Spektr missions. Some of the former socialist block countries, notably

Hungary, maintained programmes with Russia through Chibis, Trabant, *Sergei Vernov* and Obstanovka, as did Poland in the case of Koronas F. There were false starts too, with Spektr X, Laplas and Éos.

In the 1970s, the paths of cooperation set down by the 1966 Soviet-French agreement matured, with the USSR launching the French satellites SRET and Signe 3, plus the three joint satellites, Aureole 1–3, that used a Soviet satellite design to which jointly-planned Russian and French instruments were fitted. Aureole was the achievement of the leading personality of French-Russian cooperation, Yuri Galperin. The physics of the Arctic were a key area of expertise for both countries and the Aureole satellites brought them a high level of return. Galperin's laboratory convened international conferences to share the outcomes, events that later became routine.

Next, the Prognoz programme of solar observatories provided an opportunity to host French instruments in a series that ran during the Soviet period from 1972 to 1985 (Prognoz) and in the Russian period as Prognoz M/Interball (1995, 1996). French instruments were fitted to Prognoz 2, 6, 7 and 9, principally the previously developed Signe but also Gémeaux 2 and Galactika. Prognoz was not a well-known programme outside the USSR, but it provided a substantial volume of information from 20 years of observations of the Sun, the interaction of the solar wind with the interplanetary medium, the shock wave and the transition zone with the magnetosphere. This and related programmes were so extensive that the American Office of Technical Assessment (OTA) estimated that no less than a third of all French scientists and engineers were involved in joint projects with Russia, using a tenth of the CNES budget, FFR51m **[43]**. France was not the only country engaged in bilateral cooperation, as Sweden entered the picture with Interkosmos 16 and subsequent instruments flown on Prognoz. Unlike France, this did not lead to a programme of cooperation and it remains a little known part of the story.

European-Russian cooperation finally moved out of the dominant Paris-Moscow axis with the arrival of the biology and materials science programmes, Bion and Foton, first with Germany and then through ESA. Space biology had long been recognized as one of the least controversial areas of international space cooperation, even being approved by the Americans, so it was a safe starting point. The attraction of Bion and Foton lay in their providing opportunities for experiments not available elsewhere and at relatively low cost. Typically, France was first through the door (Cosmos 785, 1975), with ESA joining on Bion 8 (Cosmos 1887, 1987) and developing an increasingly sophisticated set of payloads for the Bion missions (Bion 9, 10), including its own Biobox and Biopan. ESA joining the Bion programme was an example of 'opportunistic' collaboration: the loss of the *Challenger* shuttle meant that European space science missions in key areas were grounded for some time, so the USSR quickly spotted an opening, one that was 'eagerly grabbed' as one observer put it. Europe was limited to 6 to 12 minutes of microgravity on sounding rockets from Esrange, but Russia could offer periods of up to two weeks in a proven, reliable system.

Foton used the same cabin as Bion and although the programme was dedicated to materials processing, it also carried biological cargoes and there was some interchangeability between the two series. As usual, France was first on board (Foton 2), with ESA experiments incorporated from Foton 7. Foton was the programme in which a German presence became quite significant (for example the MIRKA re-entry cabin) and it also became a substantial area of cooperative endeavour, with growing numbers of experiments. Bion and Foton improved the knowledge of radiation dosimetry, gravitational biology and radiation biology. The biggest gains in the area of cooperation were in astrobiology (the effects of spaceflight on living organisms) and meteoritics (the effects of re-entry on minerals and living organisms). These were themes never explored by the world's other space agencies. They showed how some terrestrial organisms, including multicellular macroscopic ones, could survive. The survivors were some lichens and plants seeds, bacterial spores, cyanobacteria and tardigrades, raising fundamental questions about the meaning of life, panspermia, planetary protection and meteoritic survival. This was truly cutting edge science.

Looking at the economics of scientific cooperation, Foton offered a level of microgravity better than the ISS and results that could not have been obtained on the station. Because Foton was unmanned, safety and documentation standards could be lower. Between that and currency factors, flying with Russia cost much less. If the Bion and Foton experiments had flown on the shuttle or ISS, the number flown would have been much reduced.

By the mid-1980s, cooperation between Russia and Europe reached the stage that a Soviet interplanetary mission could carry a substantial suite of European instruments. This was VEGA, which unsurprisingly had its origins in the French Éos balloon project that went through so many evolutions. The outcome fell far short of what its promoters – and in particular, Jacques Blamont – would have liked. Éos illustrated the impact of domestic politics on the space programme in France, especially after the arrival of Valéry Giscard d'Estaing as President. The Russians must have been exasperated with the French withdrawing from Éos, a second major project after Roseau. In the event, domestic Soviet politics provided a fresh injection of energy, with the combined forces of Mikhail Gorbachev and Roald Sagdeev turning VEGA into a showpiece international triumph.

VEGA set the benchmark for European-Russian cooperation as a norm for the next interplanetary missions, Phobos and Mars 8. Negotiations over the mission again became embroiled in the question of French balloons, once more leading to another unsatisfactory outcome for the aerostat campaigners. The German presence was now more evident, with Germany providing the Mars 8 camera system. Other countries also joined in, for example Finland and Austria.

The Mars 8 disaster paved the way for an extraordinary *volte face* by France, albeit led by an idiosyncratic minister, Claude Allègre. Throwing over their partners of 30 years, he instead went to the Americans to organize a joint sample return mission. Whilst questions could be asked about the reliability of their rockets, the Russians always stuck to agreements that had been made, but the Americans did not and pulled the budget for the project, leaving France with nothing. Despite the fickleness of the American budgetary process, ESA returned there to set up a similar mission for the 2020s, Mars Sample Return. In a volatile American financial situation, might it meet a similar fate?

In the last period of the space science story, the most productive area of cooperation was that of the great observatories. This involved France (Astron, Granat, Gamma), as might be predicted, but also Britain (Kvant), Germany (Kvant, Spektr RG) and some of the less prominent European players such as the Netherlands and Switzerland (Kvant). These observatories provided a substantial volume of new astronomical data. Spektr RG, with the prominent role for the Max Planck Institute, continues this line of research, its extraordinary imaging making up for the many years of uncertainty and delay before its eventual launch. Although ESA is sometimes criticized for emphasizing the space industry at the expense of space science and human spaceflight, the scientific outcomes of all of these missions can only be considered impressive.

Russia attempted to engage Europe in its Jupiter mission for the 2020s, but had to withdraw for budgetary reasons. Despite that, Europe later came aboard Russia's much delayed return-to-the-Moon programme in Luna 27, with other countries joining Luna 25, 26 and 28. Once ExoMars is concluded (see Chapter 5), these lunar missions, along with Spektr M, will become the focus of European-Russian collaborative efforts in space science in the later 2020s. The big gap here is of course the loss of Bion and Foton, which could still be continuing. One commentator pointed out that 'it took years and years to turn them into a streamlined, robust, effective and attractive programme – and then it was stopped'.

From the Russian point of view, cooperation enabled them to get instruments and expertise onto their spacecraft that they had not developed so much themselves (e.g. Prognoz, Kvant, Spektr). From the European point of view, Russia had lifting power (e.g. Proton) and spacecraft and flight opportunities not otherwise available (e.g. Foton, Bion). From the start of the Russian period, a new imperative emerged, which was the financial crisis, making the attraction of foreign resources ever more urgent. It was not just a resources issue, however, because Russian scientists found that their sources of original scientific data were drying up. INTEGRAL was a clear case of Russia spotting the opportunity to keep a source of data flowing, obtaining 25 percent of observing time in exchange for an expensive Proton launcher. These financial crises made Russia a trying partner, as projects were announced that never happened, were delayed indefinitely, were strung out along improbably lengthy schedules (e.g. Spektr, Koronas), or even disappeared altogether.

One may also notice an institutional evolution. Over the years, cooperation evolved from one of '*We are planning mission X, would you like to propose some instruments?'* to one of integrated *mission planning* (e.g. VEGA), to *joint planning* at the earliest, conceptual stage (e.g. Spektr, 1987), to the point that the latter is now the norm (e.g. Trabant). Similarly, bilateral approaches (e.g. only to France) gave way to approaches to multiple countries.

A big near-absentee here is Britain, despite its important role in the development of astronomy, astrophysics, radio astronomy (e.g. Jodrell Bank) and with centres of expertise in universities (e.g. University College London). The high point of British collaboration was Kvant, but the British government seemed to want such participation shielded from view. If the government was not impressed, the Russians were, as they offered the British the lead role in Spektr X, with two tonnes of instrumentation – some gift horse. That Britain's developed so little cooperation with the USSR or Russia may be attributed to deeper factors of politics and language, such as attachment to the American world of science. British space experiments, from Kvant to Foton 7, left almost no trace of a media imprint behind, although they were newsworthy stories, especially the repair of the space telescope by Titov and Manarov. After the period of Patrick Moore, the BBC's famed *Sky at Night* programme reported extensively on American scientific discoveries, but rarely travelled to Europe and even less to Russia. When the Institute for Space Research convened a 60th anniversary conference in Moscow in 2017 to mark the first Sputnik, there was an evident presence of scientists from Germany, France and Italy, even the United States – but not from Britain.

A second absentee here is Italy, which had also little cooperation with the USSR. Italy presents a contradictory picture. On the one hand, the ruling post-war party, the Christian Democrats, was strongly anti-communist – its main political opponent being the Italian Communist Party – and shared long-standing American concerns about the communist threat. On the other hand, its governments supported long-term trade relations with the USSR [44]. Most prominently, in 1966 Fiat began construction of a huge car plant on the Volga to build 660,000 Fiat 124s a year, with a new town, Togliatti, built for the car workers. It was named after the Italian communist party leader, Palmiro Togliatti. Other companies active in the USSR were Olivetti (typewriters and office machinery), Pirelli (rubber) and Snia (textiles and chemicals). These efforts were supported by Italian governments, both to make sure that Italy got there ahead of the competition and to ensure a place for Italy in the process of détente. This accelerated in the 1990s, with Italy importing energy and tourists while exporting food, textiles and *Made in Italy* products (e.g. shoes, handbags, clothes, cosmetics). From the end of the cold war, Italy acted as a 'middle power', a bridge between Russia and Europe. Italy became Russia's third largest trading partner worldwide, with 500 Italian companies trading there. In the high-tech area, Leonardo Finmeccanica had a partnership with Russian Helicopters to build helicopters, but there remained almost no cooperation in the space industry. Not until 2016 was there a joint project: an agreement between Thales Alenia Space Italian division and ISS Reshetnev to build two communications satellites, Ekspress 80 and 103, which eventually launched in July 2020.

It seems that the Italian space programme started on the American side and never really emerged from there. The founder of the Italian space programme, Luigi Broglio, started by working with the Americans on sounding rocket tests from Perdasdefogu in Sardinia, leading to a 1962 memorandum between Italy and the United States for the development of satellites and launching platforms (San Marco), with eventual cooperation on ISS. This was its chosen line of development.

The final part of this chapter was 'the Interkosmos programme after Interkosmos'. With the end of the Soviet Union, the socialist block countries turned westward to join the European Union and the European institutions, like ESA. It would be natural to assume that links to and collaboration with Russia waned, but this does not appear to be the case. To the contrary, several countries, notably Hungary, maintained a strong presence in Russian space science projects: KORONAS, KOMPAS, *Sergei Vernov*, Chibis M, Obstanovka I, with more to follow in the form of Chibis AI, Trabant and Obstanovka II. Hungary made a distinctive contribution in the form of the multi-wave plasma complex and the SAS data system. This participation may be the linear descendant of links developed during the Soviet period by Russian-speaking, now older scientists. The space agencies of several countries in eastern and central Europe were remarkably shy about reporting on their ongoing work with Russia, but they are an integral part of the story.

A final point is that such cooperation, especially across a linguistic, organizational and cultural divide, was never easy, even if it may have looked so. The Bion 10 mission illustrated the practical difficulties of cooperation and showed how they had to be sorted out and re-formalized in the new Russia. Agreement was not reached in the case of flying French balloons to Venus or Mars, one of the missions that never happened. Aerostats have still not ballooned over the deserts and valleys of Mars. From Yuri Galperin onward though, strengthened by the annual meetings, a critical mass of human resources and relationships developed to carry forward the project that dated from 1966 and which now embraces other countries. The wisdom of the 1966 decision to invest in annual reunions which cemented personal relationships was all the more apparent as the space science programme developed, because it enabled cooperation to survive inevitable setbacks and disappointments. Scientists who started with preconceptions and paranoia about the other side lost their prejudices over time and soon created a good rapport as they worked together. They 'forgot the politics and found that they ultimately shared the same dreams and fantasies'.

In the English-speaking world, the international language of scientists, the findings and outcomes of many of these missions have been little documented or disseminated, while some were hardly mentioned. Even with the cold war over, the astronomical world became more divided than ever into world dominated by American missions and discoveries, which were well popularized and the rest, which were much less well known, receding in the institutional memory. An example comes from Germany, when scientists in the Max Planck Institute that joined the Spektr RG project found themselves asking a colleague from the former GDR for help in finding documentation on Soviet and Russian space science: they simply did not know where to start, although there was plenty once they found where to look. One that stands out from this period was *The Earth's plasma-sphere*, by Konstantin Gringauz and Joseph Lemaire, a collaboration between one of the USSR's greatest space scientists, who discovered the plasmapause and Joseph Lemaire of the Université catholique of Louvain, Belgium, who met Konstantin Gingauz in 1985, with the most famous American expert in the field, the renowned Don Carpenter. It was completed posthumously after the death of Gringauz in 1993 [**45**]. It is a pity that there are not more.

References

1. Cambou, Francis & Galperin, Yuri: *Main results of the joint French-Soviet space project ARCAD 1 and ARCAD 2 for magnetospheric, auroral and ionospheric physics*. Annales de Géophysique, vol 38, #2, 1982.
2. *The ARCAD 3 project*, Annales de Géophysique, vol 38, #5, 1982; CNES: *Results of the ARCAD 3 project and of recent programmes in magnetospheric and ionospheric research*. Paris, author, 1984.
3. Lundin, Rickard; Hultqvist, Bengt; Pissareko, N and Zacharov, Alexander: *Composition of the hot magnetospheric plasma as observed with the Prognoz 7 satellite*. Kiruna, Kiruna Geophysical Institute with IKI, Moscow, 1981.
4. See this author with Zakutnyaya: *Russian space probes*, Praxis/Springer, 2012.
5. Sandalh, I *et al: First results from the hot plasma instrument PROMICS 3 on Interball 2*, Ann Geophysicae, 17, 1999; Sandahl, I *et al: First results from the plasma composition spectrometer instrument PROMICS 3 on Interball 2*, Ann Geophysicae, 15, 1997.
6. Wormbs, Nina and Källstrand, Gustav: *A short history of Swedish space activities*. Noordwijk, European Space Agency, 2007; Sheldon, Charles: *International science projects*, Global Security, undated; Zenker, Stefan: *Space is our place – Swedish Space Corporation, 25 years, 1972–1997*. Solna, Swedish Space Corporation, 1997; Grahn, Sven: *Interkosmos 16*. Sven's place, www.svengraph.pp.se.
7. Gärdebo, Johan: *Environing technology: Swedish satellite remote sensing in the making of the environment, 1969–2001*. Stockholm, Royal Institute of Technology, 2019.
8. For this history, see Facon, Isabelle & Sourbès-Verger, Isabelle: *La coopération spatiale Russie-Europe, une entreprise inachavée*. Revue Géoéconomie, 2007, #43.
9. Fournier-Sicre, Alain: *Focus on ESA's Moscow office*. On station, #1, December 1999.
10. Demets, René: *Biological experiments on Bion 8 and Bion 9*. ESA, Noordwijk, ESA, 1996. Ref ESA SP-1190.
11. Powell, Joel: *Biopan and Biobox – ESA's recoverable experiment carriers*. Spaceflight, vol 38, no 2, February 1996; Demets, René & McAvinia, Ruth: *From Eureca to EXPOSE and ExoMars – the evolving tools of astrobiology research*. ESA Bulletin, #172, Q4 2017.
12. Demets, R; Jansen, WH & Simeone, E: *Biological experiments on the Bion 10 satellite*. ESA, Nordwijk, 2002. Ref: ESA SP-1208. Mission outcomes are also covered in more detail in this author with Zakutnyaya, Olga, *op cit*.
13. Chayen, Naomi: *Microgravity crystallization of HIV reverse transcripts*. Erasmus Experiment Archive, record #7573. See also *Kashtan on Foton 7*, monograph supplied by René Demets, ESTEC.
14. Powell, Joel: *German re-entry vehicle flies on Foton 11*. Spaceflight, vol 40, no 6, June 1998.

15. Verga, Antonio; Baglioni, Pietro & Demets, René: *ESA success with Foton 12.* On station, #1, December 1999; see also Powell, Joel: *Microgravity upgrade.* Spaceflight, vol 43, no 8, August 2001.
16. Baglioni, Pietro; Demets, René & Verga, Antonio: *Payloads and experiments on Foton M-1.* ESA, 2002.
17. Sancho, LG et al: *LICHENS 2 – lichens as extremophile organisms in space.* Erasmus Experiment Archive, Experiment Record #8243.
18. Sancho, Leopold G et al: *Lichens survive in space – results from the 2005 LICHENS experiment.* Astrobiology, vol 7, #3, 2007; Rettberg, P et al: *Survivability and protection of bacterial spores in space.* ESA, Proceedings of second European workshop on exo-astrobiology, Graz, Austria, 16–19 September 2002.
19. Rettberg, P et al: *MARSTOX 2 – Martian soil, solar UV radiation and spores: protection and toxicity.* Erasmus Experiment Archive, Experiment Record #8244; Ehrenfreund, P; Peters APA & Ruitercamp, R: *ORGANICS 2 – extraterrestrial; delivery of organic molecules.* Erasmus Experiment Archive, Experiment Record #8245; Gilichinsky, DA et al: *PERMAFROST 2: the influence of the space environment on the viability of the ancient permafrost microbial communities.* Erasmus Experiment Archive, Experiment Record #8246; Demets, René & Weems, Jon: *Extremophiles – choosing organisms to study.* ESA Bulletin, #172, Q4 2017; Jonsson, K Ingemar et al: *Tardigrades survive exposure in low Earth orbit.* Current Biology, vol 18, #17, 9 September 2008.
20. Dachev, Tsvetan et al: *Preliminary results from the radiation environment observed by the RD3-B3 radiometer inside Bion M-1 spacecraft.* Aerospace Bulgaria, 25, 20.
21. Spaceflight, vol 31, no 3, March 1979; Spaceflight, vol 21, no 5, May 1979.
22. Sagdeev, Roald Z: *The making of a Soviet scientist.* New York, Wiley, 1994.
23. *Mars 94 takes shape.* Spaceflight, vol 30, no 112, November 1988; Klaes, Larry: *The rocky Soviet road to Mars.* Spaceflight, vol 32, no 8, August 1990; Bond, Peter: *Soviets revise Mars targets.* Spaceflight News, #50, February 1990.
24. CNES: *Rapport d'activité, 1998.* Paris, author, 1999; *US, France urge broader tack to bolster Mars exploration.* Aviation Week and Space Technology (AW&ST), 1 March 1999.
25. *NASA howls over massive budget cut.* (AW&ST), 2 August 1999; Day, Dwayne & Kane, Van R: *The rover returns.* Spaceflight, vol 62, #6, June 2020.
26. Evans, Ben: *Delving into Mir's astrophysical attic.* Spaceflight, vol 49, no 5, May 2007; Spiteri, George: *Mir's first module – the inside story.* Spaceflight News, #51, March 1990.
27. This writer: *Spektr and Russian space science.* Space Chronicle, vol 71, 2017.
28. Zelenyi, Lev et al: *Plasma F experiment – three years of on-orbit operation.* Solar System Research, vol 49, 2015.
29. *UK-Soviet space deal.* Spaceflight, vol 28, no 11, November 1986; Breus, Tamara: *UK projects for Mir space station.* Spaceflight, vol 29, no 3, March 1987.
30. *UK/Soviet agreement on space.* Spaceflight, vol 29, no 5, May 1987; Gavaghan, Helen: *Moscow visit signals glasnost in space.* New Scientist, October 1987.
31. eROSITA x-ray telescope captures hot, energetic universe. Sci News, 19 June 2020.
32. Japan steps in to supply key component to Russia's space programme. Space Daily, 20 August 2019; *Russia's future space observatory to be able to search for other civilizations.* TASS, 28 May 2020.
33. Blanc, Michel at al: *Laplace – a mission to Europa and the Jupiter system for ESA's Cosmic Vision programme,* Exp Astron, 2009; Zelenyi, Lev et al: *Europe lander mission – a challenge to find traces of alien life,* from Galileo's Medicean moon – their impact on 400 years of astronomy, International Astronomical Union symposium, 2010.
34. ESA: *PROSPECT project objectives and requirements document.* ESA, Noordwijk, 2015; Swiss Academy of Sciences: *Space research in Switzerland, 2012–2014.* Davos, author, 2014; Trautner, Roland: *PROSPECTing for lunar polar volatiles.* ROOM, autumn 2019.

35. Vladimir Kuznetsov (ed): The Coronas F space mission: key results for solar terrestrial physics. Berlin and Heidelberg, Springer, 2014.
36. Ministry of Education, Youth and Sports: *National space plan*. Prague, 2010; Czech Space Office: *Czech space projects*. Prague, author, 2010.
37. Dolgonosov, Maxim *et al: VHF emission from lightning discharges record by Chibis M micro-satellite*. Presentation, COSPAR, Moscow, August 2014; Klimov, SI *et al: Microsatellite Chibis M (25.1.2012–15.10.2014): results, lessons and prospects*. Presentation at 10th small satellite symposium, Berlin, 21 April 2015; Klimov, Stanislav *et al: First results of MWC SAS3 electromagnetic wave experiment on board the Chibis M satellite*. Advances in Space Research, vol 54, 2017. Klimov, Stanislav *et al: Microsatellite Chibis M (25.1.2012–15.10.2014) – results, lessons and prospects*. Small satellites for Earth observations, 10th International Symposium, Berlin, 20–24 April 2015; Pilipenkoa, V *et al: IAR signatures in the ionosphere – modelling and observations at the Chibis M satellite*. Journal of Atmospheric and Solar Terrestrial Physics.Vol 154, February 2017; Dolgonosov, MS *et al: 'Solitary' Trans Ionospheric Pulse Pairs onboard the micro-satellite Chibis M*. Advances in Space Research, vol 56, 2015.
38. Panasyuk, Mikhail *et al: Non-linear effects in electromagnetic wave activity observed in the Relec experiment on board the Vernov mission*. Ann Geophys, 15 November 2018; Panasyuk, Mikhail *et al: The global distribution of sub-relativistic electron fluxes and VLF waves in the near-Earth space as measured in the Vernov mission*. VERSIM, Apatity, Russia; Bogomolov, Vitaly *et al: Observation of TGFs onboard Vernov satellite and TGEs in ground-based experiments*. Geophysical Research Abstracts, vol 18, 2016..
39. Rothkaehl, Hanna: *Multi instrument radio diagnostics*. Warsaw, Polish Space Research Center.
40. Klimov, Stanislav *et al: Monitoring of space weather electromagnetic parameters in the ionosphere – projects Chibis M, Obstaovka I and Vernov*. Proceedings of 8th workshop, Solar influences on the magnetosphere, ionosphere and atmosphere, Sunny Beach, Bulgaria, 30 May–3 June 2016; Klimov, Stanislav: *Spatial-temporal study of plasma waves and ionospheric parameters using micro-satellites*. Fourth COSPAR symposium, Small satellites for sustainable science and development, Herzliva, Israel, 4–8 November 2019.
41. Kirov, B: *Langmuir probe measurements aboard the International Space Station*, from AB Stepanov & YA Navogitsin: Sun and Sun-Earth physics, Russian Academy of Sciences, St Petersburg, 5–9 October 2015.
42. Gramatikov, Pavlin; Nedkov, Roumen; Stanev, Georgi: *Secondary power supply for space potential monitor DP-1 and DP-2 Obstanovka project, International Space Station*, Aerospace Bulgaria, #31, 2019; Dachev, Tsvetan *et al: Description of the Liulin type instruments and main scientific results*. Aerospace Bulgaria, 26, 2014; Dachev, Tsvetan *et al: Description of the Liulin type instruments and main scientific results*. Aerospace Bulgaria, 28, 2016.
43. Office of Technology Assessment: *US-Soviet cooperation in space*. US Congress, OTA, Washington DC, 1985.
44. Sidi, Marco: *Italy: the advocate of cooperation* in Marco Siddi (ed): EU member states and Russia – national and European debates in an evolving international environment. Finnish Institute of International Affairs Report, 53, March 2018; Siddi, Marco: *Economic cooperation between Italy and the Russian Federation – history, success stories and challenges*. Finnish Institute of International Affairs Report, 2019; Siddi, Marco: Italy's 'middle power' approach to Russia, The International Spectator, 2017.
45. Lemaire, JF; Gringauz, KI, with contributions from Carpenter, DL and Bassolo, V: *The Earth's plasmasphere*. Cambridge, Cambridge University Press, 1998.

3

Human spaceflight

Given the progress made by the USSR and France in unmanned spaceflight cooperation, it was only a question of time before manned or piloted flight should
appear on the agenda[1]. By this stage, though, Presidents de Gaulle and Pompidou
had both passed on and the nature of that relationship became more complicated
under their successors, Valéry Giscard d'Estaing and François Mitterrand. By the
time of the Russian Federation and the International Space Station (ISS), human
spaceflight cooperation had broadened out to include Germany and other European
Space Agency (ESA) countries. The high-profile, high-risk nature of human
spaceflight meant new and different types of cooperation, standards, structures,
presentation and sensitivities.

First French piloted mission: PVH

Europe began to consider human spaceflight from 1970. With the Apollo Moon
landing achieved, the Americans moved quickly onward, with the Space Shuttle
programme being decided upon as the next field of endeavour. The United States
invited Europe to join the programme, the agreement being that ESA, led by
Germany, would build a laboratory called Spacelab to fly on week-long missions
in the shuttle's large cargo bay, on the understanding that there would be abundant

---

[1] The terms 'manned', 'piloted' and 'human' spaceflight are used interchangeably in this and
other chapters, with preference given to the gender-neutral second and third forms.

The original version of this chapter was revised. The correction to this chapter is available at
https://doi.org/10.1007/978-3-030-67686-5_7

opportunities for Europeans to fly on board. Spacelab was hugely expensive for Europe – mostly so for Germany, where it was built – but the prospect of Europeans flying on it remained vague. France made a pitch for a French astronaut on the first Spacelab, which was not agreed, so as these prospects receded, the idea of getting a lift into orbit elsewhere became more attractive. The shuttle experienced lengthy delays and would not get airborne until 1981. By contrast, the USSR had constructed the world's first orbital space station, Salyut, in 1971 and this was the main line of development of its piloted programme. From 1978, guest cosmonauts from eastern and central Europe, starting with Czechoslovakia, had flown to Salyut 6, which with some justification could be called the 'first international space station'.

It was only a question of time before a west European flight to Salyut would arise. Soviet cosmonaut Pavel Popovich, then chairperson of the France-USSR association, visited Brétigny space centre in April 1971, the first time a cosmonaut had visited a European space facility. Over the years, other cosmonaut visitors – Georgi Beregovoi and Yevgeni Khrunov – met French space officials. It is not known whether such a flight was discussed on any of these occasions.

Pavel Popovich in France. CNES.

Yevgeni Khrunov in France. CNES.

Although the mission was seen later as a French response to a Soviet invitation, it was in fact the French who first asked the question. CNES, through its President Maurice Lévy and on the original suggestion of balloonist Jacques Blamont, first suggested a French manned flight to a Soviet orbiting station at the annual meeting of the Soviet-French cooperation commission in Kiev in 1974. The Russians initially said no, probably the first time that a cooperation idea had been turned down. What was going on?

Here, political factors become important once again and the politics of post-Gaullist French-Russian relationships required some adjustment. Georges Pompidou died suddenly in April 1974 and was succeeded by independent republican Valéry Giscard d'Estaing. It seemed like business as normal when General Secretary Leonid Brezhnev visited France in December 1974, which was reciprocated when Giscard d'Estaing made his first visit to the USSR in October 1975. Although areas of cooperation were discussed, space was not highlighted, nor was there a visit to Baikonour and the French president's reception was even described as 'chilly'. Interestingly, congressional observers detected a loss of momentum in Soviet-French cooperation [1]. Though drawing support from the same constituency as Pompidou and de Gaulle, Giscard d'Estaing spoke on the one hand of a 'multi-polar world' that broke away from the hegemonism of the two nuclear superpowers, but on the other did not speak of a special connection to Russia, unlike de Gaulle. It is possible that the two sides were testing one another and two years later there was an improvement. Brezhnev returned to France in June 1977 and was given the red carpet treatment, with a 101-gun salute, an escort of Mirage fighters and the gift of two sports cars (Brezhnev accumulated so many cars that his disapproving mother once questioned him 'what will you do if the communists ever come to power?') [2].

This much better relationship led the Russians to change their mind and an invitation for participation in manned flight was subsequently issued by Brezhnev to Giscard d'Estaing on the next state visit (28 April 1979). The suggestion was for a French mission to an orbital station in 1982, quite a short time in which to put a piloted mission together. The French formally agreed in principle on 6 September 1979 when the first reports found their way into the French press. A possible programme was debated at the annual cooperation meeting in Ajaccio, Corsica, where a formal agreement was signed on 20 October 1979. Given that the United States was about to fly the first western Europeans on board Spacelab, Salyut gave the Russians the opportunity to do so ahead of their traditional rivals. France would pay for the transport of its trainees, their assistants and the scientific equipment for the mission, estimated at FFR14bn, but there was no cash payment for the mission itself. This meant that France would be the first west European country to get a person into orbit. The first Spacelab did not fly until 1983, with German Ulf Merbold aboard.

President Valéry Giscard d'Estaing (third from left) at Le Bourget air show. CNES.

The warmer relationship brought France closer to the traditional Gaullist position. France refused to join the chorus of western condemnation of the USSR's invasion of Afghanistan that December, with the president making it clear that maintaining détente in Europe was more important than the invasion of a far-off country about which they knew little. Giscard d'Estaing visited Brezhnev in increasingly troubled Warsaw the following year and French companies took up trade contracts from those American companies now vetoed by Washington.

Preparations for the first piloted mission, in France called the PVH (Premier Vol Habité), now went ahead. Candidates for the mission were invited by advertisement within the aerospace industry and the military in September 1979. There were 430 applications, of whom 196 were selected for interview, including 26 women. Of these, 70 went to the first round of medical examinations, reduced to 30 for the second round. A group of seven, later five, was selected and announced on 1 March 1980. It is believed the French favoured a female candidate, but the Russians were insistent on a military pilot. In the end the problem was solved for them, because Françoise Varnier, an optical specialist from the University of Marseille, broke her leg during a heavy parachute landing. Eventually, Jean-Loup Chrétien was selected, with Patrick Baudry as his backup, the announcement being made on 11 June 1980. Both had over 3,000 hours of flying time, with Chrétien being head of the Mirage F-1 fighter test programme, so it is no surprise that he became the top candidate.

They then headed to Moscow for training, where they had to learn Russian and familiarize themselves with Russian space systems. The French prepared a set of experiments for their week on board the station, with 30 hours allocated to their operation. CNES chose 37 experiments, which cost €6 m. The Russians gave no details of the space available for them, so the French scientists had to ask, '*Is this too big?*' and worked at the problem until the right dimensions were achieved according to whether they got a 'yes' or 'no'.

Jean-Loup Chrétien and Patrick Baudry. CNES.

Well in advance of the PVH, two French experiments had already been operating on the previous Soviet orbital station, Salyut 6. Cytos, brought up on 10 January 1978, was an experiment to test the effects of spaceflight on simple biological objects, especially the cellular proliferation of micro-organisms compared to an analogue on Earth. The second was Elma, a joint programme of cooperation in materials processing experiments in the station's Splav and Kristall furnaces. Here, Salyut gave France unique access to one of many areas where the USSR was at the cutting edge: materials processing, where sophisticated furnaces (e.g. Kristall) had been operating for a number of years. Russia had pioneered materials processing in space ever since the first such instruments were installed on Salyut 5 in 1976 and had one of the top plasma scientists in the world, Vladimir Fortov, who received his degree in the subject back in 1971. For Elma, samples prepared in laboratories in Bordeaux, Grenoble and Meudun were transported to the station by Progress 5 on 25 March 1979. The experiments covered crystal growth in solutions, the solidification of alloys, vapour phase crystal growth and the production of magnetic materials. They were tested during the long-duration mission by Vladimir Lyakhov and Valeri Ryumin that year.

Splav.

PVH in trouble

The politics of Europe experienced a seismic shift in May 1981 when François Mitterrand was elected to the presidency. A visibly worried American vice president, George Bush, rushed to Paris to express American governmental alarm about the socialist-communist programme for government and the fear that Italy, too, would soon 'go communist'. The French were predictably furious at the attempted American reprimand but these developments soon impacted on the PVH for a set of convoluted political reasons.

Since 1959, France had been ruled by two Gaullist presidents (the general himself and Georges Pompidou) and the republican Giscard d'Estaing. All had been repeatedly challenged during presidential elections by the socialist party candidate, François Mitterrand, in 1965, 1969 and 1974. The traditional outcome was anticipated for the next presidential election in 1981, with Giscard d'Estaing expected to win a second term.

Strangely, Moscow had backed Giscard d'Estaing in preference to François Mitterrand in the 1974 presidential election, breaking normal diplomatic rules when the Soviet ambassador paid a well-publicized call on him during the short interval between the two rounds of voting [3]. Why would Moscow favour a conservative in preference to a socialist candidate, moreover one supported by the French Communist Party (Parti Communiste Français, or PCF)?

It appears that Moscow may have responded to the concerns of PCF leader Georges Marchais that a socialist-communist alliance in government would benefit the socialists more than the communists (he was right about that). He considered it better to wait until declining economic and social conditions made the triumph of the PCF on its own inevitable (wrong so far). Another theory lay in the Kremlin's fear that a socialist-led government with communists might show the voters of western Europe that a parliamentary road to socialism might be possible, a treasonable notion for Leninist purists. Such was the complex, conspiratorial, calculated nature of the Marchais-Mitterrand-Giscard-Moscow relationship. Mitterrand was a sophisticated political operator, experienced as a government minister in the Fourth Republic. He needed the help of the traditionalist PCF and the vote of its supporters to get into power, but he also needed to show the larger number of moderate left voters publicly that he would not be a prisoner of the PCF once in government, an ambition in which he succeeded better than anyone might have anticipated. He was not called 'The Sphinx' for nothing.

Earlier, in April 1975, the USSR cancelled a visit to Moscow by Mitterrand at two weeks' notice, probably because Mitterrand had attacked both human rights issues in the USSR and the actions of the Stalinist Portuguese Communist Party. The cancellation got a bad press in France and the Soviet Union quickly changed its mind before further damage was done. Mitterrand touched down in

Moscow on 24 April 1975 and was received by General Secretary Leonid Brezhnev. Mitterrand was quick to notice Brezhnev's physical deterioration – he had suffered a stroke the previous year – but ever the diplomat, kept that knowledge to himself, as he had with Georges Pompidou's long illness. Mitterrand, though, was still sore about losing the presidency by a small margin to Giscard d'Estaing when Moscow backed Giscard the previous year, although Soviet backing was unlikely to have weighed heavily on the minds of the average French voter.

Once elected in May 1981, Mitterrand found that his relationship with Moscow had become increasingly difficult. The French stumbled across a Russian espionage plot. A Russian official, code named Farewell (real name Vladimir Vetrov), had passed a vast amount of intelligence information to France with the identities of almost 400 Soviet spies, mainly in industrial espionage. Martial law in Poland that December created further tension.

Even before that martial law, new American President Ronald Reagan had been arguing for restrictions on trade with the USSR. Mitterrand would have none of it, pointing out that Napoleon's blockade of Russia in the 19th century had failed and asking why, when the Americans were exporting vast amounts of sanctions-exempted grain to the USSR, Europe should not export machinery. The United States warned Europe against participating in an oil pipeline project with Russia, which greatly riled Mitterrand, who not only signed the deal but agreed to buy 8bn m^3 of gas. The US responding by embargoing gas pipeline machinery built by American subsidiaries in Europe. A diplomatic war of words followed, with Mitterrand insisting that Europe would not accept subordination to American orders 'at whistle point'. Mitterrand found himself caught between a Soviet side that had been breaking the rules by supporting his political opponent, spying and had reacted badly to his disapproval of their policies in Afghanistan and Poland; and the Americans who regarded his independence as insubordination. The French intelligentsia backed the Poles, its new heroes being Solzhenitsyn and Sakharov, who Mitterrand also supported.

It is no surprise that the president also came under pressure to cancel the upcoming PVH flight. Strangely enough, when Mitterrand was elected, the Russians presumed that the mission would be cancelled because it was associated with his predecessor, although the two astronauts explained to their colleagues in Star Town that French politics did not necessarily work like that. In France, leading members of the Academy of Sciences wrote a letter to the president urging the flight be called off and 4,000 scientists signed a petition. CNES was so worried that its president, Hubert Curien, went to prime minister Laurent Fabius to stress the long-term importance of the French-Russian space alliance. Mitterrand was unshaken and never interfered with the Soviet-French institutions or ties that dated to 1966, which remained stable and unaffected. Mitterrand agreed to the mission, but to cool political criticism ordered 'minimal symbolism': no hugging, kissing, medals, music nor ministers. The highest representation was that of the ambassador. The Russians then said there could be no CNES representation at the launch, which caused another row. Now the PCF was not pleased and insisted on its own representation, some sympathetic scientists and veterans of the Normandie Niemen squadron.

In the meantime, preparations for PVH had also become problematical. There was a first clash at the annual reunion in October 1980, in Baku, Azerbaijan. The French had devised an ambitious scientific programme, which they expected to be lifted into orbit by a Progress freighter dedicated for them. A military officer explained that there was no way the French could have a Progress to themselves and that their scientific allocation was 100 kg only. The French threatened to pull out of the mission completely, but some long-distance phone calls were made and the mission was back on again. In the end, 80 kg went up aboard the new orbital station, Salyut 7, in April 1982 and a further 230 kg on a Progress in May.

PVH Russian language training. CNES

PVH team training. CNES.

On arrival in Star Town, Patrick Baudry and Jean-Loup Chrétien were greeted by cosmonaut Boris Volynov who explained the rules: no phones, no cars, no holidays, no newspapers, no letters, no photos. The French were glad that they had selected military astronauts who could cope with barracks life, which revolved around what they called their 'Bermuda triangle' of the classroom, the canteen (*stalovaya*) and their flat. Classes were dictated, with no photocopies, exams every three weeks and the professor under evaluation as well to make sure he was sufficiently rigorous. There was a library, but books could only be read on site rather than borrowed and this was checked on exit. They had some time off to visit Moscow during the weekends, when they tried to get away from their minders. Their spartan, simple lifestyle was outweighed by the warm welcome they received from the cosmonauts with whom they trained, their colleagues and an often demanding social programme around emotional departures and returns from missions. The simple fare on offer included the regular cabbage and cucumber menus in the *stalovaya*, but Baudry arranged for some Bordeaux (wine) to be delivered by diplomatic bag. Their sounding off about the food in their flats – or rather, to the microphones in them – did lead to some improvements in the *stalovaya* and recorded wishes like *'If only the canteen would serve some...'* were often met remarkably quickly. The two Frenchmen had been learning Russian for two months by the time they arrived, but nonetheless found mastering the language the most difficult part of the training. One of their biggest challenges was to not make any unguarded political comments, for fear that this would damage the already difficult political atmosphere around the flight. They had contact with the French cultural attaché at the embassy and went there once a week to pick up their post. It was all quite a contrast for the Europeans training indefinitely for Spacelab in Houston, who had plenty of personal freedom but had to organize their own accommodation, sometimes some distance from the training area. In Star Town, they walked to work nearby. By the time the Americans flew on Mir, new houses had been built for them in Star Town, but this was not the case for the French.

PVH flies

In September 1981, two crews were formed: Jean-Loup Chrétien with Yuri Malashev and Alexander Ivanchenkov; and Patrick Baudry with Leonid Kizim and Vladimir Soloviev. In February 1982, Malashev had a medical issue and was replaced by Vladimir Dzhanibekov, though a later account suggested that it was a personality clash. The Chrétien crew was selected as the prime crew in September 1981. Dzhanibekov was a veteran of the Soyuz 28 flight which docked with Salyut 6 in 1978, while Ivanchenkov had flown 139 days on Salyut later that year [4].

French join Soviet space mission

TWO FRENCH test pilots have started training with Soviet cosmonauts for a manned space flight to take place next summer.

One of the two will fly on a three-man mission to an orbiting space laboratory of the Salyut type, aboard the recently tested Soyuz-T three-seater space vehicle.

It has not yet been decided whether the flight will be made to the veteran Salyut-6, which has already accommodated a dozen missions, or whether a new Salyut is to be orbited.

Two complete crews are already in training for the mission, the final choice of main crew and back-up to be made when training is completed.

Commanding the first crew is Yury Malyshev, who tested the three-man Soyuz-T-2, when it docked with Salyut-6 last June.

Flying with him will be Alexander Ivanchenkov, who has already spent 140 days on Salyut-6 during an earlier mission, and French test pilot Jean-Loup Chretien.

The other crew is headed by Leonid Kizim, who took part in the first flight of a Soyuz-T with a three man crew, and includes Vladimir Solovyev with the other French participant, Patrick Baudry.

French participation in this flight represents a new stage in a long-standing co-operation between France and the Soviet Union in unmanned space exploration, which includes the carrying of French equipment on flights to the Moon, Mars and Venus.

Above, the two crews, left to right: Vladimir Solovyov, Leonid Kizim, Patrick Baudry, Alexander Ivanchenkov, Jean-Loup Chretien and Yury Malyshev. Below, Patrick Baudry (centre) and Vladimir Solovyov (second from left) on a visit to the Aivazovsky Gallery in Feodosia with Leonid Kizim (right) and Alexei Leonov.

PVH crew selection.

Chrétien and his colleagues boarded Soyuz T-6 on the evening of 24 June 1982. French teams in Baikonour were astonished at how close they could be to the rocket, from the hangar to right up close before it was fuelled, a complete contrast to Cape Canaveral. The night before the mission, the crew members bathed in an alcohol bath and slept in beds sterilized with ultraviolet light, a procedure designed to minimize infection.

The launch was broadcast live on Russian and French television, which followed the swing arms falling back, the ignition and the rocket's ascent into the summertime night sky. A day later, Dzhanibekov orientated the Soyuz for docking 900 m from the station as they overflew a tracking ship positioned near Gibraltar. After a hair-raising approach in which the computer broke down and the spacecraft began to gyrate, Dzhanibekov regained control and steered the craft in manually from a record 1,000 m out and they docked with the Salyut 7 orbital space station over Libya. They were welcomed aboard by the resident crew of Anatoli Berezovoi and Valentin Lebedev. French embassy officials had been invited to watch the docking in the mission control centre, the TsUP, in Kaliningrad (since renamed Korolev). When his diary was published later, Berezovoi particularly remembered this visit for breaking the tedium of Russian space food with French delicacies of crab, cheese, hare and lobster followed by strawberries.

However, they were not there for the menus. The experiments illustrated the benefits that an orbital station could offer to a European country, starting with the medical. The most original French instrument in the scientific package was the 85 kg *Echographie*, which provided sectional images of the heart, a concept that was later used in hospitals the world over, accompanied by a sound system of earphones to listen in to heartbeats. The idea that doppler ultrasound could follow internal human body movements in this way was first put forward by Léandre Pourcelot in the mid-1970s. In orbit, it was possible to see arteries and veins, heart cavities and the speed of blood flow. The experiment was used three times, on days 3, 4 and 5 and again after landing. Originally, it was to be one-off experiment during PVH, but the Soviet cosmonauts were so impressed that they kept it on board the Salyut and used it in particular for Oleg Atkov's record 237-day mission in 1984.

Second, there were biological experiments, starting with Cytos 2 which studied the behaviour of the reaction of bacteria to anti-bacteria in weightlessness. Cytos 2 contained microflora in little glass ampoules inside a thermostat called Biotherm, which kept them at 4°C until the glass was broken and they were exposed to anti-bacteria, the cosmonauts taking note of the subsequent changes in colour and form. Bioblock 3 examined the effects of heavy ion radiation on six sets of samples, of which two were brought back during the French expedition and the others at later stages. *Bracelet* concerned the normalization of blood circulation during adaptation to zero g, whilst *Neptune* concerned the adaptation of the retina to different levels of light. *Échange microbien* analysed the hygienic and sanitary condition of the station.

Third, there were astronomical experiments. PIRAMIG (Photography Infra Red Atmosphere, Interplanetary Medium, Galaxies), developed by the CNRS aeronomy service at Verrières-le-Buisson and the space aeronomy laboratory in Marseille, was a photographic system designed to take long-exposure nighttime pictures of Earth's atmosphere, as well as astronomical objects in the near infrared and visible wavebands, through 20 filters, with 350 pictures taken. It studied sources of weak luminosity in the atmosphere and interplanetary and galactic space. Each image had an automatic marker to note the precise time it was taken to the second. For this, the crew switched off the lights and used only torches, with Berezovoi aligning the station to within 1 arc second of such objects as the Crab Nebula. The French equipment was left behind and later used extensively by subsequent Salyut 7 crews, some of it eventually being shipped across space to the Mir station. PIRAMIG enabled astronomers to make a double-sphere map of the galactic plane and made possible an estimation of the population of supergiants. PIRAMIG subsequently took 600 images to map the Magellanic clouds.

PVH: Jean-Loup Chrétien (2nd left) on Salyut 7.

Photographie du Ciel Nocturene (PCN), developed by the Institute of Astrophysics in Paris, took pictures through a porthole and was a highly sensitive instrument to take views of the night sky, especially galactic dust, zodiacal light, Gegenschein, nebulae, subtropical arcs, polar aurorae and noctiluminous clouds. PCN even identified a band of volcanic cloud around the Earth from the volcano El Chichon. *Sirène* was a spectrometer developed by IKI and CESR in Toulouse to study cosmic x-rays from 2−600 keV.

Finally, there was materials processing. The *Magma* experiment, carried out on the Krystal furnace on board, had 14 thermocouples to measure the power of the furnace to heat a cartridge of samples up to 1,000°C. *Accelerometer* measured the level of gravity in the furnace, with a sensitivity up to 5 x 10^{-6}, important for assessing the degree to which experiments might be disturbed by movement in the station. *Diffusion*, developed by the University of Grenoble, studied the influence of microgravity on atomic transport and the speed of the distillation of solid polycrystalline Pb-Cu. *Immiscibles,* devised by the Centre of Nuclear Studies in Grenoble, studied the role of capillary forces and thermal convection in the evolution of liquid structures, their solidification and sedimentation.

The cosmonauts returned to Earth 65 km from Arkalyk in Kazakhstan on 2 July, only 4 km from the target point. Chrétien volunteered that everything had 'gone like clockwork' during the 189-hour 59-minute mission. His cabin was put on display at the Musée de l'Air et de l'Espace at Le Bourget, together with his spacesuit. Baudry later got to fly in space too, on the shuttle. With the mission safely over and visibly successful, Mitterrand softened his 'minimal symbolism' rule and spoke glowingly of the prospects of future cooperation. Chrétien was awarded no less than three medals.

As a postscript to the mission, a French experiment called COMET (Collecte En Orbite de Matières Extra Terrestres) was flown up to Salyut 7 to collect extra-terrestrial material for subsequent retrieval on a spacewalk. They were retrieved on 28 May 1986 by visiting cosmonauts Leonid Kizim and Vladimir Soloviev during their flight from Mir to Salyut 7 and were brought back to Earth in July 1986 for subsequent electron microscopic analysis in Toulouse.

Second French piloted mission?

The chorus of opposition beforehand notwithstanding, PVH attracted substantial media attention in France, with French television providing wall-to-wall coverage. Once again, France had demonstrated its leadership role in the European space field, while other European would-be astronauts languished in Houston.

France had the unique opportunity to perform an entire week of experiments with half a tonne of equipment. Soon after the return of Jean-Loup Chrétien, thoughts turned to a more ambitious successor mission, one that might be longer and involve a spacewalk.

Once again, however, the political background set the context and it was a discouraging one. Six months after Chrétien touched down, the French found out that the Russians had been successfully bugging French diplomatic cables for the previous seven years, so Mitterrand ordered the expulsion of 47 spies on Farewell's list. Farewell himself (Vladimir Vetrov) was discovered and shot for treason in January 1985.

Mitterrand then supported the NATO decision to deploy 108 Pershing missiles and 464 nuclear-armed cruise missiles in Germany, Britain, Belgium, the Netherlands and Italy, as a response to the USSR's deployment of 225 SS-20 missiles. Although not part of NATO's operational command, such French support was quite a change from the times of de Gaulle and Giscard. The Americans were delighted and the Russians furious and once again, France found itself in the middle of these east-west tensions. In March 1984, Mitterrand argued with Ronald Reagan, making the case that the US should treat the USSR with more respect. Mitterrand was aware – which possibly Reagan was not – that Russian attitudes were informed by their experience of being encircled and attacked, from the American and European invasion of Russia in 1918 onward.

It was three years before a first state visit, when Mitterrand arrived in Moscow in June 1984 with Charles Fiterman, his PCF Minister for Transport. Before arriving, Mitterrand had raised the situation of Andrei Sakharov and his wife Yelena Bonner and there was a flurry of publicity in the Soviet press to show their robust good health. The programme centred on Moscow and Volgograd, where Mitterrand laid wreaths at the war memorial. His meeting with President Konstantin Chernenko was a short one, but we know now that Chernenko's health was already poor and Mitterrand dealt mainly with foreign minister Andrei Gromyko. The purpose of the visit was seen by both sides as an attempt to restore the better relations of earlier years. Even so, Mitterrand publicly criticized the treatment of Andrei Sakharov, with the interpreter asking him whether he really wanted his words translated (he did) [5].

Mitterrand flew to Moscow again early the following year, formally to attend Chernenko's funeral, but in reality to meet his successor, Mikhail Gorbachev. The United States, though, moved quickly to prevent too close a Russian-French relationship. Within two weeks of Gorbachev's appointment, the CIA leaked the story of Farewell in a move clearly designed to undermine any Russian-French accord. The CIA was either authorized to undermine that relationship, or, in taking the initiative to do so, was unrestrained by its political masters. The Americans then applied pressure on Europe to support President Reagan's Strategic Defence Initiative ('*star wars*'), so insistently that an angry Mitterrand threatened to take France out of the meetings of the world's leading western economic powers, the G7.

The arrival of Mikhail Gorbachev promised an all-round improvement of east-west relationships. It was hardly a surprise that Gorbachev's first visit to the west as Soviet leader, on 3 October 1985, was to France. Mitterrand told him that France was presently an ally of the United States, but looked forward to a Europe which was not an auxiliary of the US and had a stronger, better, converging relationship with the USSR. Gorbachev would propose the idea of the 'common European home', one which won enthusiastic support among people in continental Europe and saw delirious crowds in Germany chanting 'Gorby! Gorby! Gorby!'. It was hardly a new idea, since de Gaulle had quoted it earlier and attributed it in turn to Napoleon. Both Gorbachev and Mitterrand criticized '*star wars*'.

President François Mitterrand visits CNES. CNES.

Second piloted flight: Aragatz

The idea of a second piloted mission, this time of long duration, was proposed at the meeting of the Grande Commission in Samarkand in October 1984, agreed during the Gorbachev visit to Paris a year later and signed on 7 March 1986, just two weeks after the new Mir space station had entered orbit. The details were ironed out at their annual space cooperation meeting in Yerevan, Armenia, that October. It was given a code-name of *Aragatz* (Mount Aragatz being Armenia's highest summit). The mission would break new ground, not only because of its length (a month) but because it would include the first spacewalk by a third country. For France, which had already begun development of the Hermes spaceplane project, this would expand its flight experience. France reportedly invested over €18 m in the mission and left behind equipment valued at over €2 m on the station. A grumpy Chrétien later complained that the mission was originally to have been three months, but was halved to 45 days and then reduced to a month [6].

Planning for the mission proceeded at a rapid pace. The menus for the mission were again taken very seriously, with CNES inviting proposals and 15 companies suggesting 46 different products. For the record, the menu was pigeon with dates and spices, sauté of veal marengo, boeuf bourguignon, fondue of oxtail and cheeses (gruyère and chantal), Breton and campagnard paté.

France had by now expanded its astronaut group in anticipation of American shuttle missions, the new group comprising three test pilots (Jean-François Clervoy, Jean Pierre Haigneré and Michel Tognini) and four scientists, one of whom was female, rheumatologist Claudie André-Deshays. From this group, four were assigned to train for the second Soviet mission: Chrétien, Tognini, Haigneré and Clervoy. Chrétien was selected for the mission in June 1986, with Baudry again as backup. Even as they did so, the residual problems of the cold war never seemed to go away completely. In 1987, France expelled three Soviet diplomats for trying to steal the secrets of the Ariane rocket, but the USSR responded that it had no need to do so, as it was much less advanced than its own Proton rocket (that was certainly true with regard to its engines).

President Mitterrand and his minister responsible for space, Paul Quilès, flew into Baikonour from Moscow on the Concorde to attend the launch of Soyuz TM-7 on 26 November 1988, accompanied by Soviet foreign minister Edouard Shevardnadzhe. Despite the inconvenience of rescheduling orbital manoeuvres in advance, the mission was even delayed for five days to fit in the French presidential timetable, evidence of the political priority given to the mission. Flying in and out on Concorde – a French display of *panache* – was another departure, as up to that point only Soviet planes had been permitted there.

Ten days earlier, Chrétien and Baudry, who were already there, had witnessed an astonishing spectacle. Roused from their beds during the early, dark morning of 15 November, they were told to put on heavy clothes to guard against the biting cold, wind and snow. They saw the space shuttle Buran turn the inky blackness into a bright disk of light as it took off. They were even more astounded three hours later to see Buran, which in that time had twice circled the Earth, break out of low clouds and snow flurries, line up with the runway and, undeterred by crosswinds, land right in the middle of the centre line with perfect, automated precision. President Mitterrand was given a tour of Baikonour, which included the returned Buran shuttle and signed a protocol for new missions. His arrival there was heralded in the Soviet press as putting behind the 'wasted decade' of the post-Gaullist period (hardly entirely Mitterrand's fault). At a press conference, the chief of Star Town, cosmonaut Vladimir Shatalov, told correspondents that the French might fly on Buran as practice for flying Hermes.

Two days later, Chrétien was on board Mir for almost a month of experiments. This time he flew with Alexander Volkov and Sergei Krikalev. Unlike the previous mission, when his crew members returned with him, the two Russians would remain aboard while Chrétien flew home with the previous Mir resident crew, Vladimir Titov and Musa Manarov. His arrival on Mir was televised live, as was a press conference from orbit on 4 December in which he outlined his forthcoming spacewalk and the expandable truss platform that he was to assemble.

On 9 December, together with Volkov, Chrétien became the first non-American, non-Russian to make a spacewalk. The spacewalk began over Japan, with Chrétien emerging to grasp handrails from which he could deploy the ERA, a carbon-plastic truss structure made by the French space agency CNES in Toulouse. ERA was an FFR50m, hexagonal, 1,049 kg, 3.8 m carbon-fibre structure to pioneer large framework structures in orbit. It stubbornly refused to open, with the cosmonauts expressing increasing levels of frustration. It had still not been done when Mir passed out of the range of ground control. Out of earshot, though, Volkov did what he had been told not to do and gave the structure some improvised well-aimed kicks with his boot. It duly deployed, saving the mission. The delay caused an extension of the spacewalk from 4 hours 20 minutes to 5 hours 57 minutes and it included the installation of an exposure facility of paints, plastics and micrometeoroids. Samples were deployed on the outside of the station. *Échantillons* was a 15.5 kg container with four sets of samples: *Comes* (space materials, such as paints, reflectors, adhesives, filaments), *Mapol* (polymers), DIC.DMC (dust detectors) and MCAL (white paint). Five of these tested technologies for the upcoming European spaceplane, Hermes. The *Échantillons* samples were brought inside by the crew of Soyuz TM-8 in January 1990.

Training to assemble girder structures in orbit.

The principal experiments followed similar lines to the 1982 mission, but were heavier this time (580 kg). A new echograph was developed, *As de Cours* (ace of hearts), an ultrasound system for the cardiovascular system showing blood circulation. There was also *Super-pocket* (neuro-sensorial system); *Physalie* (vestibular system), which involved donning a substantial body of equipment (75 kg); *Kinesigraphie* (posture); *Viminal* (visual stimuli and a test of commanding Hermes); *Circe* (effects of harmful ionized radiation on human tissue); and an x-ray scanner to measure bone loss. *Posture, Physalie* and *Viminal* were the first joint neuroscience experiments in orbit. They would be followed by *Illusion* (Antares and Altair); *Synergies* (Altair); *Posture/Movement* (EuroMir); STAMP (EuroMir); *Cognilab* (Cassiopée, Pégase, Perseus); *Orientation* (Mir) and *Cogni* (ISS). There were tests of a new 28 kg solar panel design (*Amadeus*) and the effects of heavy ions on computer circuits (*Ercos*). As part of a space education programme, France 3 television station brought a young astronaut club together to ask Chrétien some questions during a live hook-up.

Chrétien returned to Earth on 21 December. After undocking, the landing was delayed for three hours because of a computer software problem and this also led to a readjustment of the target landing area. Five helicopters filled with French journalists had been ready to fly to film the landing, but because of thick cloud, a 200 m low-cloud ceiling and the fear of collision, they were grounded and never reached the landing site. The cosmonauts landed in a temperature of -14°C, wind speed 6 m/sec. They were quickly airlifted 180 km to Dzhezkazgan and then onward three hours to Moscow.

It was a historic homecoming, not so much for Chrétien but for his two companions Manarov and Titov, who were coming back after a record 366 days – one year – in space. All the attention was on them because the year-long mission was a real marker in the extension of long-duration flight. As they had been in space for so long, they were supposed to be carried out of the cabin on stretchers, but thanks to the bad weather the stretchers did not arrive, so they just had to get into the plane unaided – which they did without trouble. Chrétien's colleagues Volkov, Poliakov and Krikalev, returned to Earth the following April. Chrétien stayed in the USSR where he qualified to fly Buran, but the programme was eventually cancelled in 1993.

French missions to Mir

The two flights of Chrétien were one-off, pioneering French missions. Possibilities now opened up for regular flights to the Mir space station but financially, things had changed. The increasingly cash-strapped Russians saw guest missions as a means of maintaining their space programme in general and its flagship, Mir, in particular, so free or low-cost missions were over. From now on, flights would not be based purely on goodwill, but on contractual obligations. For the new flights, the Soviet agency responsible – Glavkosmos – insisted on a €10 m payment for the first mission, €12 m for the second. CNES tried to talk this down on the basis that the Russians could use their equipment afterward, but that argument did not get very far. Contracts even brought a certain advantage for the French government, which was now able to portray cooperation with the USSR as a commercial, technical arrangement with no particular political significance. Having said that, the missions were well publicized, with CNES issuing colourful and informative packs in English for the international media.

The 1988 protocol was formally initialled by the two leaders in Paris on 4 July 1989, leading to a ten-year agreement signed at government level by Paul Quilès, socialist French minister for Posts, Telecommunications and Space on behalf of Michel Rocard's government and by Soviet Vice President Lev Voronin. This provided a framework for no less than four French flights to Mir, in 1992, 1994, 1996 and 1998, specifying flight durations and the weight of experiments that could be carried, up to 300 kg. Reflecting the change of government in Russia, a fresh agreement was signed on 28 July 1992 between CNES, the new Russian space agency (RKA) and Energiya, the design bureau responsible for piloted flights and into whose coffers the French money would flow.

France's early astronauts. CNES.

Two-time backup Patrick Baudry would have been the obvious choice for the next French mission, but he had already flown on the shuttle by then. Michel Tognini was next in line and he was duly named in autumn 1990 for the 12-day Antares mission. France was allocated up to 400 kg for the experimental package, which was sent up ahead of the mission on Progress M-10, M-11 and M-12. The new commercialism was soon evident when Energiya – either in desperation or greed – asked for €2,000 each from French journalists wanting to go to Baikonour to cover the launch, but the system was becoming more open, as the French got to see the Mir mock-up in Star Town. Life in Star Town had also relaxed a bit and it was now possible for the Antares trainees to travel in a car without an escort, though the car had to be booked in advance and there was no guarantee it would turn up. When he could, Tognini used the car to explore the countryside around Moscow. Not only that, but there was contact home by phone and fax, with visits to France every 45 days, officially for 'familiarization with the experimental programme'. The team was in Russia for the coup in August 1991, about which Star Town residents now spoke openly.

Antares duly went ahead on 27 July 1992, when Tognini and colleagues Anatoli Soloviev and Sergei Avdeev launched from the summertime steppe on Soyuz TM-15, boarding Mir two days later. A commercial, contractual flight or not, it was still of sufficient interest to be carried live on French TV. The flight also had a personal angle, as Tognini had married his physical education instructress, Elena Chechina, during training. Eleven experiments were carried, some being developments or repeats of the 1980s missions, mainly biomedical and technological. Perhaps the most forward-looking was *Alice*, which concerned the behaviour of liquid at its critical point and a heat transfer effect, called the Piston mechanism. Installed in the Kristall module, the *Alice* experiments discovered a new mode of heat transfer in critical liquids, earning its authors an award from the science academy and putting France ahead in the field of critical fluid hydrodynamics. Tognini installed *Nausica* to measure radiation dosage, important for determining the need for future protection on orbit. The South Atlantic Magnetic Anomaly, it was found, gave out 50 times more radiation than the rest of the orbital path, at 1.5 millisivert/hr, or 30 percent of the daily dose. *Nausica* measured the impact of the solar eruption of 31 October 1992, which raised radiation levels by 30 percent and even more at polar levels as the particles from the Van Allen belt funnelled downward. Tognini returned after 12 days with Alexander Viktorenko and Alexander Kaleri.

The second mission, Altair, was originally scheduled for July 1994, but was brought forward by a year when other foreign missions did not arrive and Russia needed the cash. On 26 October 1992, Jean-Pierre Haigneré was selected, with Claudie André-Dehays as backup and with only nine months to go before launch. Haigneré flew to Mir with Vasili Tsibliev and Alexander Serebrov on Soyuz TM-17. Shortly before take-off, he received a sharp reminder of the difficulties facing the new Russia when there was a power blackout at the pad. An hour after take-off, the electricity supply in all the adjoining city of Leninsk failed. Aboard Mir, six medical and two technological experiments were carried out using equipment brought up during the Antares mission and there were six new life sciences experiments and two technological ones. Haigneré also spoke to French radio amateurs. He came back with the returning Mir crew of Gennadiy Manakov and Alexander Poleschuk, with a light wind of 3 m/sec during the return and the impact on the ground being a 'gentle shock' of 15 km/hr. Interviewed about the mission afterward, Haigneré described the Soyuz as 'small and not very comfortable' but much easier than doing combat manoeuvres in the Mirage 2000. He had not been space-sick, although his face had swollen in weightlessness at first and he felt disorientated only when lifted out of the cabin on his return. He expressed dislike for the food due to the lack of choice and too much tinned fish, 'which was not my favourite for breakfast'.

Jean-Pierre Haigneré. CNES.

Soon afterwards, another French instrument arrived on Mir, a LIDAR (LIght Detection And Ranging) called ALISSA (Atmosphère par LIdar Sur SAliout), to study cloud systems and measure their vertical heights. It was installed on the Priroda ('nature') module which arrived at Mir on 23 April 1995 and began operations on 2 October. It was built by the Service d' Aéronomie of CNRS with the Russian Institute of Applied Geophysics. The purpose was to pave the way for a permanent observing system for high rainclouds. This was a project of Jacques Blamont of the Service of Aeronomy and the Institute of Applied Geography in Moscow that went back to 1986. The instrument worked for the rest of the life of Mir, providing new information on the nature of clouds and proving its feasibility. The principal problem was one of pointing, in that Priroda was required for other experiments so ALISSA could not be pointed at Earth continuously.

Two missions had now been carried out and had gone so well that discussions opened at the 30th cooperation meeting in Toulouse in April 1994 to add a fifth mission to the scheduled four. The next mission, Cassiopeia ('Cassiopée' in French) with Claudie André-Dehays, duly took place on 17 August 1996, launched aboard Soyuz TM-24. It was inevitable that much of the public attention would focus on the astronaut, an exceptional person who became one of France's best known spacefarers. She had won her place in the French space corps during the

1985 competition, the only woman in the final selection of seven from the 700 applicants. She had resolved to become an astronaut the night she watched Neil Armstrong walk on the Moon and the following day traded in her bicycle for a telescope. André-Dehays achieved her baccalaureat at 15 and the first of several university degrees at 24 (medicine, followed by biology, sports medicine, aeronautical and space medicine and rheumatology). From 1990, she had taken part in the regular zero-gravity tests carried out for potential cosmonauts on board France's old Caravelle airliner, their 'vomit comet'. She combined ambition (always top of her class), determination (essential to handle cosmonaut training) and a love for contemporary art and Mozart with personal charm (she was very capable of dealing with the French media). Aged 39 at the time of her mission, she was the 31st woman in space.

Preparation of the experimental package took three years, involving a team of 20 people at CNES and drawing in 40 laboratories and industrial partners. The focus was on physics and life sciences and these were substantial experiments. *Physiolab* weighed 40 kg and involved the cosmonaut wearing a measuring jacket for two periods of 24 hours. *Cognilab*, 57 kg, was a supportive seat in which the astronaut used graphics to respond to visual, auditory and muscular stimulation. In *Fertile*, 28 kg, three salamanders fertilized at the beginning of their flight would give birth and their eggs would all be brought back to Earth, with some having been in zero-g all the time and some in a centrifuge, an experiment that required a high level of monitoring. *Alice 2*, which weighed 57 kg, tested fluids and the piston effect in the Kvant module (in the late 1990s, *Alice 2* was brought back by the American Space Shuttle, which had sufficient space to do so and it continues to be used at the Institute of Fluid Mechanics in Moscow). *Dynalab* sensors were placed on the extremities of the Mir station to measure vibration, while *Trellis* tested how vibration could be dampened to keep telescopes focussed on their targets. Some student experiments were also carried. The results of the mission were presented at a colloquium in Paris on 19 December, opened by the Secretary of State for Research, François d'Aubert.

There were significant pre-flight problems that exemplified the financial issues with the new Russia. Because of their financial difficulties, the Soyuz U2 launcher had to be replaced with the less powerful Soyuz U. This meant that the cosmonauts could not take personal effects and had to lose weight, while Mir had to drop its orbit so that they could reach it. A week before lift-off, Gennadiy Manakov failed his final medical (doctors diagnosed a possible heart irregularity) and he was taken off the crew, together with his colleague Pavel Vinogradov, so André-Dehays flew with the backups – Valeri Korzun and Alexander Kaleri – instead. The Soyuz TM-24 reached the orbital complex on 19 August for 14 days of medical and biological experiments. Already on board Mir were Yuri Onufrienko, Yuri Usachov and the American spacewoman Shannon Lucid, who was on the first long-duration American mission by a woman. André-Dehays brought up some welcome menu items: Basque swordfish, duck, quail and energy bars. She returned to Earth after a 16-day mission with Onufrienko and Usachov, whose mission had lasted 195 days, to be met by a recovery team of two rescue vehicles, 11 planes and 18 helicopters. The most vivid and only anxious moment of the flight, she recalled, was when the parachute opened, as the cabin was abruptly pulled up and swayed to and fro for 90 seconds before stabilizing itself.

Claudie André-Dehays made a number of presentations of her mission afterwards and drew huge press attention. She flew back to Paris with her lizards and presented the results of her experiments to a symposium presided over by the secretary of state for research on 19 December 1996. *Fertile* confirmed that fertilization in space is possible, while *Physiolab* made a record of the relationship between blood volume, arterial pressure, the nervous system and vascular resistance. *Cognilab* provided information on the role of the right hemisphere in the process of perception and *Alice* found out-of-equilibrium states of vapour balls in liquids, while *Castor* recorded vibration in orbit and its effect on spacecraft structures. Following her visit to Mir, she married veteran Mir cosmonaut Jean-Pierre Haigneré and took his surname. She spent the better part of ten years in Star Town.

For the fourth mission in this part of the series, Léopold Eyharts was ready to fly to Mir in August 1997 on Soyuz TM-26. Eyharts was a fighter pilot and subsequently test pilot and had already learned Russian in France. His mission, Pegasus, was designed to continue the experiments undertaken by André-Dehays on Cassiopeia. Unfortunately, the mission coincided with a period of peril on Mir, the summer starting with a fire, followed by a near collision and then a real collision between Mir and a Progress freighter, the depressurization of the Spektr module of the station and an extended period of emergency repairs. Given the problems on Mir and the lack of sufficient electrical energy, it was unlikely that Eyharts would have been able to carry out any useful scientific research. Instead, Soyuz TM-26 carried up emergency equipment in his seat for repairs, Pavel Vinogradov and Anatoli Soloviev leaving Earth without him.

Eyharts did get his chance six months later when he flew on 1 February 1998 with Talgat Musabayev and Nikolai Budarin aboard Soyuz TM-27. He spent nearly three weeks on Mir before returning to a snowy landing on 19 February. His experimental package was largely a repeat of Cassiopeia, including *Fertile*, with six salamanders whose eggs he brought back, *Physiolab* and *Cognilab*, *Castor* (technology) and *Alice*. The situation on Mir had now stabilized and no longer attracted much media interest, but his landing with Anatoli Soloviev and Pavel Vinogradov was far from routine. The temperature was already -30°C and blizzards hit the landing site as the cosmonauts re-entered the Earth's atmosphere. Fearing a collision, only one Mil-8 helicopter was dispatched to recover the crew. Kazakh air control was unable to maintain a radar image of the descending Soyuz TM-26, but thankfully Mil captain Anatoli Mikhalishev picked them up on his radar as they came through 4,500 m. The Soyuz landed in snow, the Mil close beside an instant later. Normally, the rescue crew erects an inflatable field hospital for the crew's medical examinations, but conditions were so poor that Soloviev,

Vinogradov and Eyharts were brought directly to the door of the Mil, which had kept its engines revving to stop them freezing from the extreme cold. The medical rescue team, led by Oleg Fyodorov, had to give peremptory physical examinations as the helicopter whisked them away to the nearest landing point at Kustanai [7].

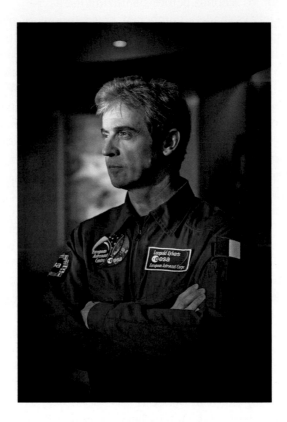

Léopold Eyharts. CNES.

By the time Eyharts returned, funding for the Mir station had begun to run out as Russia increasingly invested its diminishing resources in the ISS, whose hardware was now nearly completed. The discussions about a fifth mission to add to the series that opened in 1994 led to a 1996 agreement for a final 120-day mission called Perseus in 2000. Partly, this was because Russia needed a final crew to decommission the Mir station – and the money to pay for it. The French mission was then radically shortened to 35 days, but with a spacewalk to recover French instruments still on the outside of Mir. The figures for the cost to CNES were never officially disclosed, but various estimates of between €18 m and €30 m appeared. In a further timetabling amendment that reflected the rapidly changing fortunes of Mir, Perseus was then brought forward six months. The crew was announced as Viktor Afanasayev and veteran Jean-Pierre Haigneré, though it was an assignment which no cosmonaut ever wanted, as no one wished to be on 'the crew that killed off Mir'. Due to the complexities of crew changes, the length of Haigneré's mission was extended from 35 days to 99 days, near to the original 120, but the French saw no reason to pay more as a result.

The mission then came under threat from an unexpected quarter – at home. That summer, on 4 June 1997, Claude Allègre had been appointed minister responsible for spaceflight (see Chapter 2). He announced that he would re-orientate French space policy away from piloted flight ('I'm not crazy about manned flight') toward the planets and applications and ordered the suspension of work on the mission to Mir. France's astronauts could join the European astronaut team and take their chances there. He also decided on a geographical re-orientation away from Russia and toward the United States, with whom he planned a Mars sample return mission. Jean-Louis Cuche of the Inserm laboratory in Paris wrote to Allègre, appealing for the fifth mission to fly, outlining the importance of protecting French expertise in such areas as space medicine and warning that countries such as Germany would overtake France [8]. Such a prospective outcome hit a raw nerve and the mission was allowed go ahead. In advance of the flight, a Progress carried up 18 lizards for the *Genesis* experiment.

Star Town in winter.

Haigneré duly went into orbit on Soyuz TM-29, with Russian commander Viktor Afanasayev and Slovak cosmonaut Ivan Bella, on 20 February 1999. Despite it being a decommissioning mission, Afanasayev, Haigneré and a cosmonaut from the previous mission, Sergei Avdeev, had a busy time. Experiments covered a broad range, from the salamanders to the cardiovascular system, fluids, meteors, the human spinal column, metal alloys and bone density. In addition to the French experiments, there were also two for Germany (metals alloys and the spinal column) and two for ESA (bone density). Haigneré planted his own wheat seeds in the Svet greenhouse and wrote a weekly newspaper column. He and Afanasayev made two spacewalks together. On the first, in April, they installed experiments on the outside of the hull and retrieved the Leonids comet collector, left outside the previous November. *Exobiology*, a French device to expose organic matter on the outside of the station, was also installed. On the second spacewalk, in July, they retrieved many of the packages. The other experimental highlight was the continued study of salamanders (*Genesis*), which arrived at the station on 4 April on Progress. They also launched a micro-satellite, Sputnik 42 and the crew observed that summer's solar eclipse – a ball-shaped cloud of darkness crossing from Ireland to Bulgaria.

At 6 pm on 27 August, Afanasayev closed the hatch on Mir for the last time. 'We have grief in our hearts,' he told television viewers. The crew gathered in as much as they could of the scientific results and equipment and programmed the computer to run the complex without a human presence. Then they closed off the fans and the ventilators and the station fell silent. Separation took place at 9.17 pm and Mir could be seen receding in the distance over Petropavlovsk. Soyuz TM-29 came down two orbits later 80 km east of Arkalyk. The solid rockets fired, cushioning the final descent but also setting some steppe grass on fire. The cabin turned over and Haigneré was violently sick on landing, the symptoms persisting for an hour, but he was quickly helped to recover by Claudie with drinks of water. Haigneré's total time on board turned into 189 days, the longest non-Russian mission to the station, so the French got their long-duration mission in the end and more. His colleague Sergei Avdeev held the accumulated record of 748 days.

In the event, their grief was premature, because this was not the final mission to Mir after all. Russia attracted funding for the first-ever commercially purchased spaceflight, which did turn out to be the final visit in 2000 (MirCorp), but the inevitable could not be delayed much longer. The great space station Mir finally de-orbited in a ball of flames over the Pacific in March 2001. Mir left a permanent legacy in France itself. A full-scale model of the station, with its modules, was bought by the city of Toulouse in 1996 and shipped from the Khrunichev centre to the port of Verdon, 100 km from Bordeaux, at the end of the following year. These were working models that had been used for dynamic, electrical, vacuum and equipment tests, so were as close to the 'real Mir' as one could be. The road convoy was slowed by Christmas decorations in towns on the way – which had to be carefully raised – but the exhibit arrived on Christmas Eve and was placed in a hangar pending the arrival of 30 Russian specialists in May 1998 to help construct the display, with all the appropriate mountings, attachments and a repainting, all to ensure complete authenticity – interior and exterior – and viewing points. The specialists, many of whom had been involved in the construction of the real Mir, stayed there until the public unveiling on 10 July 1998, in the presence of nine cosmonauts and astronauts. Within less than ten years, the display had attracted two million visitors [9]. Real Mir veterans could only compliment the specialists on the quality of the finish, though Claudie Haigneré commented that the real Mir was never so tidy. In 2006, there was a reunion of all the French and Russians who had flown together on orbit on Salyut and Mir and those who had worked most closely with them, to mark the 40th anniversary of the 1966 agreement. It was a big group, presided over by Claudie Haigneré.

German missions to Mir

The dominant position of France in the European space field is not difficult to understand, motivated as it was by technological modernization and being the driver of European integration and its independent political role, the latter especially contributing to cooperation with the USSR and Russia. What was more surprising was the late arrival of Germany, which had modernized rapidly from the 1950s and had become, alongside France, Europe's industrial powerhouse.

Post-war German reconstruction had been very much American-informed and financed, with the Americans maintaining substantial armed forces there. Under the Hallstein doctrine, Federal Germany would have no dealings with the GDR, nor its friends, although there had been diplomatic relations between Moscow and Bonn since 1955. This began to change in the late 1960s, when chancellor Willi Brandt's *Ostpolitik* opened cooperation with the east. As chancellor (1969–74), Brandt visited the USSR in 1969, which was reciprocated by Leonid Brezhnev in May 1973, but he never visited space facilities there. The explanation may lie in the slow pace of development of the post-war German space industry. Because of the A-4 rocket, German rocketry was a sensitive subject and all its rocket engineers had dispersed abroad, some to Arab countries. Germany did not resume spaceflight development until 1955 and set up an engine testing station in 1959. American Vice President Lyndon Johnson (1961–3) had offered American launchers and an interplanetary programme to Germany, taking the form of the Azur mission on a Scout rocket in 1969 and the Helios missions toward the Sun. Germany was happy to be a founder member of the two European space organizations, the European Space Research Organization (ESRO) and the European

Launcher Development Organization (ELDO) in 1962 **[10]**. According to Jacques Blamont, observing from France, Brandt wanted to bury the idea of a European launcher when the Blue Streak-based Europa launcher failed and saw the European space programme as an annex of Germany's powerful American protector. The French were quite struck by the lack of German-Russian cooperation, which they attributed to the cold war and the low level of German funding **[11]**. With the formation of ESA (1975), Germany effectively terminated its national space programme, deciding to participate only in ESA projects and in building Spacelab for the Americans, re-iterating its preferred cooperation with the US.

Germany eventually reversed its 'ESA-only' policy in 1989 by setting up its own national space agency, then called DARA, but later and better known as Deutsches Zentrum für Luft- und Raumfahrt (DLR). This came at the worst possible time, however, with the federal budget devoted to completing the process of reunification with the former GDR.

The only cooperation with the Soviet Union appears to have been a 1973 agreement for information exchange between the DLR Institute for Aerospace Medicine and the Institute for Bio Medical Problems (IBMP). Cooperation with Russia was formally approved through the intergovernmental agreements of 22 July 1986 and 9 November 1990 (USSR), which were revised on 10 April 2001 (Russia), the current framework. This was a 27-page text with 14 articles and two annexes in German, with a Russian text alongside. It specified cooperation in a list of fields: astrophysics; planetary exploration; remote sensing; materials; medicine and biology; communications; navigation; launchers; technology; and spinoff. Cooperation with Russia up to then was acknowledged by the government many years later as being 'comparatively recent' **[12]**.

The idea of a German spaceflight to Mir arose in the course of a meeting between the Soviet Academy of Sciences and research and technology minister Heinz Riesenhuber, on the fringes of a meeting between Chancellor Helmut Kohl and General Secretary Mikhail Gorbachev in Moscow on 25 October 1988, when a preliminary agreement on principle was made. The general impression is that the offer was made from the Soviet side. A decision to go ahead and make technical preparations was agreed when the two heads of government met again in Bonn in June 1989. A formal agreement was signed on 18 April 1990. On the financial side, Russia originally proposed DM21m for a week-long flight with 100 kg of experiments up and 5 kg down. The hard-nosed Germans thought this excessive but the Russians stuck to the price and they ended up on DM20m (€10 m). Russia offered a date of March 1992, which was not ideal for Germany because it was heavily committed at that time to the D-2 Spacelab mission on the shuttle. DARA officials made it plain that Spacelab, with 90 experiments on a nine-day mission, should be the priority. There was a brief discussion of short-circuiting the process by putting Eberhard Köllner into orbit for a month-long mission to Mir. He was already trained for such a mission, having been backup for the first German to go into space, Sigmund Jähn of the GDR, to Salyut 6 in 1978. The German government made it clear that it wanted a 'west' German mission **[13]**.

President Gorbachev discussed the mission further with Chancellor Helmut Kohl in the Caucasus in July 1990, formally stating that a joint spaceflight would be a visible way of underlining German-Russian friendship. Thankfully, Germany already had five astronauts in training for the American-German Spacelab D2 mission and, to resolve the problem of overcommitment on the German side, the Russians said they were quite happy to accept the level of proficiency those five had attained in America without the need for additional general spaceflight training, except of course to familiarize themselves with Soyuz and learn Russian.

Two German astronauts were selected to train for the mission in October 1990, Dr Reinhold Ewald and Klaus-Dietrich Flade. They were presented in Dresden on 8 October 1990, the time of German reunification. Both were photographed with Gorbachev in Bonn. Mission planning and training began on 12 November and the two sides quickly agreed an experimental package of 14, of which five, focussed on microgravity, sleep rhythms, radiation protection and materials science, were provided by DLR and developed at some speed. A nicely illustrated media pack was presented, compiled by DARA, DLR and Energiya.

Reinhold Ewald and Klaus-Dietrich Flade. DLR.

Mission preparations took place against the heady political background of reunification and were facilitated by Germany's first cosmonaut, Sigmund Jähn, who acted as liaison, interpreter and advisor. The German astronaut Ulf Merbold, who had left the GDR at 19 before the wall went up, persuaded his colleagues to bring in Jähn to assist in planning the German piloted missions to Mir and later the ISS. His role was understated but critical. Those who knew Jähn described him as a quiet, modest and measured man who spoke to adoring crowds in the GDR not from a script, but in his own words. Highly committed to the GDR, he nevertheless worked with the reunification process and became the bridge between the two Germanies right up to his retirement in 2001. He personally accompanied German astronauts to Star Town and to the launch pad, providing personal and moral support. According to Beate Fischer, head of astronaut training, Jähn was invaluable in navigating Germany's astronauts through everything from border checks, to minders, to Russian food and how to survive vodka toasts [14]. Cinema goers will remember his role in the bittersweet comedy *Goodbye Lenin!* Jähn was not the only gain for the united Germans, as the federal republic absorbed the GDR's space agency, the Institut für Kosmosforschung and the many centres of technical excellence in the GDR, benefitting from their many, lasting contacts and colleagues in the Russian space programme.

Sigmund Jähn many years later. Bernardt Tiedt

Sigmund Jähn with Paolo Nespoli. Bernard Tiedt.

The mission, called Mir 92, duly took place in March 1992 and was the first piloted spaceflight by Russia, as distinct from the Soviet Union. Klaus Dietrich Flade was selected with Alexander Viktorenko and Alexander Kaleri. As a sprinkling of snow lay on the ground, their rocket took off from Baikonour. Down its side, it was decorated with the bars of the flag of Germany and, for the first time ever, the red, white and blue flag of Russia, marking the passing of the Soviet Union. Once on orbit, Flade spoke to German science minister Heinz Riesenhuber. While on Mir, he carried out 13 medical and one materials science experiments. Amongst the medical studies were HSD (skin tissue), TON (intraocular eye pressure), *VestABrille* (space sickness) and VOG (eye movement). Flade reported on his experiments on vegetable and animal cells; could be seen wearing goggles for the OVI experiment in the Kristall module; carried out materials sciences experiments (TES); and used dosimeters to measure radiation levels. He brought 10 kg of samples back when he returned to a snowy Earth a week later with the previous Mir crew of Alexander Volkov and Sergei Krikalev, the latter still with his party membership card in his pocket. He had been launched 312 days earlier by a country which no longer existed.

Mir 92 departs – the German flag at Baikonour. DLR.

Agreement was soon reached between Germany and Russia for a second mission in 1996, though it slipped into early 1997 and was called Mir 97. The two sent to Moscow were Flade's backup, Reinhold Ewald and Hans Schlegel. This was a more ambitious and much more costly mission (€50 m), with 27 experiments (twice as many as Flade) and double the duration at 20 days. Ewald, from Mönchengladbach, was a scientist who had written a thesis on the spectroscopy of interstellar matter, studied medicine in his spare time and became a research assistant at the University of Cologne's 3m radio telescope. His principal experiments were MEDES (medical diagnostic device), investigating oedema, the behaviour of metal alloys and fluids, ultrasound tests of the body and bone loss, investigating changes to the spinal column and how the human body stored salt in microgravity (this was found to be substantial, to the point that it may be a causative factor in bone loss on orbit).

Ewald arrived at Baikonour cosmodrome for his mission at the end of January 1997, on a Tupolev 134-A2 with *Yuri Gagarin Cosmonaut Training Centre* painted on the side. He watched the Soyuz TM-25 roll down to the pad on the railway system, the *Mir 97* logo displayed on its side, in a temperature of -15°C, something that seemed to bother no one. On 10 February, he took the bus to the pad with Vasili Tsibliev and Alexander Lazutkin. In a dramatic contrast to Cape Canaveral, anyone was free to walk around the fuelled rocket, from VIPs, to press, to launch officials and their children. It was so bitterly cold that ice had formed on parts of the rocket. Because the Soyuz U version was now being used, personal effects could not be carried, but they still found room for 12 paintings by young German artists. Ewald brought with him a laptop computer, an advance on the handwritten notes of Mir 92.

Little can Ewald have known that he had joined the most troubled, difficult and dangerous period of the entire Mir operation, though he would escape the worst of it. When he arrived at Mir on 12 February 1997, he was greeted by the resident crew of Alexander Kaleri, Valeri Korzun and American Jerry Linenger.

On 24 February, a fire broke out in the Kvant module in the station, caused by an oxygen candle. These are small, 20 cm long, 8 cm diameter blocks inside a small oxygen generator and are used by the cosmonauts periodically to renew the air supply, a process which takes 20 minutes. This time, instead of releasing additional oxygen into the cabin, the burning oxygen candle acted as a blowtorch, discharging flame and thick smoke into the station which filled the cabins in under a minute. The crew quickly donned masks and tried to extinguish the fire, but without any effect. The next minute was tumultuous, as some cosmonauts raced through Mir to fetch extinguishers while others tried to unload the return-to-Earth schedule from the computer printer. Eventually, after 90 seconds though it must have seemed much longer, the fire burned out, but smoke continued to fill the station for the next six or seven minutes, while the cosmonauts coughed and spluttered and the master alarms blared. The problem in a space station of course is that smoke has nowhere to go − one can't just open a window − although the ventilation system will eventually remove it. The smoke had even spread to the descent craft, making the use of Soyuz for evacuation problematic. The station was over Africa at the time and once things had calmed down, the crew radioed in news of the incident to ground control over a crackly air-to-ground link. One of the NASA engineers in mission control noticed the sudden flurry of activity among the controllers and tried to find out what had happened. Eventually, one picked up the alarming word *pozhar*, mentioned several times − the Russian word for fire.

Ewald survived the scare to return on 2 March. For his colleagues Vasili Tsibliev and Alexander Lazutkin, this was the beginning of a nightmare summer, in which there was a collision, the loss of the Spektr module, repeated computer crashes and moves by the nervous Americans to pull out of the joint missions that had brought the Russians precious funding.

Ewald's flight had a postscript. Progress M-37, arriving in December 1997, brought up a small, German-made fly-around robotic spacecraft called Inspektor, built by Daimler Benz. The aim was to fly the Inspektor at close range around the station to investigate collision damage and take continuous high-precision television pictures. It was a test of a system that could later be developed for the ISS, possibly saving spacewalk time. Inspektor was prism-shaped, 930 mm long,

560 mm wide and weighed 72 kg, with three solar panels, batteries, antennae, gyros, video cameras and tiny nitrogen gas jets for manoeuvring. It was intended that Inspektor would fly around Mir in a number of circles and then move across to the now undocked Progress, testing inspections over a 29-hour period. However, Inspektor lost its star lock which inhibited its engines from firing and it soon posed a new collision threat to the station. Accordingly, Progress fired its engines to move Mir out of Inspektor's way. The concept was good one, but it has not since been repeated.

The 25th anniversary of the Mir 92 and Mir 97 missions were celebrated by the German space agency, DLR, in March 2017, the party taking the view that their success was the basis for the subsequent expansion of German-Russian cooperation across numerous fields. One regret was that they passed up a chance to fly Germany's first female astronaut (30 years later, Germany has still to fly one).

Mir 92, 97 reunion. DLR.

Although German space budgets were stretched by the cost of reunification, an overlooked aspect of this period was that Federal Germany acquired the considerable assets of the GDR's space programme. This comprised the human resources of scientists and engineers who had trained in the USSR; specialists in optics and cameras (they made the cameras for Salyut 6); and institutionally, the little documented and likely undervalued Space Research Institute (IKF) from Berlin. Because they had all learnt Russian and understood the Russian system, they were able to act as a bridge between their new western German colleagues and Russia, 'making Russia understandable' to them.

The other Europeans: Austria and Britain

Germany's neighbour, Austria, was an early beneficiary of invitations for missions at the end of the Soviet period. This should not have been surprising, since Austria had already contributed instruments to Venera 13/14, VEGA, Phobos and Interball. The mission arose from a visit by the Soviet prime minister Nikolai Ryzhkov to the Austrian chancellor Franz Vranitsky in July 1987, during which a surprise invitation was issued that led to the *AustroMir* mission. The Austrians were offered a scientific payload of 150 kg with 42 hours of crew time to work on experiments. The Austrians had no previous headline involvement in ESA and for €20 m, this was a chance to make up lost ground [15].

The Austrians wasted no time, beginning astronaut selection in April 1988, with 200 candidates from whom 50 finalists were selected in February 1989. This was reduced to 15, then six, then two by 10 March 1989. The government funded the Joanneum research centre in the city of Graz, Styria, to manage the mission, including the choice of experiments. Of the 34 proposed, there were 15 selected, of which 11 were medical (e.g. blood, motor systems, the heart, eyes, body fluids), three technological and one remote sensing. They were sent up in advance by the Progress M-9 freighter. Not slow to pass up a commercial opportunity themselves, the Austrian government attracted €9 m in sponsorship from 54 companies, including Austrian Airlines, Z-länder Bank, Bundesländer insurance, Vöslauer mineral water and Kodak. The Russians made it clear that the Austrian would have to be able to fly the Soyuz if necessary and would therefore have to spend at least 500 hours in language instruction.

Two Austrians trained for the mission – Dr Clemens Lothaller, a 27-year-old anaesthesiologist from Vienna and Franz Viehböck, a 29-year-old electrical engineer from Peichtoldsdorf. They arrived at Star Town early in 1990, where they posed beside Vöslauer mineral water crates and a distribution van with the slogan *Vöslauer – mit der Kraft der Erde* (*Vöslauer – the power of the Earth*). The mission took place against a background of some uncertainty, as the coup had taken place in Moscow only a few months earlier and Soyuz TM-13 turned out to be the last *Soviet* piloted space mission.

It was Viehböck who flew to the Mir space station between 2–10 October 1991, launching on Soyuz TM-13 with Alexander Volkov and Toktar Aubakirov. He was seen off by Chancellor Vranitsky, who travelled to Baikonour for the occasion.

Viehböck could be seen waving a small Austrian flag as, *2001 A space odyssey* style, he floated into Mir to the amplified taped music of the Strauss waltz *The blue Danube*. On a personal level, it was a dramatic time for Viehböck, as his daughter was born on his launch date and the first pictures of her were relayed up to him by television. Viehböck issued reports on his progress in carrying out the scientific experiments, such as LOGION (ion emissions from liquid metals); MIGMAS (materials analysis); blood and body fluids; *Monimir* (eye, arm and head coordination with spinal reflexes); *Motomir* (muscular system); and *Cogmir* (the brain). He used the MKF6MA camera already on board to survey Austrian territory. Viehböck landed near a chilly, cloudy Arkalyk a week later. From the point of view of the Austrians, it all seems to have gone very smoothly, with the scheduled experiments carried out, but *Austromir* was not very popular within ESA and the Austrians were accused of making a solo run. Why not do piloted flight within ESA? Their reaction may have subliminally discouraged Austria from some form of follow-up mission.

Although Britain was a founder member of ESA, its level of financial support was far behind the big three of France, Germany and Italy (£112 m a year, compared to France, £700 m; Germany, £500 m; then Italy). Its withdrawal from Blue Streak left a permanent mark and Britain was rarely prominent in ESA decision-making. As for cooperation with Russia, individual research establishments could cooperate if they wished (see Kvant, chapter 2), but the government did not encourage them to do so. When they did, it was little advertised, such as a British experiment on board Salyut to combat space sickness. Eladon, in Bangor, Gwynedd, created the Elagen supplement, harvested from Eleutherococcus shrubs in eastern Siberia. Vladimir Lyakhov and Valeri Ryumin took 4 ml every day, the adaptogen having the advantage of being neither addictive nor a stimulant, without side effects and strengthening the immune system. It has since been used routinely **[16]**.

Things began to change when the British National Space Centre (BNSC) was established in 1985, led by ESA's first director general, Roy Gibson. On the cooperation side, a key moment was a visit by a parliamentary delegation to Moscow in May 1986, led by deputy prime minister William Whitelaw on the government side and Denis Healy for the Labour party opposition. This followed the ground-breaking visit made by President Mikhail Gorbachev to Britain in November 1985. Denis Healy expressed the wish that a Briton might fly in space, but, when asked, made it clear that he did not intend to volunteer himself. The visit included a tour of Star Town guided by cosmonaut Georgi Beregovoi, in the course of which he twice made an offer on behalf of the Soviet government for a Briton to fly on a Soviet spacecraft. This was followed up by a formal letter of invitation to the minister responsible for trade and industry, Geoffrey Pattie. The delegates saw the mock-up of the Mir space station and clambered on board.

Despite France's positive experience of cooperation with the USSR, domestic reaction was suspicious that the offer was a 'propaganda ploy'. According to Denis Healy, 'Willie Whitelaw would have mentioned [the offer] to Mrs Thatcher, but that was never taken up by the government' [17]. The BNSC took the decision to put the offer on hold while a more structured cooperation agreement was put in place. To do so, the BNSC visited Moscow between 29 September and 1 October, led by Roy Gibson, who met Roald Sagdeev, the director of the Institute for Space Research (IKI) as well as the Interkosmos council. IKI proposed joint activities in the areas of sub-millimetre astronomy, life sciences, radio telescopes and material sciences and an initial agreement or protocol was signed between Gibson and Sagdeev [18].

Roy Gibson. ESA.

An intergovernmental accord was the next step and a full-up, ten-year agreement was signed in March 1987 by the foreign secretary Geoffrey Howe and Soviet foreign minister Eduard Shevardnadze. The agreement set down cooperation in the areas of space science (x-ray astronomy, high energy astrophysics, solar and terrestrial physics), life and material sciences (but the only specifics were commitments to the Kvant mission, already underway in the University of Birmingham) and the Phobos project, with three scientists invited to participate (David Southwood of Imperial college; Greville Turner of University of Sheffield and John Guest of University of London). Britain was specifically invited to participate in a project then called Spektr X (see Chapter 2) but there was no commitment to a piloted spaceflight, though the issue was unlikely to go away.

Then it all went horribly wrong, with the government decision of 24 July 1987 to reject the BNSC's plan and withdraw from frontline space research. Trade and industry minister Kenneth Clark called ESA 'a hugely expensive club' and deplored European manned space activities as 'an expensive frolic' [19]. Roy Gibson resigned in protest. The BNSC continued in existence, but instead of acting as a space agency it retreated into a minor role of coordination and exhortation. The idea of a British astronaut to Mir, discussed in Moscow the previous autumn, never seems to have won support at political level, although this was not stated explicitly. The ten-year agreement was effectively gone with the BNSC plan, with Spektr X a collateral casualty. British hopes of flying an astronaut on the American Space Shuttle had already faded when the *Challenger* exploded.

Then, out of nowhere, came probably the most bizarre of the cooperation stories in piloted spaceflight. In April 1989, up popped the British Astronaut Project (BAP), connected to the Soviet space agency Glavksomos. This involved a consortium of British companies undertaking to raise £16 m to fly the first Briton into space. Consistent with its decision not to fund European piloted spaceflight, the government made it implacably clear that it would not provide any funding, nor would it allocate responsibility to an agency or public-private partnership, unlike Austria. Of course, this did not stop ministers availing themselves of photo opportunities or attending signing opportunities in both London and Moscow on 29 June 1989. BAP was then renamed Antequera, then Juno and was given the task of raising the funds, banked by Moscow Narodny Bank. Dr Heinz Wolff was appointed science director for the project, for which he had an allocation of 300 kg and he quickly devised, from 60 proposals, a set of 26 experiments in medicine and biology. Contributors included Imperial College London and Surrey University, which provided a miniaturized furnace, with the experiments ranging from protein crystals, to plants to be housed in an incubator with a centrifuge. Many prospective funders were mentioned, such as Memtek (video and audio tapes); Zeon (watches); Linguarama (languages); British Airways (BA would fly them back on Concorde after the mission); ITN (television rights for £3.2 m); Memorex (tapes); British Aerospace; Interflora (flowers); and British Telecom; with talks of a toy company and the company that made Mars bars.

There was no difficulty in finding would-be astronauts, with 12,000 people responding to the call and 3,000 completing applications. Of these, 150 were selected for medical tests, with the final selection cut to 16. These 16 participated in gravity training in Britain's only centrifuge at the RAF Institute of Aviation Medicine in Farnborough in October 1989, the only government facilities to be made available for the mission. This had been arranged by an Air Vice Marshall on the selection panel. After this, four candidates were left, from whom two finalists were selected on 25 November 1989 – Helen Sharman and Maj. Timothy Mace. Sharman was chemical engineer who had heard the recruitment advertisement by chance on her car radio.

Despite a flying start, the project lurched from one uncertainty to another, with many phases of being on and then off. Whereas Austria had no trouble attracting commercial sponsorship, this was not the case in Britain, where sponsorship revolved around the television rights. One of the hopes of the original organizers was to sell television coverage of the launch: after all, there was wall-to-wall coverage of the French missions by their TV, so there was clearly a public appetite. The Independent Television Network (ITN) appears to have been the media of choice and did much filming of the mission preparations. *Soviet Weekly* hounded the mission promoters to name confirmed sponsors but got excuses about the Gulf War, economic recession and 'over-hyped media'. The one company that did appear to come on board was Suttons Seeds, which sent up 125,000 seeds on the mission. At the end of the mission, having been exposed to weightlessness and radiation, the seeds were brought back to be distributed to, grown and analysed by schoolchildren.

Despite the uncertainty, Mace and Sharman left for Moscow for 18 months of training. Two days before their departure, they were presented to a typically well-briefed and personable Prime Minister Margaret Thatcher in Downing Street, who demonstrated a thing or two about how to handle the media. After only a short time in Moscow, though, the fundraisers called time, with little, if any, money having come in and the government still ruling out support. It seems that ITN pulling out was the crux of the matter and it is not known if it ever paid for its earlier filming. The astronauts were instructed to shun the BBC because ITN was a potential sponsor. This was likely a blunder in media management, because when ITN pulled out, the BBC got its own back by largely ignoring the mission, pushing the launching well down the news during a quiet news weekend. *The Mail on Sunday* organized a competition for five children to go to Russia, but at least part of the trip was paid by a charity, the Science and Technology Trust. There was the promise of a set of stamps by the British Post Office.

This should have been the end of the matter, but Mace and Sharman stayed on in Star Town. A deal was agreed in which the costs would be met by the Moscow Narodny Bank — though everyone knew where the funding really came from. The unpublicized consequence was that Britain lost the equipment package. Instead, whoever flew would oversee experiments sent to the station by the Energiya design bureau on Progress M7 in spring 1991 and the two trainees abruptly had to adjust their training to suit the Energiya experiments. They comprised experiments in biology (behaviour of newts, growth plants in magnetic fields, growth of cells (*Vita*)); physiology (noise on the space station and its effect on hearing; blood analysis using the Austrian *Reflotron*; astronaut response times (*Pleven*); Earth observations (equipment with different wavelengths); materials processing (*Elektropograph*); and heart monitoring (*Cardiorecorder*). One British experiment that did survive was the Applied Potential Tomography unit for non-invasive fluid distribution studies built by the University of Sheffield, though this was sold to the German space agency and flown as part of Mir 92 [20].

Helen Sharman later provided her insights into the practical aspects of cooperation with the USSR in its final days, the culture shock experienced by westerners and the spartan accommodation, from basins with no plugs to meat-based diets ill-suited to vegetarians. She was struck by the endless medicals (no one wanted to be the doctor or nurse who missed something), the chain of command, control over their movement and the isolation (there were only 13 phone lines between the USSR and Britain and mission managers did not want them using them anyway), but also the superb physical training facilities in Star Town and the challenge of learning Russian [21]. Whatever the attitude of the government at home, the British embassy at least helped out in small but important ways and invited the pair to receptions.

The selection of Helen Sharman was made on 22 February 1991, to fly with Anatoli Artsebarski and Sergei Krikalev. Sharman herself was somewhat surprised, since helicopter pilot Tim Mace better fitted the profile of the 'right stuff', but she reckoned that her even-tempered disposition counted in her favour. Her mission duly lifted off on 18 May 1991. Only one television station in Britain, BSkyB, televised the launch, at a time when satellite television was in its infancy and had few subscribers. All went smoothly and there were telecasts from space showing Sharman carrying out experiments. When on board Mir, it was her with whom President Gorbachev wished to speak, making an issue of the importance of good relations with Britain, but she had to make do with a 'good luck' message from her own prime minister, John Major. She also spoke to school students during Mir passes over Britain. Sharman returned to the Earth with Viktor Afanasayev and Musa Manarov on 26 May [22].

Helen Sharman — first Briton in space.

There were several postscripts to the Juno mission. On arriving back in Moscow, Sharman was greeted by the USSR's female cosmonauts, Valentina Tereshkova and Svetlana Savitskaya, while she and Tim Mace were both awarded the title Order of the Friendship of Peoples. There were some catty exchanges in the Soviet press as to why the USSR had earned little or no money from the flight, but both Energiya and cosmonaut Alexei Leonov defended it on the basis of gains in 'neighbourly relations' with Britain. After her return, Sharman made a science-promoting tour of schools, one encouraged by John Major [23]. As for the postage stamps, she got them in the end, but in Madagascar and the USSR. She kept a low profile for many years, resuming her career out of the public eye, but returned many years later to give talks and interviews, providing a fresh and welcome perspective of the events of many years earlier. On the 25th anniversary of her mission, Sharman again met her colleagues from the mission, cosmonauts and scientists, at the British Interplanetary Society in a celebratory reunion.

Helen Sharman stamp: USSR

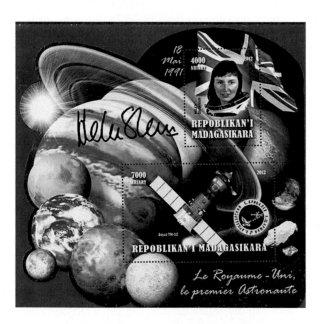

Helen Sharman stamp: Madagascar.

The unsuccessful candidates from the Juno mission were given the opportunity to have their names forwarded to the ESA astronaut corps, but in the absence of British participation in ESA human spaceflight – Kenneth Clark's 'an expensive frolic' comment had not been forgotten – they had no hope of flying. British space fortunes did not improve until 2010, when the UK Space Agency was established, Britain increased its support for ESA to surpass Italy and the country finally decided to participate in piloted flight, being rewarded with the assignment of Tim Peake to Soyuz in November 2015.

This, though, re-awakened some of the issues from the time of Helen Sharman's flight. Over the intervening years, British-born astronauts flying as naturalized Americans on the Space Shuttle (Michael Foale, Piers Sellers, Nick Patrick) had received enthusiastic press, film and documentary acclaim, in stark contrast to that given to Sharman during her mission. When Tim Peake flew, he was described as 'Britain's first astronaut', while the country's new space agency called him 'Britain's first official astronaut', which was hard not to see as a pointed put-down of the unmentioned Sharman. Tim Peake was referred to as 'the UK's only professional astronaut', implying that Sharman was somehow amateurish (her Russian cosmonaut colleagues did not think so). In renewing its identification of Peake as the first official astronaut, the BBC acknowledged that although there had been other British astronauts before, they were not paid for by the taxpayer nor were they 'wearing our flag', although 'our flag' must have been hard to miss on the photograph of Sharman's spacesuit [24]. It is hard to avoid the conclusion that Juno never shook off the default lines of the cold war, old or new. In the end, Helen Sharman's dignified, personable, cheery manner was the part of the story that ordinary people would remember.

Helen Sharman spacesuit.

Helen Sharman and Tim Peake. ESA.

EuroMir

Until 1994, all the European visiting missions to Salyut and Mir were organized on a national basis, (France, Germany, Austria and Britain). Their experiences suggested that there should be opportunities for the wider family of nations within ESA to participate in piloted flights. In April 1990, ESA and the USSR set up a number of joint working groups, including human spaceflight and in 1992, the Granada ESA council meeting approved ESA's participation in the Mir programme, with a view to flying European astronauts (accordingly called EuroMir, also written 'Euromir'). On 12 October 1992, ESA signed an agreement with the Russian Space Agency for two ESA flights to Mir (30 days, 135 days with spacewalk and 100–450 science hours respectively); cooperation on Mir 2; and a spaceplane project to fly in 2003.

This was a rapid pace of development, possibly explained by Russia's urgent need for cash. Fortunately, trained European astronauts were already available and required only instruction in the Russian language and Russian space systems.

They offered the prospect of a significant expansion in Europe's human space-flight experience. The cost to ESA is believed to have been between €45 m and €100 m per mission.

From 1992, at the time of the German missions, contacts between ESA and Russia intensified, with ESA visits to Moscow, Star Town, Baikonour, the industrial plants of Energiya and Molniya which designed spaceplanes and the Soyuz production plant in Samara, which hugely impressed the European delegation. There was discussion of Russia joining ESA and even a united Eurokosmos. The poor shape of the Russian space programme was already apparent in the form of contracting programmes, a shrinking workforce and declining conditions and salaries. Unlike Germany, which had access to Russian-speaking colleagues from the GDR, ESA had almost no Russian speakers and had to bring in translators. Things improved when ESA opened a Moscow office and from the mid-1990s, travel arrangements became much easier.

An important practical change for EuroMir was the arrival of CADMOS (Centre d'Aide au Dévelopment des activités en Micropesanteur et des Opérations Spatiales). The French were struck by the huge effort involved in decamping all their personnel to Moscow for the duration of a mission, reliant on negotiating access to their astronaut at the TsUP and having poor contact with France, so – with Russian agreement – they built their own mission control in Toulouse. CADMOS did more than provide a direct television link to their astronaut, it was a facility for overseeing all the mission experiments and taking in the data and results.

Four astronauts were assigned to EuroMir: Pedro Duque of Spain, Christer Fuglesang of Sweden and Thomas Reiter and Ulf Merbold from Germany. They arrived in Star Town for training in August 1993. The preparations for the mission involved the astronauts in 160 hours of language instruction, 175 hours of space-craft familiarization, 315 hours of technical instruction, 325 hours of biomedical training and 530 hours of mission-specific training. There was also water survival training in case they came down in the sea or in a lake. The original assignment was for Duque and Merbold for the 30-day EuroMir 94 and Fuglesang and Reiter for EuroMir 95. With Germany already in the Mir programme, EuroMir made it possible to introduce astronauts from Spain and Sweden.

For the first mission, ESA had just over a year to prepare a scientific package of 29 experiments. It was therefore decided that the first mission should mainly use equipment already brought up to Mir by European astronauts and that ESA should prepare more original experiments for the second. The main focus was on life sciences (cardiovascular system, neuro-sensory system, muscles, bone density) and materials sciences (e.g. the Czech CSK-1 furnace), some of which would be brought back (e.g. 100 bags of frozen blood and urine samples). Most of the experiments came from Germany and France, but there were also some from the Netherlands, Italy, Belgium, Denmark and Austria. In advance, 150 kg of equipment for the 29 experiments was ferried up by the Progress M-24 freighter craft. Each astronaut was allowed to carry 10 kg personally. This was one of the main contrasts to the US Space Shuttle and Spacelab. With the Americans, one could bring up and return a much greater range and heavier weight of experiments, but the missions were much shorter, lasting only a week or two. For EuroMir, one experiment was ferried up in advance: a Swiss Radiation Environment Monitor, similar to one fitted on the British STRV-1b satellite, to count electrons and protons penetrating its shielding. It was installed on the outside of Mir during a spacewalk on 9 September before EuroMir arrived [25].

As Soyuz TM-20 was wheeled out for the first EuroMir mission, the ESA logo could be seen stamped on its side for the first time. Clanking along the railway to the pad, the rocket passed the accumulated rusting debris of an increasingly poorly maintained cosmodrome. The flight began at night time on 4 October 1994, a historic date at the Baikonour cosmodrome. On board Soyuz TM-20 were Germany's Ulf Merbold, mission commander Alexander Viktorenko and flight engineer Elena Kondakova. Merbold joined the very small club of people to fly on both American and Russian spacecraft. Before that, he worked in the Max Plank Institute for Metals Research in Stuttgart, his main fields being solid state and low temperature physics.

The mission began a month of experiments on Mir. Normally, Mir had a crew of either two or three cosmonauts, more during the week-long handovers, but this was the longest continuous time six people had lived there. One of the long-duration Mir crew was Dr Valeri Poliakov, coming to the end of the longest-ever spaceflight of a year and a quarter (438 days). Merbold reported on the progress of his experiments through 20-minute daily video broadcasts to the main ESA space facilities in Paris, Cologne and the Netherlands. On 19 October, he connected with 45 schoolchildren from ESA's 13 member states in a 30-minute *Classroom in space* event in Noordwijk, the Netherlands. While in orbit, he communicated with the scientists via DICE, a multi-way teleconferencing system developed by Matra Marconi that had been brought up on Austromir and was designed to enable four-way communication between TsUP, CNES, DLR and the European centres in Cologne and Noordwijk. They also had a hook-up with Chancellor Kohl. ESA provided Merbold with a freezer based on Biorack for medical samples.

The mission was not entirely trouble-free, with Mir experiencing power and battery problems, but there were no emergencies. Toward the end of his mission, 3 November became a red-letter day for ESA, when Merbold's fellow astronaut Jean-François Clervoy blasted off on the Space Shuttle, meaning that two European astronauts were in orbit at the same time.

On 4 November, Merbold returned with the previous Mir crew, Talgat Musabayev and Yuri Malenchenko. They had a rough landing, with strong winds blowing them 9 km from the designated landing point, 79 km north east of Arkalyk. Malenchenko recalled later that he had not buckled up during the re-entry – he thought he would do that when the parachutes came out. The gravity (G) forces built up to 3G, which was nothing for a jet fighter pilot, but after several months of weightlessness was almost impossible to bear and Merbold could not move his arms to attach his seat belt. When the parachute did come out, the cabin swung so violently that he was nearly thrown out of his seat and could barely see his shaking control panel in front of him. The wind caught the capsule as it came into land and it bounced three times, 'like a blow to your head with a piece of firewood,' he recalled afterwards. Merbold brought back with him 16 kg of samples from his medical experiments, including saliva, blood and urine (the rest had to wait for later returning missions). Soon after, he was appointed head of the European Astronaut Centre in Cologne.

Later, Merbold was able to contrast the shuttle experience with the Mir experience. He noted that the pace on Mir was more relaxed and that there was always time on its lengthy missions to do an experiment the next day. On the other hand, Mir required a lot of maintenance, was underpowered, communications with the ground were intermittent and the space available for experiments was more limited. Disproving the saying 'what goes up must come down', Mir had also accumulated a huge amount of items, equipment and hardware over the years which sometimes took time to find, while the shuttle was much more ordered [26]. Merbold's mission was Europe's longest so far, while Kondakova went on to make the longest female flight to date at 174 days.

On 21 July 1995, in between the two EuroMir missions, cosmonauts Anatoli Soloviev and Nikolai Budarin installed the 235 kg, 2.5 m-long Mir Infrared Atmosphere Spectrometer (MIRAS), a Belgian instrument to observe Earth's atmosphere during solar occultations, also called a grill spectrometer. MIRAS was a cooperative project between IKI and the Belgian Institute for Cosmic Aeronomy and its director general Baron Marcel Ackerman. The purpose was to establish a profile of gas concentrations in the atmosphere from 20 km–120 km, especially 20 minor gases. MIRAS was launched on the floor of the Spektr module (20 May 1995) and then installed during a 5-hour 50-minute EVA using the 14 m Strela telescopic boom. It was the first significant foreign equipment installed on the outside of a Russian orbital station. Unfortunately, it short-circuited on power up and did not produce results [27].

The second EuroMir mission, EuroMir 95, lifted off in daylight on 3 September 1995 and reached Mir two days later. The ESA crewmember was Thomas Reiter and he accompanied mission commander Yuri Gidzenko and flight engineer Sergei Avdeev (later to hold the record for days spent flying in space (747)). Reiter was a test pilot who had qualified from Boscombe Down, England. ESA had longer to prepare for this second, long-duration flight, which focussed on changes in bone density in orbit and the reaction of biological samples to radiation. EuroMir 95 had 41 experiments, of which 18 were life sciences, five astrophysics, eight material sciences and ten technology. Experiments weighing 350 kg were delivered on a Progress ahead of the mission, with another 80 kg brought up on Progress M-29 during the flight and 70 kg during its extension. The bone density experiment was highlighted as the most important, followed by the Respiratory Monitoring System 2. The main life sciences experiments concerned blood pressure, bone mass loss, changes in muscles and the functioning of kidneys and lungs, while the materials sciences experiments used a tubular furnace. A French contribution was the Bone Densitometer (BDM) specifically designed to measure calcium in the human body, as this was now known to decrease significantly on orbit with consequences for bone strength. BDM was subsequently manufactured as an Earthbound medical device, an example of a spinoff. There was also a biokit to measure human changes in reaction to weightlessness and another machine to measure bone density.

Thomas Reiter.

For his first days aboard Mir, Reiter calibrated the experiments and set up the data relay system to download results to the ground during the night. A highlight of the early part of the mission was a 5-hour spacewalk by Thomas Reiter on 20 October, the first ESA spacewalk. In advance, he and his backup Christer Fuglesang had spent much time practising in the 12 m-deep water tank in Star Town. Reiter exited Mir with his colleague Sergei Avdeev, becoming the second European to make a spacewalk from a Russian spacecraft. By coincidence, the ESA council meeting was taking place that day in Toulouse and he made a broadcast to the gathering. The main function of the EVA was to install cassettes on the European Science Exposure Facility (ESEF) on Spektr and to bring in cartridges from a Swiss exposure experiment. They were at some distance from the hatch and the job had to be done slowly and carefully down along the Strela jib. The samples were designed to trap dust and tiny particles to assess the level of hazard to the ISS. During the mission, the Earth also passed through the tail of the Draconids, the shower created by the comet Giacobini Zinner and it was hoped to trap some of its debris. The experiments were closed over during shuttle and Progress arrivals at the station in order to avoid contamination. Reiter removed the covering on the radiation detectors and placed the exposure cassettes in position, locking them into place and then removed the Swiss package from the same area. There was also a British experiment from the University of Kent, Canterbury, to measure the impact of space dust on Mir, the dust being either man-made debris or micrometeoroids.

A big bonus for Reiter was a second spacewalk, for 3 hours and 6 minutes, this time to recover the samples put out the previous October. The spacewalk took place during an extended part of the mission. Reiter and Yuri Gidzenko exited the Mir hatch on 9 February 1996. Again, they made their way carefully along the complex to retrieve two cassettes, which ESA hoped would provide valuable data on microscopic debris in Earth orbit and also trap some cometary samples. New traps were put in their place.

As for the other experiments, Reiter took saliva samples; attached radiation sensors to himself to measure levels; used the new respiratory monitor during exercise; tested vitamin K supplements; took blood and urine samples after fasting; and attached a tourniquet to his ankle for an hour each day to measure the effect of venous pressure on bone mineral density. He tested a bioluminescence analyser to search for bacteria and operated TITUS, developed by the German space agency to achieve temperatures of 1,250°C in the area of semiconductors, alloys and glass. Technology experiments concerned radiation monitoring and the detection of microbial contamination. Transient Heel Loading was an experiment to stimulate bone growth. Reiter also tested a Video Integrated Service Controller, designed to enabled investigators to operate payloads from the ground and used the Austrian Optovert system for medical tests of the motor system. Just before Christmas, Reiter posted his opinion on the EuroMir webpage that many scientific results were now ready to be handed over to their principal investigators, the first use of the internet from space. There was an educational outreach programme, including two *Classroom in space* events with European schoolchildren and he also carried out a programme of radio hook-ups with German schools. There was a public day in which Reiter was interviewed by three German TV stations and gave an address, from orbit, to a conference on telemedicine in Nuremburg.

When the mission took off, ESA directors in Baikonour hailed it as a quantum leap forward in Europe's experience of long-duration flight, which was true. More so than they had imagined. Originally, Thomas Reiter was to have returned on 16 January 1996, but not long after he reached Mir it was clear that the rocket for the next crew switch-over would not be ready in time 'due to cash shortages'. Accordingly, his return was delayed by about 44 days, though there was no extra charge to ESA for the extension. Reiter was actually left in charge of Mir on 8 December when his two Russian colleagues left the station for a spacewalk to refit a docking cone. He became the first European to see in the new year in orbit and posted a Christmas message on the ESA internet page to tell of the work done on the orbital station so far. One of the most spectacular sights of the mission was witnessing an undocked Progress freighter burn up high in the atmosphere. First he saw flames, then explosions as the residual fuel was ignited and then the rest of the debris burning up in shreds [28].

Reiter eventually returned to Earth on 29 February 1996 with Yuri Gidzenko and Sergei Avdeev. The temperature in Arkalyk was -18°C, so they were quickly wrapped in warm clothes and put into the recovery helicopters. Reiter had been in orbit 179 days, making him the most travelled non-American, non-Russian ever. Little wonder that ESA was delighted with its progress in piloted flight. Gidzenko and Avdeev joined him in Noordwijk to present mission outcomes both to ESA and the *Classroom in the sky* group. The external samples picked up a huge amount of dust, with Mir apparently hitting a dust cloud every ten hours or so and, at one stage, encountering 5,000 tiny particles in ten minutes. The calcium experiments were important, because cosmonauts were losing 5–10 percent of bone mineral content a year, compared to someone with osteoporosis where the loss rate was 1–2 percent, so they wanted to find out why and then develop countermeasures. Reiter spoke of the potential of the Respiratory Monitoring System suit to be developed for infants to provide a warning if they stopped breathing (Sudden Infant Death Syndrome). He also described his recovery from long-duration flight, where walking normally took three to four days, blood pressure took several more days, physical fitness took up to two months and bone remineralization took longer, although he could not feel it.

During the post-flight press conference for EuroMir 95, ESA announced its intention to fly a third mission, EuroMir 98, with Reiter's backup Christer Fuglesang as prime candidate. Reiter emphasized the importance of Europe gaining operational experience in space in advance of the ISS [29]. He later returned to Moscow to undertake a 600-hour course to qualify as a Soyuz TM commander, receiving his certificate in 1997.

In the event, EuroMir 98 got no further. Mir was nearing the end of its lifetime and the first ISS module, Zarya, flew that year, 1998. Fuglesang flew later on the Space Shuttle. ISS operations were expected to be the new focus of European-Russian cooperation, while Mir was de-orbited on 23 March 2001. By coincidence, CNES organized an international colloquium *International scientific cooperation on board Mir* in the Palais des Congrès in Lyon on 19–21 March 2001 to review the achievements and scientific outcomes of the first, great, permanent orbital station. The European-crewed missions to Salyut and Mir are summarized in Table 3.1.

Table 3.1:
Piloted European missions to Salyut and Mir

Launch	Participant	Flight	Mission	Partner
24 Jun 1982	Jean-Loup Chrétien	Soyuz T-6	PVH	France
26 Nov 1988	Jean Loup Chrétien	Soyuz TM-7	Aragatz	France
18 May 1991	Helen Sharman	Soyuz TM-12	Juno	Britain
2 Oct 1991	Franz Viehboeck	Soyuz TM-13	Austromir	Austria
16 Mar 1992	Klaus-Dietrich Flade	Soyuz TM-14	Mir 92	Germany
27 July 1992	Michel Tognini	Soyuz TM-15	Antares	France
1 July 1993	Jean-Pierre Haigneré	Soyuz TM-17	Altair	France
4 Oct 1994	Ulf Merbold	Soyuz TM-20	EuroMir 94	ESA
3 Sep 1995	Thomas Reiter	Soyuz TM-22	EuroMir 95	ESA
17 Aug 1996	Claudie André-Dehays	Soyuz TM-24	Cassiopeia	France
10 Feb 1997	Reinhold Ewald	Soyuz TM-25	Mir 97	Germany
1 Feb 1998	Léopold Eyharts	Soyuz TM-27	Pegasus	France
20 Feb 1999	Jean-Pierre Haigneré	Soyuz TM-29	Perseus	France

What if? Hermes as European-Russian spaceplane?

Although the coming together of the American and Russian piloted space programmes in 1993 looks inevitable now, this was not guaranteed and another sce-
nario could well have developed. In 1987, Europe approved the development of
the Hermes piloted spaceplane, to be launched atop the new Ariane 5 rocket [30].
It made sense to explore its compatibility with Mir, so such a study was made
between October 1986 and December 1988, with working groups meeting on
16–17 October 1986 (Moscow), 14–17 April 1987 (Toulouse), 22–25 September
1987 (Moscow) and 18–22 April 1988 (Toulouse). Their work was signed off on
20 December 1988 by CNES and Glavcosmos, leading to an agreement for a flight
by Hermes to Mir in 1999, staying in docked configuration for 5–90 days and
with the flight mapped out in some detail. At The Hague summit on 9–10
November 1988, Hermes became a European programme, not just a French one.
ESA's intention was that Hermes should also be interoperable with the American
Freedom station and the European Columbus laboratory.

The second cooperative thread to the Hermes project was that Europe in general and France in particular could use Russia's existing expertise in shuttle technology and cosmonaut training. With the approval of the French minister for space
Paul Quilès, Jean-Loup Chrétien travelled to Moscow for discussions with the
head of the Buran test pilot programme, cosmonaut Igor Volk, for prospective
Hermes astronauts to undergo cosmonaut training in Russia, including shuttle-
type training. At that time, it was expected that the first Hermes astronauts would
be recruited in 1993 and that Hermes pilots would have made prior spaceflights
under either the American or the Russian programme. On the Russian side, Hermes
offered the possibility of using the technology developed during the Buran programme. The prospects for Buran had begun to dim after its first and only flight in
1988, leading to its eventual cancellation in June 1993.

Ejector seats were a key issue with regard to the technology, especially in the aftermath of the *Challenger* disaster where there was no escape system (there were ejector seats for the first four shuttles, but they were taken out at that point to accommodate larger crews). Zvezda had presented its K-36RB ejector system and the Strizh suit used for the Soviet Buran shuttle at the 1989 Paris air show. Zvezda also − unintentionally − demonstrated the K-36 basic version when a MiG 29 crashed at the air show, the pilot making a successful ejection at almost ground level, so the Europeans needed little further persuasion. Zvezda argued that the K-36RB offered ejection up to 22 km or Mach 3, providing the chance to escape an Ariane 5 fireball. Zvezda offered ESA its ejector and suit for €200,000 each, another persuasive argument when compared to the €50 m for the lighter but lower performing rival British Martin Baker system.

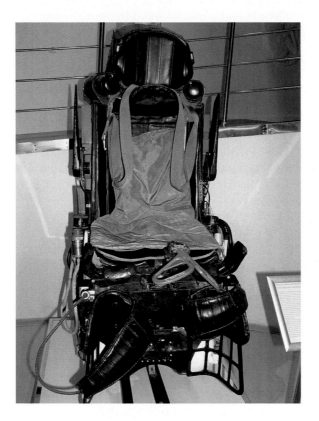

Zvezda ejector seat.

There was much discussion within ESA about the desirability of shopping for off-the-shelf technology abroad. It did little for Europe's domestic technologies, but the approach had the advantage of providing speedy access to proven technology. However, European industry preferred a joint enterprise that would bring some business Europe's way and it went to tender in 1990. Dassault was responsible for issuing the tender, bringing in proposals from Zvezda and Fiat on the one hand and Martin Baker and Aermacchi on the other. Either winner would be good news for Italy, which had lagged in obtaining Hermes contracts. Unsurprisingly, on 16 February 1990, Dassault announced Zvezda as the winner and a joint company was formed, Association des Vols Habités of Zvezda with Energiya and Molniya. It enabled the development of the ejector seat for Hermes, designed to be used to an altitude of 24 km and a speed of less than Mach 3, during take-off and for the last 30 km before landing. The European version would use European solid fuel rockets and electronics. ESA and Russia also gave attention to crew rescue. The original proposal was for the ageing French Breguet Atlantic aircraft which could drop dinghies and medical supplies, but the Russian amphibian Beriev 40 series promised to be faster and less expensive.

Beriev 40. CC Yevgeni Pashinin.

Space suits were an important related area. Hermes required both an ordinary space suit and a spacewalk suit, but Europe's experience was limited to pressure suits for jet fighters. The German company Dornier took the lead (British Aerospace withdrew when Britain decided not to participate in Hermes) and went to the world's leading experts in space suits, the Zvezda company again. In August 1992, Fiat bought two Strizh suits for testing. These suits were an important part of the process, as they needed to protect the pilots from the extreme forces and heat of an ejection, as well as offering survivability in the cold waters of the Atlantic in the event of either a launch or landing abort.

In 1992, a formal feasibility study was agreed between Dornier and Zvezda, with Dassault as sub-contractors for a spacewalk suit called EVA 2000. In 1993, Zvezda completed the new suit and argued that it should be the standard suit on the ISS, not just for Russian and European spacewalks, but the Americans too, though there was little chance of that. Unfortunately for Zvezda, this project stalled as ESA became overcommitted to the ISS and pulled out. Zvezda reverted to developing new versions of its successful existing space suit, the Orlan.

Zvezda suit for Hermes.

Also at that time, the TsAGI Aerodynamic Institute offered its test facilities, such as wind tunnels and vacuum chambers, for Hermes for heat and glass tests. Plans were set in motion for scale models of Hermes to be built in Russia for wind tunnel tests at Mach 10 and 14, as well as for Russia to transfer the test data from its four orbital BOR (*Bespilotny Orbitalny Raketoplan*, or Unpiloted Orbital Rocketplane) mini-shuttle missions in the 1980s which tested shuttle technologies. There was even discussion of Russia re-flying BOR, but with the nose and tiles from Hermes.

By the end of 1991, Hermes had made rapid early progress, with Europe buying in critical technology items from Russia in the form of ejector seats, ejector suits, space suits, hypersonic modelling and testing, heat transfer and thermal protection, windshields and rescue. The total initial cost was €603,000. These developments held the prospect of a substantial convergence of the European and Russian space programmes in human spaceflight, with Hermes/Mir operations in the late 1990s. An indicator of this convergence was Russia's close approach to membership of ESA at that time.

Hermes. ESA.

The progress proved short-lived, however, as Hermes came under increasing financial pressure. Europe became progressively over-committed to the American space station project and the Ariane 5 launcher, with Germany already burdened by bearing the costs of reunification with the GDR.

Moves were made to stretch the Hermes programme – always a worrying sign – and at the ESA Munich summit of 20 November 1991, its first flight was postponed to 2003. Early in the new year, the situation worsened, so ESA developed a second strategy of contracting out more Hermes tasks to Russia. Of ESA's €280 m budget in 1992, €5 m was allocated to 30 contracts with Russian design bureaux and technical centres (Energiya, Molniya, TsAGI, TsNIIMash) covering new areas such as landing gear, auxiliary power unit, fuel cells, lithium batteries, pyrotechnics, star sensors, altimeters and environmental controls. This was a bargain, because while €5 m was a small percentage of ESA's budget, it was a big cash injection for the struggling Russians and, moreover, one that included technology transfer to Europe. ESA opened the possibility of buying Russian docking systems to save it the cost of developing its own.

More strategies were bought into play to save the programme, namely an unmanned, stripped-down demonstrator, the X-2000, with a view to returning to a piloted Hermes early in the new century. At a French-German ministerial meeting in July 1992, Germany announced that it was prepared to pay for X-2000, but France regarded an unmanned demonstrator as a poor substitute. Eventually the French minister responsible, Herbert Curien, abandoned Hermes. At an unminuted ESA council meeting in Paris on 8 September 1992, the member states decided that there was no place in ESA's long term plan for X-2000 either, which in effect buried the whole project. The decision was not announced at the time and the official position was that the programme was still continuing in the form of the X-2000. News of its demise was, in effect, hushed up. The Hermes programme board continued to meet and figures were published indicating a €567 m budget for a European-Russian Hermes over 1993–5. The 12 October 1992 agreement for EuroMir included a shared spaceplane project to fly in 2003, immediately dubbed *Hermeski*.

The Granada ESA summit of November 1992 is seen as the point at which Hermes was finally buried, but the sequence of events is a complicated one and the fiction of the project being stretched out rather than abandoned was maintained for some time. Its precise moment of termination remains elusive, but the inner circle of the project knew what was happening. Shortly before Granada, the Hermes consortium – a roll call of Europe's leading aerospace contractors – made a public appeal to save Hermes, or at least the X-2000. There was a counterblast from the Comité National d'Évaluation de la Recherche (National Research Evaluation Committee), which lashed Hermes for over-ambition, mismanagement and outdatedness. At the Granada summit, ESA agreed on a 're-orientation' of Hermes and allocated €338 m for ESA-Hermes 'system studies', while also making the decision to study a space station lifeboat, called the Assured Crew Return Vehicle (ACRV). In the French parliament, the general's grandson, Jean de Gaulle, denounced the decision, but Curien defended it on the basis that Hermes had no clear mission.

When ESA declared for the new ISS project put together by American President Bill Clinton with Russian President Boris Yeltsin, France was an unenthusiastic participant in a project for which it had been little consulted. The trade unions in

CNES went on strike and issued a statement protesting that France, the third space power in the world thanks to General de Gaulle, would have no influence or control over the ISS, which would be at the mercy of the United States.

There was no room in the ISS project for Hermes and the drawings disappeared at this point. Like most cancelled projects, its death was never officially confirmed, but the Hermes industrial consortium was closed down in mid-1993. The 're-oriented' ESA-CNES Hermes team was not formally disbanded until 1995, when those still working on what was effectively a zombie project were told to box up their efforts for the archives and turn off the lights. Construction works that had commenced in Kourou were abandoned. The financial books on the project were not finally closed until 2014, when the project was estimated to have cost €1.2bn.

In the event, the ACRV, for which ESA member states had approved an investment of €140 m, was cancelled by NASA in 2002 and the Russian Soyuz became the lifeboat. Only one of the spaceplanes discussed here ever flew, the Russian BOR, which was re-developed by the Sierra Nevada Corporation as the Dreamchaser spaceplane for flights to the ISS from 2021. Europe has still not developed its own independent access to orbit.

What if? Mir 1.5

Hermes was one of the great *What ifs?* of European-Russian cooperation. It joined a long list of unfulfilled spaceplane projects and Japan had a similar experience with its HOPE spaceplane. Russia had substantial spaceplane experience dating to the 1960s (*Spiral*), flew BOR four times into orbit in the early 1980s and Buran into orbit in 1988. In better circumstances, such cooperation, partly based on Russian technologies, could have ensured European piloted access to space from the early 2000s. This was not the only *What if?* At this point, a successor space station to Mir was in development, as Mir had been designed with only a five-year life, not the 15 it eventually enjoyed. During the summer of 1992, Russia counter-proposed ESA participation in the Mir 2 space station that was due in 1996, with ESA not only providing a module and getting early crewed access to the station, but also with a possible view to a joint station by 2005. The official leading the negotiations, Jean-Jacques Dordain, tabled a paper to the September ESA Council in Paris – the one that quietly buried Hermes – outlining two years of system studies with Russia for 'intensified cooperation' [31]. Dordain was already trying to save something from the ruins. On the margins, there was even some discussion of Russia joining ESA as a full or associate member.

Following the Granada council of November 1992, three working groups were set up as part of the process of intensified cooperation proposed by Dordain. The systems group covered the modernization of the Soyuz spacecraft, with a new crew vehicle to replace Soyuz and Progress from 2005, while the technology group looked at space station subsystems. The third group addressed the use on

Mir 2 of the robotic arm designed for Hermes, as well as the new EVA space suit 2000 already in study between the Zvezda company and the Dornier company. Granada also established FESTIP, the Future European Space Transportation Investigation Programme, designed to look at future launchers and their engines after Ariane 5, even though the Ariane 5 had yet to fly.

However, the Russian financial climate continue to deteriorate and when its Council of Chief Designers met in November 1992, it was decided to scale back Mir 2. It would still look bigger in size than Mir because it would have an external truss structure for its solar arrays to try address its predecessor's problems of power shortages, but in reality it would be lighter. Instead of four large, 20-tonne modules attached to its node, the new Mir 2 would host four small, seven-tonne science modules built from converted Soyuz spacecraft. It would be a 'newer, smaller, better' Mir and received the light-hearted, contemporary nickname of 'Mir 1.5'.

The new proposal went through further evolutions to take into account Dordain's 'intensified cooperation'. The Russians had to reach their own painful shuttle decision to abandon Buran, but reaffirmed their commitment to 'Mir 1.5' with Europe, giving a flight date of 1997, about the same time as Mir's termination was now scheduled. For the Europeans, 'Mir 1.5' would be a collaborative effort guaranteeing them a permanent place on the station, with at least one European on board at a time, ferried up by Soyuz spacecraft. The scene was set for 'Mir 1.5' to evolve, over that year, as a joint European-Russian project.

Mir 1.5 was overtaken by events far away over that summer. The American space station, *Freedom*, announced by President Reagan as far back as 1984, had still progressed little, with even repeated re-designs and downscaling barely surviving congressional votes. New President Bill Clinton, who came into office in January 1993, worked hard to save the station by making it smaller. An interim downscaling was agreed in June 1993 but not long after, Clinton came to the conclusion that the best way to save the American station was the unthinkable: merge it with the Russian one. NASA officials met with Russian space leaders in Moscow, notably the newly-appointed head of the new Russian space agency, Yuri Koptev. This lead to a formal invitation to Russia to 'join' what was now called 'space station Alpha'. 'Join' was an understatement, since Russia would supply the first two large modules for the new space complex, which was soon re-titled the International Space Station, as well as committing a fleet of Proton and Soyuz launchers [32]. This had the side effect of making the Russian-European discussions redundant, but the benefit that there would at least still be a space station up there.

America's partners in the former *Freedom* station renegotiated their participation for ISS. Europe and Japan would have their own science modules – later called Columbus and Kibo respectively – and both would supply large refuelling spacecraft, the Automated Transfer Vehicle (ATV) and H-II Transfer Vehicle (HV) respectively. With the demise of both Hermes and HOPE, both partners would be dependent on Soyuz and the shuttle to get their astronauts to the station. Canada would build the station's large robotic arm.

Yuri Koptev. Roskosmos.

Some cooperative projects survived the sudden restructuring of *Freedom* and Mir 2 into the ISS. Russia had experience of automated orbital docking since 1967, whereas Europe had none, so it was only natural to seek help from the experts. For the ATV, Astrium contracted Energiya in Russia to build the mechanical, thermal and electrical parts of the docking apparatus and their interface with the rest of the ATV, as well as the refuelling system and the Kurs rendezvous system (although built in the Ukraine). Over 2008–2014, five such cargo ships flew to the space station: the *Jules Verne, Johannes Kepler, Albert Einstein, Edoardo Amaldi* and *Georges Lemaître*. Europe contributed the Data Management System–Russia (DMS-R) for its core module, the Zvezda. The DMS-R, made at Airbus in Bremen, was responsible for attitude control and the life support systems. ESA provided the Global Transmission Service (GTS) for Zvezda for its time signal, a project of DLR and Daimler Chrysler. This transmitted timer signals and data at 1 kps on 400.10 MHz and 1428 MHz to a distance of up to 1,200 km. The makers hoped to develop an application for tracking stolen items (e.g. cars, even credit cards) [33]. Zvezda and the ATV were a barter arrangement, with no money changing hands.

The next area of cooperation was the European Robotic Arm (ERA), an 11 m arm for which Fokker of Leiden in the Netherlands was allocated a $130 m contract on 19 March 1996. Fokker had little previous experience of space robotics but hoped to make the ERA the most sophisticated robotic arm ever, able to handle cargoes of 8 tonnes. ERA had started life as the Hermes robotic arm, HERA (Hermes European Robotic Arm) and substantial design work was undertaken during the Hermes development programme. When Hermes collapsed, HERA needed a new home. The American segment already had the main robotic arm built by Canada, supplemented by the Japanese Remote Manipulator System, so it was logical for ERA to be part of the Russian segment. Russian had abundant experience of building robotic arms for the Mir space station.

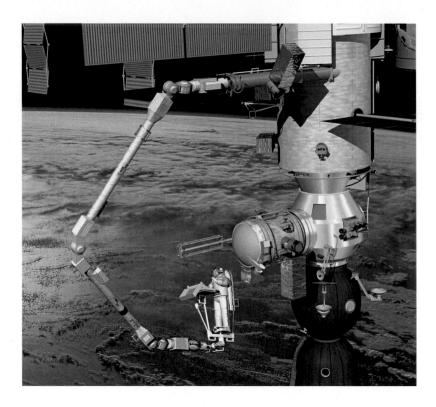

ERA robotic arm on Nauka. ESA.

ERA became a joint project between Fokker and the Energiya design bureau. The ERA is 630 kg in weight, 11.3 m in length, can lift eight tonnes and is extremely versatile, because each end can act as the base while the other is the effector, which has a tip accuracy of 5−8 mm. The effector can then moor itself on four different parts of the station. The effector has seven joints, giving a high level of dexterity, with 'wrists', 'limbs' and 'elbow joints', hand-and-foot functions, cameras, a lighting unit, a screwdriver and a computer designed for hands-on and remote use. ERA will be used for the assembly of the power systems and radiators, the movement of solar arrays, inspection of the station's exterior and the replacement of parts. The computer is equipped with a map of the station to inhibit any movement that might damage another part. ERA can be controlled either from a laptop computer inside the station, or, as displayed by ESA in 2019, by an external control unit manipulated by spacewalking cosmonauts. Its lid hinges open only when needed and is otherwise kept closed for protection from heat and cold, radiation and micro-meteorites.

ERA was scheduled for installation on the science module, also called the Multipurpose Laboratory Module (MLM), or in Russian, Nauka ('science'). Completed in 1998 as the backup Zarya module before being converted to Nauka, its launch was repeatedly delayed due to under-funding and technical problems, the worst being contaminated fuel lines which required years of repair work and refurbishment. A contender for one of the most delayed space projects ever, Nauka did not reach Baikonour until August 2020. The delay in getting the ERA into operation was one of Europe's most frustrating. One consolation was that a spare component from the project, an elbow arm, was brought up to the ISS in 2010 by the Space Shuttle on one of its last missions there, attached to the Rassvet module. In February 2005, a small robotic arm developed by DLR with Energiya and called ROKVISS (RObotics Verification on the ISS) was installed outside the space station. Unlike the other robotic arms, it was designed to be operated automatically from the ground. It worked there for five years and was brought back to ZNII RTK (Russian State Scientific Center for Robotics and Technical Cybernetics) in St. Petersburg. The results were published in a paper by 17 authors in 2016.

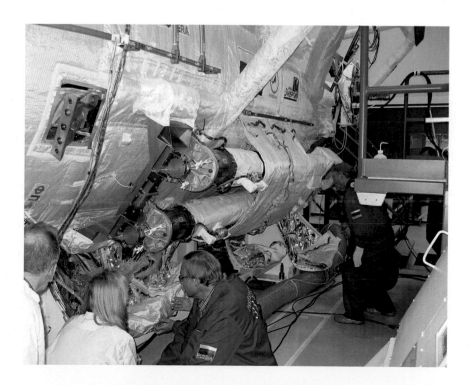

ERA robotic arm on Rassvet. ESA.

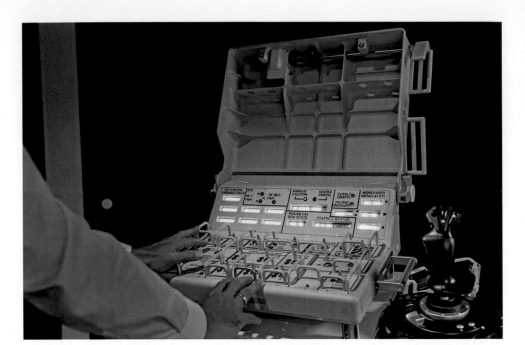

ERA controller. ESA.

What if? Kliper

One of the subject areas of intensified cooperation was a new piloted spacecraft to replace the Soyuz, which dated to 1960 and had made its first flight in 1966. Whilst rugged, robust and reliable, Soyuz was limited to a crew of three and had almost no spare room for experiments. The last chief designer, Valentin Glushko, had designed a replacement just before his death in 1989, but there was no funding to pursue it. In 2004, the Energiya design bureau suddenly presented a small lifting body-type spaceplane named Kliper ('clipper') as the Soyuz replacement, able to bring a crew of six to and from the ISS and its successor. Kliper built on all the earlier work on shuttles and lifting bodies [**34**]. A particular problem was that it did not have an obviously suitable launcher, although one, called the Onega and using existing combinations of rockets and engines, was put forward, with other possibilities appearing later (e.g. Soyuz 3, Zenit).

Although Kliper was included in the federal space plan, with the promise but not the certainty of funding, Energiya invited international collaboration, with the funding that this would hopefully bring. The United States was in the start-up phase of George Bush's return-to-the Moon Constellation project and financially fully committed, so Energiya brought the Kliper mock-up on tour to Japan and Paris, where ESA expressed its interest.

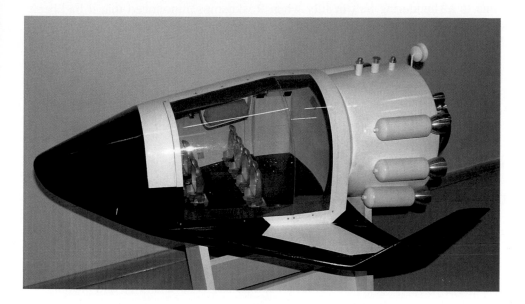

Kliper.

Europe's understanding was that it was the preferred partner and Roskosmos made a proposal accordingly. The principal benefit was in assuring European access to orbit, the goal which had eluded Hermes [35]. Cooperation got off to a flying start, with an Energiya-Astrium working group meeting in February 2005 and again in April, followed by a heads of agency meeting in June between Jean-Jacques Dordain and Anatoli Perminov. There was even a target of a launch in 2011, with both launch centres (Baikonour and Kourou) considered possibilities. Unlike Hermes, the project does not seem to have got as far as considering the *Who-does-what?* part. A formal proposal to fund cooperation on Kliper, called the Kliper Preparatory Programme 2006–7, was presented by the ESA Director General Jean-Jacques Dordain to the December 2005 ESA summit in Berlin, proposing that ESA should allocate €400,000 for Kliper. There was a suggestion that Japan would be brought in through its formally joining ESA, which was organizationally feasible because Canada was already a member. There was astonishment when the government delegates to ESA rejected the proposal. Dordain flew to Moscow immediately afterwards to try salvage something and promised Perminov fresh proposals by June 2006.

In December 2005, as unexpectedly as the Kliper project had arisen, its construction was put out to tender, with Energiya the expected winner in February 2006 although the Khrunichev and Molniya design bureaux were also invited to tender and certainly expected some sub-contracting opportunities. The project came to a shuddering halt in July when Roskosmos announced that it had been scrapped because the government was unlikely to provide any funding over 2006–2015 [36]. No full explanation has ever been given why the project collapsed, but it seems to have been a combination of financial problems, design doubt, technical prematurity and the reluctance of Europe to commit. A chance for Europe to get back into the human spaceflight business escaped a second time – but a fresh one presented itself almost immediately.

What if? ACTS

Despite this dissuasive experience, Europe and Russia then embarked on a fresh round of discussions on a new piloted spacecraft, the Advanced Crew Transportation Spacecraft (ACTS).

Attempting to salvage something from the Kliper misadventure, at the Paris air show in summer 2006, Perminov and Dordain agreed a two-year study of a new piloted spacecraft that would combine Europe's ATV as a service and propulsion module with a new Russian crew cabin. The new approach built on existing hardware, design and experience and was a much more conservative, less risky approach than Kliper. Early consideration was given to the most suitable launcher, such as a new version of the Russian Soyuz, the Zenit, or the European Ariane 5. The process appeared to be making good headway, joined by the European Union which had now begun to fund space projects like Galileo (navigation) and Copernicus (Earth resources). In March 2006 came a formal agreement for cooperation embracing Russia, the European Union and ESA, called the 'tripartite space dialogue'. This covered launchers, human spaceflight (e.g. ACTS), Earth observation, navigation, communications, technology and space science, with ESA allocating €15 m.

Progress was made and at the MAKS air show in Moscow in August 2007, the two space agencies set up a Joint System Engineering Team (JSET), the principal members coming from EADS Astrium, Thales Alenia and Energiya. Within a month it was looking at ten possible configurations, narrowed to six by October. Although their focus would be Earth orbital operations, missions to lunar and deep space destinations were also in consideration, paralleling the requirements of the American Orion. In the end, JSET came up with a similar design to Orion – a cone-shaped crew module on the ATV base – and this was presented to an ESA-Roskosmos meeting in Noordwijk, the Netherlands in December 2007. After a further JSET meeting in Moscow in January 2008, a Memorandum of Understanding (MOU) was agreed between ESA and Roskosmos that April, with Energiya as the Russian prime contractor and the European one to be chosen later. Russia drew up a more detailed design of the crew module, which it called the PTK-NP (*Pilotirumiye Transportny Korabl Novogo Pokoleniya*, or 'new generation piloted transport ship') and both sides sketched a series of mission profiles.

The first threat to ACTS began to emerge in spring 2008, but it was political, not engineering. In the Russian press, there was criticism that ACTS would make future Russian human spaceflight dependent on another country, or group of countries. In Europe, accusations began to fly that Russia was trying to dominate all the design decisions. The principal design problem focussed on the service and propulsion module, which the Russians felt could be simplified and scaled down to fit its existing launchers, though this would mean less business for the European contractors. At Kourou for the launch of the *Jules Verne*, Jean-Jacques Dordain said Europe was interested in ACTS only as a joint venture, not a project in which Europe would transfer funds to Russia over which it had no subsequent control. Despite this unsettling background noise, in May 2008 an agreement on ACTS was signed between Energiya, Thales Alenia and EADS Astrium. The basic architecture of the project was confirmed, with Russia to make a six-person crew module and Europe to build the propulsion module. Launches would be from the new base of Vostochny. Technical documentation was concluded in October 2008 with a view to an ESA decision on funding in November. Political noise in the background continued, with the Russian-Georgian war taking place that summer and Russian space officials extolling the merits of Russia proceeding independently. In the end, ESA pulled the plug at the November meeting, deciding to defer European participation in a programme that would probably have been a challenge to the tight ESA budgets. ACTS foundered on a combination of organizational problems (how to share production and funding), tight budgets and lack of political support from European ministers.

Some regard ACTS as a missed opportunity [37]. The 2006 period was an effort by Russia to secure European partnership, one which offered Europe a 'unique opportunity for its future manned space activities with important scientific and industrial consequences'. The project had many hurdles in the form of production costs, technology safeguards, funding and political support, but offered a real opportunity for Europe to be 'more than a passenger' in human spaceflight.

Russia and Europe then went their separate ways again in piloted spaceflight. Russia continued to advance the PTK-NP, which thankfully eventually acquired the less tongue-twisting title of *Federatsiya* ('Federation' for the Russian Federation) and then the shorter, less political *Orel* ('Eagle'), but with a long lead-in period and a first scheduled flight not due until 2023. As for Europe, when the ATV programme concluded with the *Georges Lemaitre* in 2014, it decided to supply its propulsion and service module to the Americans for the Orion space-craft that is intended to bring the Americans back to the Moon in the Artemis programme in the early 2020s. This was not that different to what might have been achieved with ACTS many years earlier.

Between them, the four *What ifs?* showed both the potential and limits of European-Russian cooperation. Hermes, Mir 1.5, Kliper and ACTS could have given Europe a much stronger role in human spaceflight and its related technologies in cooperation with Russia. Whereas Russian cooperation had been driven by the need for financial resources for which they were desperately short, on the European side there were clear advantages in dealing with a country that had such a knowledge and experience of human spaceflight, including shuttle-related technologies, along with access to its space station. In the event, Hermes, Kliper and ACTS were a bridge too far for the ESA member states.

Europe, Russia and the International Space Station

The first element of the ISS was the Zarya module launched by Russia in December 1998, its second the service module, Zvezda, in July 2000, with DMS-R and GTS. The first ATV, *Jules Verne*, with its Russian equipment, arrived at the station in 2008. In the event, the ISS was to be the biggest ever site of cooperation that involved Europe and Russia together. One engineer, who subsequently went on to the ExoMars project, described how, when this began, he made the first of 245 trips between Europe and Russia over 25 years.

In advance of its participation in the ISS, Europe got its astronaut corps in order, whose members would fly to the station either from Baikonour on the Soyuz or from Cape Canaveral on the shuttle. On 26 March 1998, ESA announced the formation of a 'single European astronaut corps', presenting a team of 11 in Cologne in September 1998, wearing blue suits against a background of Earth. In October 2000, the ISS was occupied by a Russian-American resident crew for the first time. The question was, how were the Europeans to get up there?

Agreements were signed between ESA and the Russian Space Agency, the RKA, to formalize European participation: an inter-governmental agreement (January 1998); DMS (March 1996); flight opportunities and ERA (July 1996); microgravity research (May 1996) and taxi missions (2001). They became a formal Cooperation and Partnership Agreement in February 2003, ratified by the Duma and Federation Council that May. The European taxi missions emerged more by accident than design and arose from the requirement that a Soyuz had to be attached to the station at all times to facilitate an emergency return to Earth should this be necessary. Because the fuel and electronics on board had a limited guaranteed life, these lifeboats had to be changed every six months, which required the delivery of a new Soyuz. The delivery crew would return in the old one a week later, essentially a form of taxi service.

The idea of using taxi missions to provide Europe with flight opportunities is attributed to ESTEC in the Netherlands, where there was a live broadcast of the launch of Zvezda on 12 July 2000. Arlène Ammar-Israël, responsible for piloted spaceflight in CNES, suggested the idea to Jörg Feustel-Büechl, head of piloted flight at ESA, of taking advantage of a taxi flight to ISS to fly an experienced European astronaut and re-establish the Europe-Russia link in piloted flight. Otherwise, ESA would have to wait until it had its own workspace on board in the form of the Columbus module in 2008.

Likely concerned that France would not fare so well in the competition for seats to the ISS, this was something of a pre-emptive strike. Following the preliminary conversation at ESTEC, the matter went to France's new research minister Roger-Gérard Schwartzenberg, the replacement for Claude Allègre who had wanted to call time on piloted flight. Schwartzenberg negotiated a week-long taxi mission to the ISS including 80 kg of experiments, with the seat costing CNES about €15.2 m and the whole mission organized in a short time.

This was formalized as an agreement between ESA and Russia in 2001, which provided for an initial six taxi missions up to 2006 by European astronauts. In practice, each mission was negotiated within ESA on a case-by-case basis. The mission costs – as distinct from the seat costs – were normally divided between ESA and the national space agency concerned, reflected in the practice, to this day, of European astronauts sporting both the ESA emblem and their national flag. But who in particular would take the available seats? Those who wanted the assignment most had their own ways of indicating why they should be chosen. Claudie André-Dehays (France) Thomas Reiter (Germany) and Christer Fuglesang (Sweden) travelled to Moscow to get a certificate of proficiency in piloting the Soyuz, so they must have topped the list for candidates. Later, those who wanted to fly on the Chinese space station made it known that they were becoming proficient in mandarin in their spare time.

Andromède

Mir veteran Claudie André-Dehays arrived to train for her mission in January 2001. Given France's early intervention and her own Soyuz commander qualification, the assignment could have gone to no one else and she flew as flight engineer (#2 position) rather than the more normal and humble title of research engineer (#3 position). Now Claudie Haigneré (she had married Jean-Pierre in May), she flew to ISS aboard Soyuz TM-33 on 21 October 2001 with Viktor Afanasayev and Konstantin Kozeev. The launch was transmitted live on the CNES website by videostream from Baikonour and was given the code-name of *Andromède* (Andromeda). Prime minister Lionel Jospin was in the TsUP control centre to see her Soyuz TM-33 dock with the space station two days later. While in orbit, Haigneré held a press conference with Roger-Gérard Schwartzenberg, between them sending the signal that France was back in the piloted spaceflight business. The station's resident crew comprised Frank Culbertson, Vladimir Dezhurov and Mikhail Tyurin.

Andromède **leaves. CNES.**

Haigneré's experiments were ferried up in advance on Progress and she brought up another 15 kg with her on Soyuz TM-33. She carried out a range of experiments involving space medicine, crystals, salamanders, frogs and computers, as well as an educational programme for schoolchildren. She made four hours of observations of sprites (optical flashes between 50 km and 90 km above thunderstorms, sometimes in a shape of a 'red jellyfish') using a lightning and sprites observation instrument devised by IZMIRAN **[38]**. French followers might have been forgiven for imagining that this was a French national mission, so well was it publicized, but it became the first of the ESA-Russia taxi missions.

Andromède **returns. CNES.**

The summer after Haigneré returned from ISS, Lionel Jospin's socialist government in France collapsed, to be replaced by Jean-Pierre Raffarin. He appointed Claudie Haigneré, out of the blue, as one of ten women in the new government he formed in June 2002, where she became the new minister for research. It was a popular choice, widely and warmly applauded. That lasted until March 2004, when she was moved to European affairs until May 2005. Her political career was less successful and she lost out in the governmental reshuffle that followed the disastrous French referendum on the European constitution. In 2009, a later science minister, Valérie Pécresse, appointed her the first President of Universcience, a merger of the City of Sciences and Industry and the Palace of Discovery. In 2015, Haigneré moved on to take a post at ESA. For the French, though, *Andromède* marked the end of an era. France had managed to build up a considerable human spaceflight experience on Soviet and Russian space stations, with 292 days on Mir alone. French cosmonauts flew once to Salyut 7, six times to Mir and once to the ISS, all at the comparatively inexpensive price of just over €110 m. France knew that it would no longer have fast-track access to the ISS and so it proved.

European missions to ISS

Over 2000–2020, there were 18 ESA missions to the ISS using the Soyuz launcher (see Table 3.2). It is not intended to describe each one individually here and there are other accounts which do so, notably that of John O'Sullivan [39]. What is more relevant is the nature of the Russian-European cooperation involved. While Europe (ESA and the member states) paid the mission costs, the 'seat' costs – getting up there and back – were covered by NASA as part of a barter arrangement, with Europe providing lifting space for American equipment on the European ATV. These costs were absorbed on a shuttle mission, but it required American cash to pay for the seats on the Soyuz, rising to €70 m by the time of the Pesquet mission. When the shuttle was grounded following the *Columbia* tragedy (2003–5) and retired (from 2011), Soyuz became the only way for European astronauts to reach orbit. On Soyuz ISS missions – when the shuttle was not flying – there was usually a Russian seat and an American seat, so the other participants (Europe, Canada and Japan) were essentially left to argue over the third seat. Normally, the seats were allocated and the teams formed about 18 months ahead of the mission concerned.

The second ESA mission to ISS was the *Marco Polo* mission by Italian Roberto Vittori aboard Soyuz TM-34 on 25 April 2002. Italy had built a Multi-Purpose Logistics Module (MPLM) which had arrived at the station the previous June and whose cargo included a Microgravity Science Glovebox. This was a large experiment box, almost the size of a small car, making possible a substantial expansion of Europe's experimental potential on the station in such areas as materials, combustion, fluids and biotechnology, with experiments studying osteoporosis being especially highlighted for this mission. Vittori flew on his ten-day flight with Russian mission commander Yuri Gidzenko and South African space tourist Mark Shuttleworth, returning on Soyuz TM-33.

The third flight was that of Belgian Frank de Winne. He had waited since 1990, having been the finalist of 526 candidates during the recruitment round for Columbus and Hermes. His flight, on a foggy 30 October 2002, marked the introduction of the Soyuz TMA, with improved electronics, solar panels and seat sizes. His *Odyssey* mission with Sergei Zalyotin and Yuri Lonchakov involved 23 experiments and he returned on Vittori's Soyuz TM-34 ten days later. Belgium contributed €15.5 m to the mission. There was a personal note too, as de Winne married a Russian woman, Lena, whom he met while training there.

Frank de Winne. ESA.

In February 2003, the Space Shuttle *Columbia* was lost, shuttle missions were suspended and the ISS became dependent on the Soyuz. Although the visiting Europeans would still fly ten-day missions, they would become part of the system of relieving crews on six-month missions, so they would fly up with a new crew and return with an old one – no longer taxi missions in the original sense. First in this system was Pedro Duque (Soyuz TMA-3, 18 October 2003, the first time the Spanish flag had flown at Baikonour) who flew to the station with Alexander Kaleri and Michael Foale. The mission was named *Cervantes* after the writer Miguel de Cervantes, author of *Don Quixote*. Duque brought up 22 experiments in physical sciences, biology, physiology and technology and there was a televised linkup with prime minister José Maria Asnar. Duque landed in a blizzard aboard Soyuz TMA-2 with Yuri Malenchenko and Ed Lu on 28 October. Press interest focussed not on him but on Malenchenko, who became the first man to be married while in orbit and whose wife was waiting to greet him on the crew's return.

By this stage, with multiple European missions coming up, a system had set in. Europeans trained in three locations: the Manned Spaceflight Centre, Houston, Texas; the European Astronaut Centre, Cologne, Germany; and Star Town, Moscow. Star Town was the most important of the three and in many cases the astronauts were accompanied there by their families. Houston was vital for learning the American side of the space station. The training had several key learning points: the Soyuz; the space station; and how to operate the experiments planned for the mission. Astronauts were expected to maintain a high level of physical fitness throughout, though they were the type of people who did not need persuading.

As with the other countries participating, the European astronauts were allocated apartments in Star Town and it became quite an international place. Star Town had a full-scale model of the Soyuz and the ISS, physical training facilities, a centrifuge and a Soyuz simulator. It was possible to get around Star Town quite easily on foot, though some used bicycles. Children were able to play around the lake and enjoy the snow in winter. Each European astronaut was assigned a crew support engineer, to help with training, science and health issues – including the food to be chosen for the mission – and, most importantly, for family liaison. This meant accompanying the family to Baikonour for the launch, but also being on hand for the landing. The support engineer would be one of the first faces the Europeans met on their return.

The principal training challenge, according to the astronauts, was learning Russian. Although the Russian cosmonauts now spoke English – it was taught extensively in the schools – few Europeans had a prior knowledge of the Russian language and they found themselves plunged into immersion language classes. Instruction methods in Star Town – in Russian or anything else – were quite traditional: classrooms, chalk and blackboard, formal exams. The dictation approach had changed little: charts were hung up on the walls and discussed, but taken away afterwards so that notation skills remained as critical as ever. Nothing was ever made easier. The Europeans were expected to familiarize themselves with everything to do with the Soyuz, right down to wiring diagrams. Tim Peake mastered the Russian language for the purposes of flying the Soyuz, but volunteered that colloquial discussion in the Moscow shops was 'a different matter'. On the plus side, there were plenty of other trainees in Star Town at various stages of mission preparation from whom one could learn.

The European astronauts had to undergo two types of survival training. The first was splashdown training, in case the Soyuz should land in the sea or a lake (Soyuz 23 came down in a lake in 1976). This involved travelling to the Black Sea to learn how to exit the Soyuz and get into the water while awaiting rescue. The more demanding was survival training, which took place in Tiksi, Siberia, identified by meteorologists as one of the coldest places on Earth, with temperatures down to -58°C. Refresher training took place in forests near Moscow. There also used to be summer training in the central Asian desert (+50°C) in a place renowned for scorpions. The Siberian training was so cold that it was even hard to sleep in an igloo, while in summer, desert temperatures fell to zero at night. In Tiksi, they would learn to bivouac, set up radio beacons and fire flares. The Soyuz survival pack included a gun. Although some western commentators put it about that this was for disciplinary purposes, the real reason went back to 1965 when Voskhod 2 landed in a snowy forest and the crew was surrounded by howling wolves.

ISS teleoperator training. ESA.

The European astronauts also did simulated gravity and weightless training. The gravity training involved being whirled around on the centrifuge in such a way as to simulate the Soyuz launch (4.2G). To do so, trainees had to learn to breathe using their stomach muscles not their chest, but had a chicken switch to hand if it was all a bit much. You had to squeeze the chicken switch throughout, which was clever, because if you became unconscious, it cut out automatically. The Soyuz launch simulation also included the boosters dropping away, so there was a brief moment of being thrown forward unexpectedly. Possibly more fun was weightless training in Illyushin 76 jets, which flew parabolic curves and in whose padded cabin one could experience weightlessness in short bursts.

There were some joint training programmes. ESA believed in the value of underground simulation training in testing teamwork, problem solving, isolation, sensory deprivation and exploration, an idea pioneered by Czechoslovakia in the 1980s (*Stola 88*). The European programme was, appropriately, called CAVES

(Cooperative Adventuring for Valuing and Exercising human behaviour and performance Skills). In September 2019, CAVES was set in the extensive Slovenian karst cave system at Lepa Jama. Those selected were European Alexander Gerst (from Germany), Takuya Onishi from Japan, Josh Kutryk from Canada, Joe Acaba and Jeanette Epps from the United States and, from Russia, Nikolai Chub, who had been selected in 2012 as one of a group whose members had yet to fly.

Obviously, Russia provided the launcher and with it the range of launch services, from flying Europe's astronaut to Baikonour on the Tupolev 134 (later the 204) belonging to the Yuri Gagarin Cosmonaut Training Centre, to pre-launch checks, suiting up and pre-launch ceremonies attended by families. Once the astronaut departed the ISS, Russia deployed a fleet of Mil-8 helicopters to fly in racetrack formation while awaiting the cabin's return; cross-country recovery vehicles; and medical teams and tents. Once landed, the European astronaut would be extracted from the cabin, put in a director chair to recover, medically examined and would be back in Cologne by evening. It remains a big, efficient, impressive operation with multiple danger points, even if it looks routine.

The fourth mission was the Dutch DELTA mission (Dutch Expedition for Life, Technology and Atmospheric research) by André Kuipers (Soyuz TMA-4, 19 April 2004), who brought up a record 21 experiments, with most of the previous missions having carried 12–20. He launched with Gennadiy Padalka and Michael Fincke. Perhaps the most unusual experiment was International Caenorhabditis Elegans (ICE), a CNES-managed ESA/Dutch project comprising four Biorack containers with 44 bags of worms! Known as *C elegans*, these worms are biologically related to humans and are one of the favourite subjects of study for biologists. Kuipers landed near Arkalyk in Kazakhstan with Alexander Kaleri and Michael Foale on 30 April 2004 and he and his worms were brought straight to Toulouse, where they proved to be no worse for wear for their flight. Detailed analysis was prepared in time for the 15th international *C elegans* convention in June 2005.

Roberto Vittori became the first ESA astronaut to visit the space station twice (Soyuz TMA-6, 15 April 2005), launching with veteran Sergei Krikalev and John Phillips. His second mission, *Eneide,* had 23 experiments on human physiology, biology and technology, ranging from plant germination to upper body fatigue, the durability of electrical components for microsatellites and parts of the forthcoming European Galileo navigation system. The Italian national financial contribution came from a research consortium of the Lazio region, the Ministry of Defence, Finnmeccanica and the Chamber of Commerce. Vittori returned early on 25 April aboard Soyuz TMA-5, with Leroy Chiao and Salizan Sharipov.

The next European mission, *Astrolab* (2006) is normally listed as an American mission because the astronaut concerned, Germany's Thomas Reiter, launched on shuttle flight STS-121 and returned on STS-116. The original Russian-American agreement for the operation of the ISS included barter provision for cosmonauts to fly on the shuttle and astronauts to fly on the Soyuz. When regular shuttle flights resumed in 2006, Russia sold one of its shuttle seats to ESA so that Thomas Reiter could make a second long-duration mission to follow his earlier one on EuroMir 95. Reiter's mission lasted almost six months, including another spacewalk and overtook the previous European long-duration record of Jean-Pierre Haigneré on Mir.

There was now a gap pending the arrival of Europe's space laboratory, Columbus, in February 2008, which would provide a substantial expansion in on-orbit experimental capacity and enable Europeans to stay on board for six months. These long-duration missions also benefitted from the new large European ATV supply ships, the first of which, the *Jules Verne*, was launched in March 2008. The first European up to Columbus was Belgian Frank de Winne's on the OASISS mission, a name chosen from 520 entries in a competition. He was launched on Soyuz TMA-15 on 27 May 2009 with Roman Romanenko and Canada's Bob Thirsk. De Winne achieved the distinction of being the first European commander of the space station. He returned on 1 December with Romanenko and Thirsk in a freezing fog that grounded the recovery helicopters, so they were brought out in armoured personnel carriers converted into rescue vehicles.

Next to launch was Italian Paolo Nespoli on the MAGISTRA mission (Soyuz TMA-20, 15 December 2010), who flew with Dmitri Kondratyev and Cady Coleman on a 159-day mission with 30 experiments, including an educational programme with schoolchildren. For the press, the high point was a video call with Pope Benedict XVI, but Nespoli built a reputation for the quality of his photography of Earth from orbit. The three landed on Soyuz TMA-20 on 23 May, Nespoli later describing the re-entry and landing as quite an experience: 'rapid, cruel and rough, like being shaken by a hammer'.

Paolo Nespoli carried away. ESA.

André Kuipers then returned to space for the PromISSe mission (Programme for Research in Orbit Maximizing the Inspiration from the Space Station for Europe) with a record 33 experiments, launching with Oleg Kononenko and Don Pettit (Soyuz TMA-03M, 21 December 2011). He was the first European to fly the latest iteration of the Soyuz, the TMA-M series. By this stage, the American Space Shuttle had been retired and the Soyuz would again be the only way to reach the ISS, the pattern for the rest of the decade. This was an even longer mission, at 192 days, with Kuipers, Kononenko and Pettit returning to Earth on a warm, sunny 1 July the following year. Kuipers described himself as much weaker than after his first flight and dizzy unless he kept his head still.

The most dangerous of the early missions was that of Italian Luca Parmitano's VOLARE (meaning 'to fly' in Italian), who launched with Fyodor Yurchikin and Karen Nyberg (Soyuz TMA-09M, 28 May 2013). Soyuz now followed a six-hour fast-track rendezvous pattern with the space station, which was much more comfortable for the crew and provided an extra two days on the station itself. Parmitano made two spacewalks (6 and 16 July), the first Italian to do so, but on the second, using an American spacesuit, there was a water leak in the helmet. Spacewalks are physically very demanding, so water is supplied to the helmet via a small tube so that thirsty astronauts may drink some water by sucking on the tube. This time, the water would not stop flowing and by the time he beat a hasty retreat back to the airlock, Parmitano could neither see nor communicate. Once inside, Nyberg and Yurchikin quickly removed his helmet and mopped up. To say that he was close to drowning would be an overstatement, but he was saved by his own quick action and those of his colleagues, a reminder of the ever-present dangers of the spacewalk. The three returned on Soyuz TMA-09M on 11 November after 166 days.

Parmitano was followed by 37-year old German geophysicist and vulcanologist Alexander Gerst, for whom 40 experiments were carried up by the final European ATV, the *Georges Lemaitre*. He was launched with Maxim Surayev and Reid Wiseman on Soyuz TMA-13M on 28 May 2014. The 166-day mission was named *Blue dot* in honour of the description of Earth by Carl Sagan as a 'pale blue dot' in the darkness of space. Gerst's most important task was perhaps the installation and commissioning of the German-built 400 kg MSL-EML electromagnetic levitator to test new alloys for such areas as turbines and aircraft engines. *Magvector* was another futuristic test of an electrical shield to protect a spacecraft from radiation, an idea already pioneered by Russia in the Cosmos programme. Gerst made a thankfully undramatic spacewalk on 8 October, while his colleagues Surayev and Alexander Samokutyayev spacewalked to activate the Russian-European EXPOSE R experiment. Gerst, Surayev and Wiseman returned at dawn on 10 November.

The *Futura* mission saw the first European woman to set foot in the station since Claudie Haigneré. Samantha Cristoforetti was another exceptional person, with qualifications from the Technical University in Munich, the École Nationale Supérieure de Aéronautique et de l'Espace in Toulouse and the Mendeleev University of Chemical Studies in Moscow, where she wrote a thesis on solid rocket propellants. She earned a masters in aeronautical science (Naples) and was also a fighter pilot in the Italian Air Force. Her spare time activities included scuba diving, caving and mountaineering. Cristoforetti brought with her a palm-size computer with her favourite works, including Tsiolkovsky's formula for rocket propulsion. She was launched into a misty sky on 24 November 2014 with Anton Schkaplerov and Terry Wirts [40]. She clearly relished her experience on the mission and was often pictured with her fellow female cosmonaut, Russia's Elena Serova. Together they tested SPHERES, sending them up and down the inside of the station. SPHERES (Synchronized Position, Hold, Engage, Reorientate Experimental Satellite) were coloured hexagonal droids based on a *Star Wars* drone, which used ultrasound detectors to know their position, carbon dioxide gas to manoeuvre and eight AA batteries for power. They were later intended to fly outside and inspect the station.

Samantha Cristoforetti working on Thermolab experiment. ESA.

The most talked about experiment was undoubtedly the last: the first coffee machine on an orbital station. Developed by the Italian coffee brand, Lavazzo, its purpose was to enable the crew to enjoy standard Earthly coffee while in orbit. In reality, constructing the 200 kg coffee machine to operate safely in zero-g was quite a challenge because of the pressures (10 atmospheres) and temperatures involved. Cristoforetti returned with her colleagues after 199 days, just before sunset on 11 June 2015. She went on to train with Chinese astronauts, or hangtianyuan, in the first-ever joint training between ESA and China, a sea survival and rescue training course off the Chinese coast at Yantai in summer 2017. She was photographed posing with two hangtianyuan, all in Chinese spacesuits. Although it was a fun-photo, it also indicated a serious purpose and brought the prospect of European-Chinese piloted missions closer.

Two new, hitherto unrepresented countries now flew to the ISS: Denmark and Britain. Although the taxi missions were over, the long-duration, year-long flight of Mikhail Kornienko and Scott Kelly meant that there was no return-to-Earth at the six-month half-way point, but the Soyuz still needed to be replaced, meaning a taxi flight. The original proposal was that one 'small' nation would get a standard, long-duration flight, but not a 'small' nation that had already flown to the station (i.e. Spain, Sweden, the Netherlands, Belgium, Switzerland). There would be short-duration flight for Britain, a 'big' country, but one which had historically contributed remarkably little money. In the end, the mission roles were reversed: Britain increased its European funding considerably, making a long-duration flight either a condition or at least an expectation, whereas Denmark was slow to come up with a large financial contribution. Between the two, Denmark's Andreas Mogensen was allocated the taxi mission (Soyuz TMA-18M) and Britain's Tim Peake got the long-duration flight (Soyuz TMA-19M).

Mogensen was originally due to fly with a tourist, singer Sarah Brightman, but she withdrew. Her seat was then allocated to Aidyn Aimbetov of Kazakhstan, with Russia always keen to maintain goodwill with Kazakhstan from whom it leased the Baikonour cosmodrome. They launched with Sergei Volkov on the IRISS mission on 2 September 2015. The title IRISS combined the idea of Iris, the Greek goddess of the rainbow and the link between humanity and cosmos; and ISS itself. Mogensen worked overtime during his week on board, on experiments as varied as sprite lightning, backpain and controlling robots. He returned with Aimbetov, but Sergei Volkov stayed on board and the spacecraft brought back Gennadiy Padalka.

Andreas Mogensen of Denmark. ESA.

A British candidate, Tim Peake, had been successful in the ESA 2008 astronaut competition, but with Britain so little involved in ESA, he was unlikely to ever get a flight. This suddenly changed in 2010, when Britain established its own space agency and, having hitherto eschewed piloted flight, now put £49.2 m into the ISS programme. The mission was called *Principia*, named after Isaac Newton's text on the laws of motion and gravity, *Naturalis Principia Mathematica*. The contrast with Helen Sharman's mission could not have been greater. The UK Space Agency invested £2 m in outreach for the mission to inspire interest in STEM (Science, Technology, Engineering and Mathematics) in schools. Students were invited to design applications and experiments to go on two Raspberry Pi mini-computers that Peake would bring on board, while the mission patch was put out to competition, which was won by a 13-year old. The BBC previewed his mission in an hour-long documentary, showing the personable astronaut clearly enjoying his training in Moscow, Cologne and Houston, but well aware of the dangers and challenges involved.

British television stations covered the launch live (Soyuz TMA-19M, 15 December 2015), from the walk-out to the bus, with Peake accompanied by his parents, wife and two young boys. Television from the cabin showed Peake clearly visible in the right seat, giving a cheery thumbs up to viewers, especially the many enthusiastic Union Jack-waving young people who had assembled at both the science museum in London and his old school in Chichester along the south coast. With him were Yuri Malenchenko and Tim Kopra.

Tim Peake departs. ESA.

During his mission, Peake made amateur radio calls, described his activities on tweets and blogs, made educational films, participated in air-to-ground events – especially with schools – ran the London marathon and controlled a robot from orbit. On 15 January 2016, he became the first Briton to walk in space. He had trained to spacewalk, something he always wanted to do, but none had been scheduled, so he got lucky. His NASA colleague Tim Kopra was not so fortunate, having had to cut short the spacewalk when his helmet began to leak – it was the same suit that had caused Luca Parmitano such problems.

Peake returned on 18 June 2016 with Malenchenko and Kopra after 186 days in orbit, a dramatic and triumphant reversal of Britain's fortunes in human space-flight. He toured schools, while his cabin was put on display in Britain and Northern Ireland and attracted large crowds. Peake was an ideal ambassador for Britain in space and he expressed the hope of returning there on a future mission in the 2020s.

The next mission marked the return of France to the ISS. This was the *Proxima* flight of Thomas Pesquet, continuing the French tradition of naming missions after stars or constellations. The name was chosen from 1,300 entries, with the winner being another 13-year old, from Toulouse. Pesquet was the first Frenchman aboard the ISS since Léopold Eyharts had launched aboard the shuttle and helped to set up the Columbus laboratory nine years earlier. The reason for the long gap in French participants on the ISS was not clear, although several had flown earlier on the shuttle: Jean-François Clervoy (STS-66, 1994; STS-84, 1997; STS-103, 1999); Jean-Jacques Favier (STS-78, 1996); and Jean-Loup Chrétien (STS-86, 1997). In the 2000s, Philippe Perrin flew in 2002 (STS-111) and Eyharts in 2008 (STS-122). Pesquet flew with Oleg Novitsky and Peggy Whitson, launching at night on 17 November 2016 on Soyuz MS-03, Europe's first experience of its newest version. He worked on 50 experiments, made two spacewalks and returned to Earth with Novitsky on 2 June 2017 after 179 days, quickly receiving a con-gratulatory phone call from newly-elected president Emmanuel Macron.

ESA decided to offer a second flight opportunity to the already-flown astro-nauts from the 2009 selection (Gerst, Cristoforetti, Mogensen, Peake, Parmitano, Pesquet). First to benefit was Paolo Nespoli (Soyuz MS-05, 28 July 2017), flying with Sergei Ryazansky and Randy Bresnik on a 139-day mission. His late evening launch took place in exceptionally clear conditions, so that the rocket's trail was backlit, its plume billowing out in the upper atmosphere. Called VITA (Vitality, Innovation, Technology and Ambition, but also meaning 'life' in Italian), the 60-experiment mission included study of the effects of gravity on middle-aged people (Nespoli was 60). He put his photographic skills to further use by shooting film for the National Geographic Channel. The three came down in light snow on 14 December.

Alexander Gerst was the second to get another mission (Soyuz MS-09, May 2018). He was originally scheduled to fly with Jeanette Epps and Sergei Prokopyev, but Epps was mysteriously dropped from the mission at short notice and replaced by Serena Aunan-Chancellor. Gerst called the mission *Horizons* and contributed to the design of a logo showing the horizon of Earth receding into infinity. His mission covered medicine, the use of the electrostatic levitator, commanding a humanoid robot and using the joint Plasma Kristall furnace. He became the sec-ond European, after Frank de Winne, to command the ISS. They returned on the morning of 20 December in cold conditions (-20°C) but Gerst was back in Germany that evening. There was a moment of history on 26 August 2018 when Gerst called in by television from orbit to a party being held in Morgenröthe Rautenkranz to mark the 40th anniversary of Sigmund Jähn's first German flight into space. Gerst had been trained by Jähn and together they hailed his pioneering role. Sadly, Jähn died unexpectedly the following year.

Alexander Gerst training. ESA.

The third repeat flyer was Luca Parmitano (Soyuz MS-13). He flew with Alexander Skvortsov and Andrew Morgan and was launched, fittingly, on 20 July 2019, marking the 50th anniversary of the Apollo 11 landing on the Moon. Parmitano called the 50-experiment mission *Beyond*. In October, he became the third European commander of the space station.

At this stage, European missions to the ISS reached a hiatus, with quite a number from the 2009 class still awaiting their second opportunity. Essentially, the American arrangements to buy Soyuz seats from the Russians ran out, while their new piloted spacecraft – the Boeing Starliner and SpaceX Crew Dragon, originally due in 2017 – had yet to fly. ESA began to worry when European astronauts would next fly again. So did the astronauts, who, conscious of China's upcoming space station, started to learn Chinese (Thomas Pesquet, André

Kuipers and 2015 selection Matthias Maurer) or took themselves to China for a joint sea training programme (Samantha Cristoforetti and Maurer). There were no more training in Star Town.

The first of the new spacecraft, the SpaceX Dragon, did not make its first operational flight until the summer of 2020. Thomas Pesquet was the first to be offered the opportunity to fly one, being assigned to the second operational mission in 2021. Whether anyone would fly on the Soyuz again was not clear, but in September 2019, the director of ESA, Jan Wörner, stated that Europe would be quite happy to continue to do so. There was no formal agreement in place, nor financial arrangements, but he argued for such discussions to begin. One would-be European astronaut, Johanna Maislinger, took matters into her own hands and tried to get sponsorship for a Soyuz tourist mission. She was dual nationality Austrian and German, a Boeing 777 captain, doctor, engineer and extreme sports enthusiast, who had taken part in *Die Astronautin*, a campaign to send the first German woman into orbit. Although it is a country known for an interest in women's equality and with a long-serving woman chancellor, Germany has yet to put a woman in space [41]. For the time being, the principal European-Russian spaceflight would be fictional, thanks to *Proxima*, an acclaimed French film by Alice Winocour about Sarah Lareau (Eva Green) training in Star Town for a space mission.

Table 3.2:
Piloted European missions to ISS from Russia

Date	Astronaut	Flight	Mission	Nationality
21 Oct 2001	Claudie Haigneré	Soyuz TM-33	*Andromède*	ESA/France
25 Apr 2002	Roberto Vittori	Soyuz TM-34	*Marco Polo*	ESA/Italy
30 Oct 2002	Frank de Winne	Soyuz TMA-1	*Odyssey*	ESA/Belgium
18 Oct 2003	Pedro Duque	Soyuz TMA-3	*Cervantes*	ESA/Spain
19 Apr 2004	André Kuipers	Soyuz TMA-4	DELTA	ESA/Netherlands
15 Apr 2005	Roberto Vittori	Soyuz TMA-6	*Eneid*	ESA/Italy
(4 July 2006	Thomas Reiter	Shuttle STS-121	*Astrolab*	ESA/Germany)
27 May 2009	Frank de Winne	Soyuz TMA-15	OASISS	ESA/Belgium
15 Dec 2010	Paolo Nespoli	Soyuz TMA-20	MAGISTRA	ESA/Italy
21 Dec 2011	André Kuipers	Soyuz TMA-03M	PromISSe	ESA/Netherlands
28 May 2013	Luca Parmitano	Soyuz TMA-09M	*Volare*	ESA/Italy
28 May 2014	Alexander Gerst	Soyuz TMA-13M	*Blue dot*	ESA/Germany
24 Nov 2014	Samantha Cristoforetti	Soyuz TMA-15M	*Futura*	ESA/Italy
2 Sep 2015	Andreas Mogensen	Soyuz TMA-18M	IRISS	ESA/Denmark
15 Dec 2015	Tim Peake	Soyuz TMA-19M	*Principia*	ESA/Britain
17 Nov 2016	Thomas Pesquet	Soyuz MS-03	*Proxima*	ESA/France
28 Jul 2017	Paolo Nespoli	Soyuz MS-05	VITA	ESA/Italy
6 Jun 2018	Alexander Gerst	Soyuz MS-09	*Horizons*	ESA/Germany
20 Jul 2019	Luca Parmitano	Soyuz MS-13	*Beyond*	ESA/Italy

Joint experiments: Matroshka, Plasma Kristall, EXPOSE, ICARUS, Kontur

As this narrative has shown, each mission had its own research programme, normally devised by the country concerned in association with its research institutes. In addition, there were some joint Russian-European experiments that crossed these missions, operating even when European astronauts might not be present: Matroshka, Plasma Kristall, EXPOSE, ICARUS and Kontur.

Matroshka was a 'human torso phantom' (or 'dummy' in the popular press), without arms or legs. Such phantoms had been used in medical research for over 30 years and this one was bought off-the-shelf and re-worked, with more than 6,000 radiation sensors — many very small — at 1,634 measurement points of vital human organs, such as eyes, colon, lungs, stomach, and kidneys. Twenty research institutes were involved, with Germany the most prominent on the European side. Its purpose was to measure radiation levels in the different parts of the body in detail, from the head to the lower insides **[42]**.

A circular Austrian-Russian precursor Matroshka had flown on Mir from May 1997 for 555 days and returned to Earth in February 1999. Now, on the ISS, Matroshka I was initially installed in one of the cosmonauts' cabins (*katuyas*) and then fitted externally on Zvezda on 26 February 2004 to measure radiation dosage outside. It was brought back in on 18 August 2005 after 18 months. Matroshka R, also circular, was on board the station from June 2007 to December 2011 and was especially important for Bulgaria, whose Liulin 5 instrument found increases in dosages as the ISS flew at higher geographic latitudes **[43]**. Next up were the anthropomorphic Matroshkas developed by ESA and DLR in Cologne: 1 (ISS), 2A (Pirs), 2B (Zvezda) and 2 (Kibo). The principal findings of Matroshka 2 were that radiation dosage rates were twice as high outside the station as inside (539 vs 337), with highest recordings on the skin, eyes and chest and lowest on the lungs, kidney and bladder. There were no surprises, but the precise levels were now known. Recovered from a Soyuz on landing, Matroskha 2 returned to the DLR Radiation Biology Laboratory in Cologne, while a ground model went on display in the Museum of Cosmonautics in Moscow. His successors are two daughters, Helga (DLR) and Zohar (Israeli Space Agency), with distinct female torsos, that have been given seats on the first, unpiloted Artemis mission to the Moon on the American Orion spacecraft with a European service module. Between them, the Matroshkas have added considerably to our knowledge of the detailed incidence of radiation on the human body.

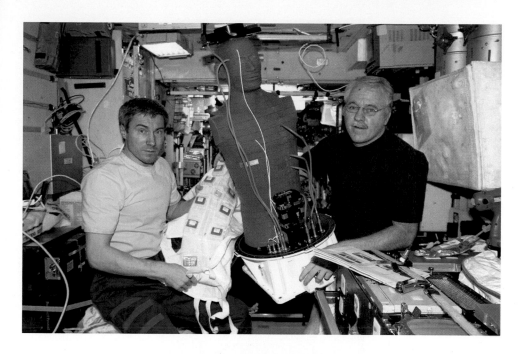

Matroshka inside. ESA.

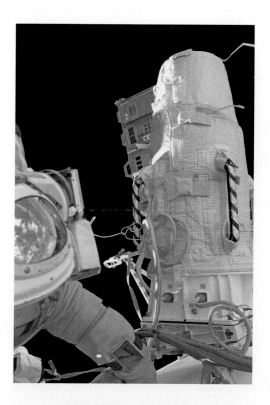

Matroshka outside. ESA.

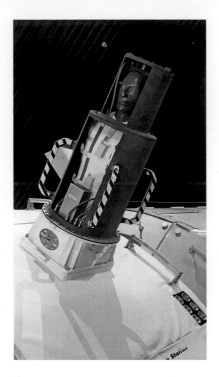

Matroshka outside Zvezda. René Demets.

Matroshka in the Museum of Cosmonautics, Moscow. René Demets.

The second cooperative area was plasma research, the German-Russian Plasma Kristall experiment (PK, or *Plazmenniy Kristall*), which evolved numerically as each new generation followed on (PK-3, PK-4, with PK-5 due in the 2020s). This was one of the most important lines of Russian-German research on ISS, the two respective institutional leaders and funders being the Joint Institute for High Temperatures of the Russian Academy of Sciences and the Max Planck Institute for Terrestrial Physics, with 12 German and seven Russian institutes also involved.

Plasma physics is one of the most exotic of the modern sciences, one which goes beyond the normal world of matter, time and space to address what Vladimir Fortov – the director of the joint institute – called 'the extreme states of matter'. Within that, there is a sub-discipline of 'dusty plasma physics'. Pure plasma is called the 'fourth state of matter' (after liquid, solid and gas), best known for the aurorae and, in practical terms, in fluorescent light tubes. Complex plasma includes dust and its nearest equivalent, in the traditional three states, is in what is called 'soft matter', like milk foam. Plasma crystals have important applications in areas such as electricity, energy transmission, solar energy cells, computer superconductors, the removal of dust from industrial units, food, medicines and clothing. Experiments on Earth are limited by gravity, whereas zero gravity enables three-dimensional studies of the behaviour of dusty plasma. On the ISS, electrodes convert inert gas such as argon or neon into plasma of particles only a few microns across, which become electrically charged and form a crystal structure (plasma crystal). Their behaviour and formation can be varied by changing the electrical field. On Earth, the particles fall to the bottom of the chamber, but not in zero gravity.

Vladimir Fortov speaking in Moscow.

PK-1 and PK-2 had already flown on the Mir space station (1998–9). PK-3, also called PKE Nefedov, was installed on the ISS in 2001 and operated until 2005, followed by PK-3+ from November 2009 to the end of 2011. Originally, the experiment operated in the Zvezda and Zarya modules, but it then moved to a more permanent home in the Poisk module. Typically, two to three 90-minute runs of experiments were conducted on three successive days each year, the first being from 27–29 January 2010 by Oleg Kotov. It was quite an intensive operation, with the tubes, fittings and connectors requiring protection from wear and tear. The chamber must be made a vacuum, which takes two days to set up. The instrument uses a pumped gas supply, has six micro-particle dispensers and two cameras and uses 3kV of electrical supply. The experiment is filmed and the videos sent back with the next returning Soyuz.

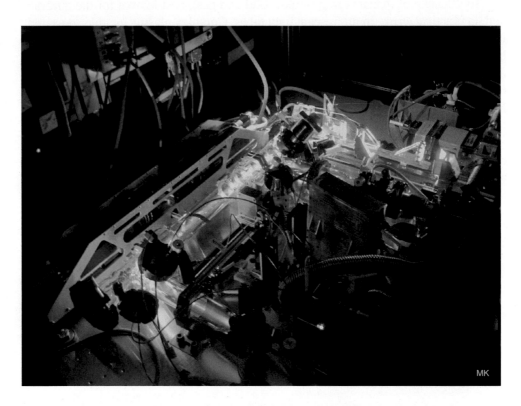

PK-4 cold plasma experiment. DLR.

PK-4 was made by OHB for the Max Planck Institute for Extraterrestrial Physics in Garching. It arrived on the ISS on 29 October 2014 and was installed in Columbus the next month by Alexander Samokutyayev and Elena Serova. PK-4 was governed by a 2013 agreement between ESA and Roskosmos, in which ESA had to build it and Roskosmos deliver it. Operations are managed by the DLR's Complex Plasma Research Group which has held conference events on the research outcomes, with scientific supervision by an International Coordination Group.

PK research generated over 100 scientific papers during its first 15 years on the space station. The big surprise in plasma experiments was that cold plasma acts as a sterilizer and can be used to treat chronic antibiotic-resistant wounds, bacterial infections and ulcers. At present, 37,000 Europeans die annually from infections picked up in hospital. Cold plasma sterilization could end that and, in developing countries, sterilization could be applied gently, with no residues or wastes, using a handheld device as small as a torch or electric shaver. It could be a lead instrument against epidemics, fungal infections, eczema, acne, herpes and skin ailments, as well as protecting plants, with further applications for mechanical objects, buildings and surfaces [44].

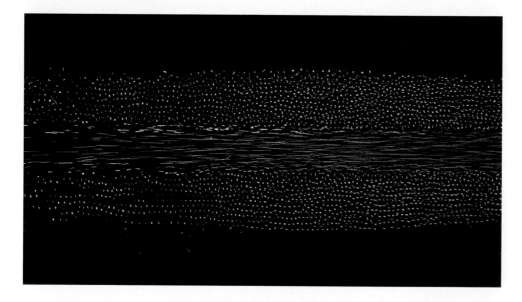

PK-4 plasma circulating. DLR.

The third joint experiment was EXPOSE. This was an experimental package developed by ESA with prime contractor Kayser-Threde, deployed or 'exposed' on the outside of the ISS to test the degree to which biological samples were affected by the vacuum, temperatures and radiation of space. This built on the experience of the Foton programme, but those missions were limited to two weeks while the ISS provided the opportunity to expose samples for up to two years and offered the ability to fly samples under Martian-like conditions. The first, EXPOSE E, launched and returned on the shuttle, was installed in 2008 on the European laboratory Columbus. EXPOSE R (R for Russia) and its successor R2 were overseen by the Energiya design bureau and installed on Zvezda in 2009 and 2014 respectively. Its trays included lichens, archaea, bacteria, cyanobacteria, algae, black fungi, mosses, liverworts, amino acids and spores, with varying levels of protection, vacuum and light under different windows. The periods of exposure were 17 months (EXPOSE R2), 18 months (E) and 22 months (R). When recovered, one lichen and one black fungus survived completely unprotected, while others survived with protection, indicating the possibility of life forms travelling distances across space to reach other planets. Many tardigrades and cress and tobacco seeds survived [45].

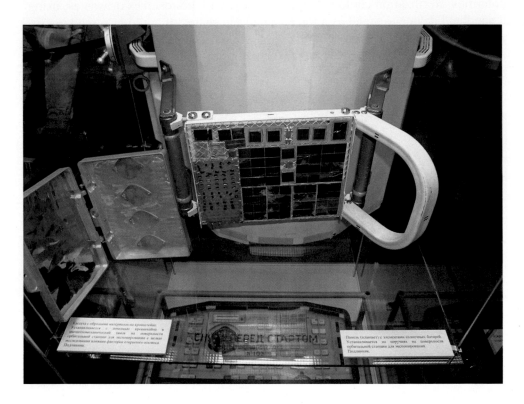

EXPOSE display.

After EXPOSE R2 was installed on Zvezda in August 2014, the samples were opened to the elements on 22 October by spacewalking cosmonauts Maxim Surayev and Alexander Samokutayev. They comprised six experiments which included BIOMEX, BOSS, PSS, Biodiversity (devised by IBMP) and two for radiation measurement: R3D-R2 and DOSIS R2. BIOMEX (BIOlogy and Mars EXperiment) involved the prior collection of cyanobacteria, such as bacteria, algae, lichens, fungi and mosses, from Antarctica and many other places, while BOSS (Biofilm Organisms Surfing Space) involved biofilms, the type of organisms that survive in pipes and shower heads.

EXPOSE outside the station. DLR.

The EXPOSE R2 samples were brought back to Earth in 2016 and transferred to Cologne for examination by 30 research institutes in 12 countries. By 2019, the BIOMEX experiment had generated 42 articles, while the results of BOSS and PSS were published together in a special issue of *Astrobiology*. Many of the samples survived the 533 days in the vacuum of space, its harsh temperatures and irradiation. This was an important finding, because vacuum and extremes of temperatures can be simulated on Earth, but solar ultraviolet radiation cannot and that is what EXPOSE tested. Some lichens were able to survive for up to two weeks. One of the survivors was archaea, single-cell micro-organisms that have lived in salty sea water on Earth for 3.5 billion years. According to astrobiologist Jean-Pierre Paul de Vera of the DLR Institute of Planetary Research in Berlin Adlershof, 'these organisms and biomolecules showed tremendous resistance to radiation and returned to Earth as 'survivors from space'.' Single cell organisms able to do this were candidates for the kind of life forms that might be found on Mars. According to Petra Rettberg of the DLR Institute of Aerospace Medicine, biofilms were some of the oldest communities of organisms on Earth, creating extracellular envelopes to survive. Some of the specimens collected from Antarctica were put in soils, clays and sediments designed to match those of the old Mars, complete with some volcanic ash that must have been present then.

EXPOSE samples back on Earth. DLR.

EXPOSE E, R and R2 included the R3D Radiation Risks Radiometer Dosimeter, called respectively R3D-E, R and R2. They were managed by the University of Erlingen in Germany and were installed on the ISS over 2008–2010. The R3D contained a Bulgarian Liulin spectrometer to monitor incoming cosmic particles and four solar sensors to measure solar light over four ranges of wavelengths. They took five million measurements of radiation levels and enabled the calculation of a dose rate for any station flying at around 360km altitude, including mapping those points over the Earth's surface where radiation was more and less intense **[46]**. A follow-up EXPOSE has been in preparation for incorporation into the Bartolomeo installation on Columbus, but awaits completion by the industrial supplier. Called the Exobiology Facility, it does not look like EXPOSE, but offers the possibility of monitoring test samples *in situ* to observe changes *during* the flight (EXPOSE offered a before-after comparison only).

The fourth joint experiment was ICARUS (International Cooperation for Animal Research Using Space), developed by Roskosmos, Energiya and the Institute of Geography of the Academy of Sciences; and in Germany, by the DLR and the Max Plank Institute of Ornithology in Konstanz. The concept was that hundreds of migratory birds would be fitted with miniature transmitters (weighing 5g), whose signals would be transmitted upward to an antenna on the ISS, which would collect signals on every pass and relay data on the passage of migratory birds back to the ground. The ICARUS antenna was brought up to the ISS on Progress MS-08, arriving on 15 February 2018. It was installed outside the ISS by cosmonauts Sergei Prokopiev and Oleg Artemyev during an eight-hour spacewalk on 15 August 2018 and activated on 10 July 2019. However, the ICARUS computer inside the station failed immediately and was sent back to Earth on the returning unmanned Soyuz MS-14 in September. The fault was identified and fixed and a new computer sent up to the station on Progress MS-13 in December. It was reinstalled and re-activated on 10 March 2020 for a test phase designed to ensure clear, reliable and noise-free signals, supervised by SpaceTech GmbH in Immenstaad on Lake Constance.

ICARUS antenna. DLR.

ICARUS installation. DLR.

ICARUS transmitter on bird. DLR.

The fifth and final joint Russian-German experiment is Kontur 2, brought up to the station in July 2015 to test the ability of someone in orbit to tele-operate a robot on Earth. Kontur 2 was developed by the Institute of Robotics and Mechatronics of the DLR in cooperation with the State Scientific Centre for Robotics and Technical Cybernetics in St Petersburg and was first operated in the Russian segment on 25 August by Oleg Kononenko. On 15 December 2015, Sergei Volkov demonstrated that it was possible for a cosmonaut to use a joystick to command a robot – called Justin – on the ground in Oberpfaffenhofen, Germany and achieve fine motor precision. The same experiment was applied to two Russian robots, Yula and Surikat, in St Petersburg. Dealing with the delay between the command being issued and it being received on the ground was one of the big challenges and they managed to reduce the delay to 30 milliseconds over 400 km. The experiment had the potential to enable cosmonauts to control a robot working elsewhere in space, on Earth or on the surface of another planet, from a space station. There were 23 sessions that autumn and the experiment was declared a success. Although it might look easy, in fact the technologies involved in what is called 'force feedback' were challenging. Kontur was the least expensive experiment of the five. Kontur 3 is now in design and anticipates a situation in which astronauts orbit Mars but, in advance of landing there, control robots to explore the surface. Kontur 3 will simulate control a long way from Earth, making accuracy more demanding and challenging.

Kontur 2 operated by Oleg Kononenko. DLR.

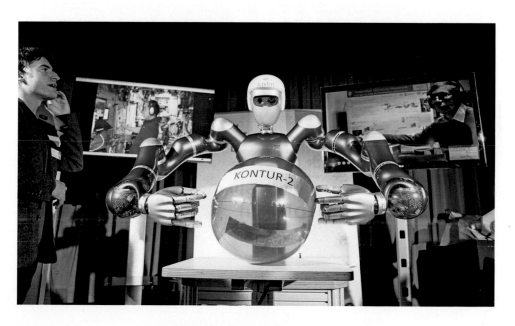

Kontur 2 Justin controlled from orbit. DLR.

By 2020, the DMS-R equipment was ageing. Its lifetime was 15 years, but by 2014 there had been a number of failures of its working memory (Random Access Memory, or RAM). DMS-R is so important that a breakdown could lead to the station going out of control. Over the years, the spares were used up or fitted to the Nauka module. Eventually, spare and replacement parts were put back into production under the supervision of a joint team between Airbus and the Energiya design bureau. Cosmonauts were very much involved in the process, as they were the people who would be changing out the computer boards. They switched out the old ones with the new, a bit like open heart surgery and by the end of 2019, the system had been fully updated. It is intended that it will be replaced from 2024 by DMS-R Next Generation (DMS-R NG), designed to last past 2030 **[47]**.

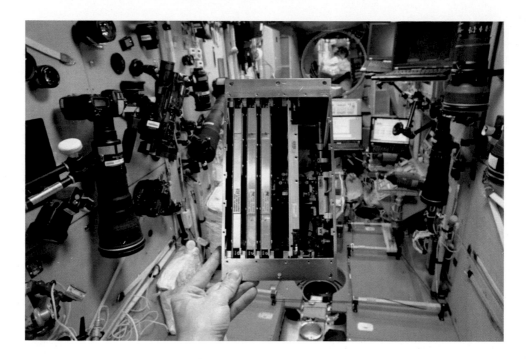

DMS installation. ESA.

European astronauts land on Mars

The final area of European-Russian cooperation was in simulated missions to Mars. Simulation missions were a long tradition in the Soviet space programme, being divided into experiments designed to simulate future space missions from the point of view of endurance and psychology (one in 1967−8 lasted a year) and closed life support systems, where the emphasis was on the regeneration of air, water and food. Facilities for simulated missions were built in Moscow at the Institute for Bio Medical Problems (IBMP) and were crudely called the 'botchka' (the box). Simulated missions had a certain advantage in a cash-strapped programme, as they enabled experimental work to be carried out when resources for the real mission were somewhere in the distant financial future. Undertaking the project with Europe would also bring in some funding. The Mars 500 project was initiated by director Anatoli Grigoriev but managed by cosmonaut Boris Morukhov.

Europe participated with Russia in two simulated Mars missions. First mention of the mission was recorded in summer 2004 as a European-Russian project to test for a long-duration mission to Mars of about 500 days. Apparently, it was underfunded and the decision was taken to make a 105-day test rather than nothing at all, even though this duration would not even get a real crew to Mars. Accordingly, it was called Mars 105 and got under way on 31 March 2009. There were two ESA participants, Cyrille Fournier (France) and Olivier Knickel (Germany), who had been chosen following a public call, similar to an astronaut selection. They joined commander and cosmonaut Sergei Ryazansky, cosmonaut Oleg Artemyev, physician Dr Alexei Baranov and sports psychologist Alexei Shpakov. The simulation involved 20-minute communication delays, medical experiments and growing food (lettuce, radish and cabbage). The mission concluded on 14 July 2009.

There was a positive reaction to the mission, if you discount catty comments from the British media ('Why don't they simulate the Olympics too?'). In reality, these tests had a serious purpose: how would the body respond − psychologically and physiologically − to the type of confinement represented by a Mars mission? Although the experiment could not simulate weightlessness, nor the absence of a capability of rescue, it was otherwise as realistic as possible. Ultimately, a sick participant − for example with a burst appendix − could leave, which would be impossible on a real mission, but there was otherwise no face-to-face contact with the rest of the world. In the event of someone leaving, the rest of the team agreed that they would carry on. Samples (e.g. blood, urine) were placed in a dumb waiter in a hatchway with an on/off red light. The exercise regime was almost identical to the space station, with expanders, a bicycle and a treadmill.

Mars 500 crew. ESA.

Following the success of Mars 105, preparations were made for Mars 500, a 500-day mission that would match the timeline of a real flight to Mars. ESA was fully involved in the mission definition – the steering committee, medical boards and operations – and was able to contribute experiments, which it invited in such areas as medicine (adaptation, stress, immune system) and psychology (confinement, cultural difference, mood). Indeed, in some places Mars 500 is described as a 'joint' ESA/IBMP mission, funded by both. ESA was allocated two places and made another public call for participants. Eleven candidates were sent for ten days of psychological and medical tests at a hospital in Moscow, following which the final six were selected. The selectors appeared to be most interested in compatibility, teamwork, forming stable relationships and some creativity. The schedule specified a typical day of eight hours sleep, eight hours work and eight hours free time, so filling the free time was important (e.g. learning Russian). Once selected, there was a period of training, which involved bonding (survival tests), mission preparation and, most importantly, preparation of 105 psychological, physiological and medical experiments. Each experiment had its own protocols and they were grouped under ESA experiments and IBMP experiments, with the ESA outcomes shared within its group, but each such experiment having an IBMP contact.

The two Europeans selected were Romain Charles (France) and Diego Urbina (Italy) who joined Russians Alexander Smolevsky, Mikhail Sidelnikov, Sukhrob Kamolov and Alexei Sitev.

The surprise was a late addition from China: Wang Yue. This was important for the development of long-duration Chinese human spaceflight, but the financial support must have been welcome too. By taking Chinese participation, though, the Russians effectively ruled out the Americans, because the US let be known that American astronauts would not be permitted to participate alongside the Chinese. China's first astronaut, Yang Liwei, appeared at IBMP during the experiment, an indicator of its importance to them.

The hatch was closed at 11.49 am European Time on 3 June 2010. The experiments were mainly in psychology, physiology and microbiology. ESA experiments covered such areas as the impact of inactivity, cardiovascular de-conditioning, the immune system response, the effects of exercise, the value of dietary supplements, blood levels, the value of blue light on circadian rhythms, skill tests and stress tests. A Dutch team had an overall responsibility for the evaluation, led by Bernadette van Baarsen [48].

Mars 500 made a simulated course correction on 24 December 2010 to put it into a spiral orbit that descended to Mars. On day 244, Mars Orbit Insertion was carried out. Three crew members made a simulated descent to Mars on 12 February 2011: Alexander Smolevsky, Wang Yue and Diego Urbina, leaving their colleagues Romain Charles, Alexei Sitev and Sukhrob Kamolov still in orbit. The landing module was quite small at 6.3 m by 6.27 m and became their home for 16 days. The landing crew made three 'spacewalks on Mars' which included planting the Russian, European and Chinese flags (the car park was commandeered and converted into a Martian surface). They left the Martian surface on 23 February, re-docked with their colleagues on 27 February and left Mars orbit to return to the Earth on 1 March.

The hatch opened on 4 November 2011, the six men stepping cheerfully forward in their blue track suits to the warm applause of flower-bearing relatives and friends. They spoke briefly to the press before going for debriefing. At 520 days, it was the longest simulation mission ever. Romain Charles spoke of how he and his colleagues followed a day of three equal eight-hour shifts (experiments (mainly medical), sleep, leisure). He found the return trip the most tedious, the best antidotes being games and e-mail contact with the outside world. Post-flight analysis provided a rich field for researchers and psychologists as to how best to support crews on such challenging missions, address conflicts and re-motivate a crew during whatever low points might emerge. Small things mattered. Although the crew could indicate food preferences in advance, Wang Yue's menu did not have any rice, which did not appear until half way through the mission. Being confined indoors was less of a struggle than anticipated (one of the crew practised in advance by not leaving his flat for a whole weekend) but they had cravings for natural sound, like wind. They missed water and longed for a swim.

Diego Urbina and Romain Charles later wrote an account of Mars 500, giving the European perspective on the experiment **[49]**. The mission was considered successful, with many turning points (the Mars landing), high points (e.g. birthdays), low points (monotony on the way home), stress points (two blackouts), learning points (other languages) and talking points (food). Six started and six finished, so it showed that a team *could* ensure a 500-day mission in confinement ('you can do it'). Of course there were ups-and-downs and good days and not-so-good days, but there was 'no hardship'. In effect, the simulation was a statement that 'the human factor matters. If you don't ask the question about how confinement works, you won't know until you try to do the real thing'. One of the participants was very sceptical of one of the Mars colonization plans, which did not involve a return flight but an indefinite stay − 'you need some light at the end of the tunnel.' Post-mission analysis, based on observations and the diaries of the participants, shed new light on how to manage extreme, confined environments and there were many important lessons on how to achieve a sense of teamwork among a culturally diverse group. Overall, though, international differences were not high on the problem list.

Romain Charles.

For the French, Mars 500 did not make sense without a follow-on in orbit on the ISS. The Americans had made a year-long flight with Scott Kelly, but that was still far short of Valeri Poliakov's 438 days back in the 1990s. The principal focus of American interest was on how Scott Kelly's experience compared to his twin brother Mark on the ground. Not only did there seem to be no American appetite for a 500-day flight by a team, but the Americans did not seem interested in staying long on the ISS either.

Mars 500 showed the Europeans that the Russian space programme of the late mid-2010s had still not emerged from the trauma and dislocation of the 1990s. The age profile was of a lot of older people who had been in the space programme all their lives, with some young people now appearing, but remarkably few in the middle age group (by contrast, the average age of workers in the Chinese space programme was 27). Many had left in the 1990s and that generation was missing. Most of the young ones held several jobs to be able to bring in household income: 'they are struggling and it's not really sustainable. It must be tough to be a scientist there now'. The lack of maintenance or new facilities was evident everywhere. European visitors to Baikonour felt that it still had 'a very Soviet feel to the place', though the ever-increasing number of tourists meant improved facilities. Russian facilities, like mission control, looked they had not changed since the 1970s 'with long, winding, quickly-get-lost corridors and big amphitheatre control room. They are still using radiograms for communicating with their cosmonauts on the ISS – have they not heard of e-mail?'.

Despite this, the Russian space programme was still impressive. One commented on how the timelines of projects differed between Europe and its partner countries: 'European projects have very long timelines, with everything planned out months and years ahead'. They were much shorter in Russia and, for that matter, China. 'In advance of our mission, the Russian institute appeared laid back, even chaotic to our European eyes. But as it got closer, everything came together in a very disciplined, orderly way.' National pride – even if Russia's greatest space achievements were some time ago – was an important driver: 'that is why they stay on and work crazy hours. They recognized that Russia was not doing all the things that they would like. They would still like to make the country proud, like in the 1980s, with Russia doing 'big things' in space'.

Human spaceflight: conclusions

Piloted flight was the next logical development for European-Russian cooperation, with France as the driver. When the USSR began to fly guest cosmonauts from the socialist block countries to the Salyut 6 station in 1978, extending the guest list was likely to be only a matter of time. Such an invitation was proposed by France in 1974 and formally issued to President Valéry Giscard d'Estaing in

April 1979. It might have happened sooner had there not been a period of many years in which the Soviet leadership tried to get the measure of the new, non-Gaullist, French president and his view of France's role in Europe.

The French were given a generous equipment load for their first mission, the PVH (Premier Vol Habité). All went well in preparation for the mission until the French presidential election of 1981, when international and partisan political complications and calculations entered the Franco-Soviet relationship. The Soviet leadership had backed Valéry Giscard d'Estaing against the winner, the socialist candidate François Mitterrand, the outcome of the labyrinthine nature of the politics of international socialism at the time, compounded by the discovery of a big Soviet spy ring and then the rise in tensions because of Afghanistan, Poland, SS20s and cruise missiles.

Mitterrand found himself man-in-the-middle between a problematical Soviet leadership and American attempts to subordinate Europe 'at whistle point'. Mitterrand, badly underestimated as a result of his three previous unsuccessful presidential campaigns, took his own decisions and stood firm and the flight went ahead, making Jean-Loup Chrétien the first western European in orbit. With the election of Mikhail Gorbachev as Soviet leader in 1985, the tensions that had dogged the first four years of the Mitterrand presidency ebbed away, with France quickly offered a second, long-duration mission with a spacewalk. Gorbachev was keen to build much better links to western Europe, with the ultimate aim of a 'common European home'. Twenty-two years after the general was there, François Mitterrand flew into Baikonour on the Concorde and made an agreement for French flights to the Mir space station every two years, an indicator of the progress, indeed the acceleration of that relationship. Jean-Loup Chrétien was the first non-American, non-Russian to make a spacewalk.

A schedule of French missions to Mir followed, in 1992 (Michel Tognini), 1993 (Jean-Pierre Haigneré), 1996 (Claudie André-Dehays), 1998 (Léopold Eyharts) and a bonus mission in 1999 (Jean-Pierre Haigneré). The ground rules had changed, with the financially stricken Russian space programme requiring ever-larger amounts of cash for flights. Given the volume of experiments the French were able to conduct and the on-orbit time accumulated, they would probably be regarded now as astonishingly good value. The experiments were ground-breaking, especially those with medical applications.

A French monopoly on piloted flight was never going to last and Germany finally made its entry to the field of European-Russian cooperation. Germany's late arrival was a function of the slow start to the German space programme (subsumed into ESA from 1975–89); lack of interest by successive chancellors; and the absence of a political relationship between Moscow and Bonn to match the Moscow-Paris axis, even during the time of peace-maker Willi Brandt. It was Mikhail Gorbachev who invited the Germans in, leading to the missions known as Mir 92 (Klaus Dietrich Flade) and Mir 97 (Reinhold Ewald). Neighbouring Austria also benefitted from an invitation (Franz Viehböck, 1991). Between them, these three countries (France, Germany and Austria) were able to get a substantial experimental payload aboard Mir, operated by their own astronauts. These missions evolved into the EuroMir, joint missions with ESA, in both cases with German astronauts: Ulf Merbold (EuroMir 94) and Thomas Reiter (EuroMir 95). By the end of the 1990s, Europe had built up a large volume of experience and experiments in space, unimaginable 20 years earlier [50]. Germany's much later start was explained by one engineer, who contrasted the slow but more structured and long-term German political decision-making process with the quicker, more impulsive decisions of the more flexible French.

The Juno flight by Britain's astronaut (Helen Sharman, 1991) is an outlier. Britain had been invited to send an astronaut to Mir as far back as 1986 but the following year, in an unheralded upheaval, the British government withdrew from a significant role in space research and would have nothing to do with the frolicking folly of human spaceflight. Two years later, the British Astronaut Project appeared out of nowhere to add Britain to the top class of European nations flying to Mir. When the fund-raising package collapsed, it seems that the Soviet government decided to fund it anyway to promote good relationships with Britain. Here, diplomatic imperatives appear to have been uppermost, though it is not evident that Moscow got much return or goodwill on its investment. Certainly it did not lead to column inches. Whereas Austrian companies like Vöslauer had rushed to sponsor the Austrian flight to Mir, British sponsors and television gave Juno a wide berth, the commercial lack of interest reflecting that of the government. Despite such an unpromising backdrop, Helen Sharman accomplished her mission with distinction.

The early 1990s were a formative period for European human spaceflight and cooperation with Russia, full of might-have-beens and potential *what if?* alternative histories. The first was a project to develop Europe's spaceplane, Hermes, in cooperation with Russia, one that intensified as its development programme became increasingly difficult. This project would have led to an unprecedentedly high level of integration between Europe and Russia in human spaceflight. The second, Mir 1.5, did not reach such an advanced stage, but was overtaken by the decision of President Clinton to abandon the American *Freedom* station in favour of a joint space station with Russia, in which Europe would play a part. The third and fourth were Kliper and ACTS, which ultimately failed to gain traction in Russia and Europe respectively. Although these turned out to be blind alleys, some ideas were salvaged, such as the European Robotic Arm. The Hermes cooperation showed the lengths to which both countries were prepared to go in cooperation and, more than that, integration. On two occasions, there was discussion of the possibility of Russia joining ESA, either as a full member or associate member, like Canada. Yuri Koptev, then head of the Russian space agency, wanted to go beyond a funding-driven relationship to a higher, more organic level of integration.

In the event, these false trails (Hermes, Mir 1.5) became part of the ISS project in different form and ensured practical cooperation. Europe provided two key parts of the ISS Zvezda service module: the Data Management System (DMS) and the Global Transmission Service (GTS), while Russia provided key technologies for the ATV (rendezvous and docking systems). Once the ISS was crewed, a system was set in place for regular European visits to the ISS. Although it was expected that Europeans would fly even-handedly aboard the shuttle and the Soyuz, in reality the shuttle served as the launcher for only brief periods. Europe's route to the ISS went through Star Town, Moscow and Baikonour. From 2000–2020, there were 19 European missions to the ISS and these involved a systematization of European-Russian training, with an almost permanent European presence in Star Town. These Soyuz seats enabled Europe to operate its Columbus laboratory, host to 230 experiments in its first decade.

On a typical long-duration mission, the ESA astronaut could expect to carry out 50 major experiments, sometimes more (63 in the case of Alexander Gerst). Europe accumulated considerable time on-orbit: 366 days for Luca Parmitano; 363 for Alexander Gerst; 351 for Thomas Reiter; 314 days for Paolo Nespoli and 210 days for Jean-Pierre Haigneré; with individual mission totals of 201 days (Luca Parmitano); 200 (Samantha Cristoforetti); 197 (Alexander Gerst and Thomas Pesquet); 193 (André Kuipers); and 186 days (Tim Peake). Luca Parmitano led the spacewalk table at 33 hours 9 minutes. Flight opportunities opened up for Spain and Denmark. The Russians gained the cash that kept their space programme going, paid via NASA, so the Europeans were left only with their mission costs. Beyond this transactional relationship, it is important not to overlook the joint experiments on the ISS: Matroshka, Plasma Kristall, ICARUS, Kontur and EXPOSE, all cutting-edge science; nor, on the ground, Mars 500. Given that Europe's contribution to the ISS was estimated to be in the order of five percent, this was a significant and cost-effective outcome.

Between them, the American and Russian connections ensured European access to human spaceflight for over 20 years and, from 2008, its own laboratory in Earth orbit, a substantial win from cooperation. The future of European-Russian piloted spaceflight cooperation is uncertain and in 2020 there were no more Europeans training in Star Town, as they will now fly to ISS aboard the American Dragon and the Starliner. However, it is hard not to imagine Europeans flying up to a future, smaller Russian Orbital Station, or indeed the Chinese station. Europe now has a solid information and experiential base in human spaceflight.

References

1.   Senate Committee on Aeronautical and Space Science: *Soviet space programmes, 1971–5 – goals and purposes, organization, resource allocations, attitudes toward international cooperation and space law.* Washington DC, 1976, p123.
2.   *Brezhnev gets elaborate welcome in Paris on first trip as President.* New York Times, 20 June 1977.
3.   For an account of the politics of the Mitterrand period, see Short, Philip: *Mitterrand – a study in ambiguity.* London, Bodley Head, 2013.

4. Kidger, Neville: *Salyut mission report. Part 13.* Spaceflight, vol 25, no 3, March 1983.
5. *Both sides are wary as Mitterrand arrives in Soviet Union.* New York Times, 21 June 1984.
6. Burgess, Colin & Vis, Bert: *Interkosmos – the eastern block's early space programme.* Praxis/Springer, 2016.
7. Mellow, Craig: *Aiming for Arkalyk.* Air & Space, August/September 1998.
8. France re-iterates space station attack. AW&ST, 13 October 1997; Taverna, Michael: *Pegasus may herald end of an era.* AW&ST, 9 Feb 1998.
9. Gracieux, Serge: *Une station route pour la ville rose.* Espace, July/August 2007.
10. Reinke, Niklas: *History of German spaceflight.* Cologne, DLR, 2010. For a history of spaceflight development in the GDR, see Hein-Weingarten, Katharina: *Das Institut für Kosmosforschung der Akademie der Wissenschaften der DDR.* Berlin, Verlag Duncker & Humblot, 2000.
11. *Cinquante ans de coopération France-URSS/Russie.* Paris, Institut Française d'Histoire de l'Espace, Êditions Tessier et Ashpool, 2014.
12. Federal Ministry of Education & Research: *German space programme.* Bonn, BMBF Publik, 2001.
13. Burgess, Colin & Vis, Bert: *Interkosmos – the eastern block's early space programme.* Praxis/Springer, 2016; Hooper, Gordon: *West Germany gets the best of both worlds!* Spaceflight News, July 1990.
14. Kowalski, Gerhard: *Sigmund Jähn's great journey.* DLR magazine, 163, December 2019.
15. Bond, Peter: *Austromir – Austrian to fly aboard Soviet station.* Spaceflight News, §56, August 1990.
16. Seedhouse, Erik: *The use of biological response modifiers by astronauts.* Spaceflight, vol 30, no 10, October 1997.
17. *Soviet space offer to Britain.* Spaceflight, vol 28, no 7, July-August 1986; Spiteri, George: *Denis Healy – my trip to Russia.* Spaceflight News, #40, April 1989.
18. *UK/Soviet space deal.* Spaceflight, vol 28, no 11, November 1986; Breus, Tamara: *UK projects for Mir space station.* Spaceflight, vol 29, no 3, March 1987.
19. *UK/Soviet agreement on space.* Spaceflight, vol 29, no 5, May 1987; *Space plan flounders.* Spaceflight, vol 29, no 9, September 1987; *ESA forges ahead – Thatcher government opts out.* Spaceflight, vol 29 no 12, December 1987.
20. Salmon, Andy: *Science on board the Mir space station.* JBIS, vol 50, #8, August 1997. For other coverage of the period, see *Juno gets Soviet go-ahead.* Spaceflight, vol 33, no 2, February 1991; Burnham, Darren: *From race tracks to ground tracks.* Spaceflight, vol 39, no 1, January 1997; *Postal tribute for Juno flight.* Soviet Weekly, 8 March 1990.
21. Sharman, Helen & Priest, Christopher: *Seize the moment – the autobiography of Britain's first astronaut.* London, Victor Gollancz, 1993.
22. Kidger, Neville: *Helen's 8-day mission.* Spaceflight, vol 33, no 7, July 1991.
23. *Lunch with Helen Sharman.* Emirates, 2016.
24. Drury, Colin: *Blastoff! Why has astronaut Helen Sharman been written out of history?* The Guardian, 18 April 2016; Rigby, Jennifer: *Tim Peake isn't first Brit in space – don't forget ballsy Yorkshire woman Helen Sharman.* Telegraph, 16 December 2015; Space:UK, summer 2013, #38, pp2, 18, 19; Spall, Nick: *Reality check.* Spaceflight, vol 61, #10, October 2019; Edward Stourton (ed): *Today – a history of our world through 60 years of conversations and controversies.* London, Cassell, 2019.
25. *European astronauts for Mir.* Spaceflight, vol 36, no 3, March 1994.
26. *Astronaut tells both sides of the story.* Spaceflight, vol 38, no 6, June 1996.
27. Audenaert, Benny; du Brulle, Christian; Laureys, Dawinka; & Pirard, Théo: *Belgians in space.* Brussels, Êditions Racine, 2004.

28. Werner, Marius: *EuroMir 95*. Spaceflight, vol 38, no 8, August 1996.
29. Van Rooij, Ton: *ESA-Mir operations prepare for ISS*. Spaceflight, vol 38, no 9, September 1996.
30. This writer: *Europe's space programme: to Ariane and beyond*. Praxis/Springer, Chichester, 2003; and Van den Abeelen, Luc: *Spaceplane Hermes: Europe's dream of manned spaceflight*. Praxis/Springer, 2017.
31. *ESA to propose euro-Russian spaceplane and space station*. Spaceflight, vol 34, no 10, October 1994.
32. Johnson-Freese, Joan & Moore, George M: *Space station reconceptualized*. Spaceflight, vol 36, no 2, February 1994.
33. Salmon, Andy: *Global Transmission Services*. Spaceflight, vol 42, no 1, January 2000.
34. Hendrickx, Bart: *In the footsteps of Soyuz*, in Brian Harvey (ed): Space exploration 2007. Praxis/Springer, 2007.
35. ESA: *Toward a long term strategy – the future of European space exploration*. ESA, Noordwijk, 2005.
36. Zak, Anatoli: *Russia in space – the past explained, the future explored*. Apogee, 2013.
37. Mathieu, Charlotte: *Assessing Russia's space cooperation with China and India – opportunities and challenges for Europe*. Vienna, European Space Policy Institute, 2008.
38. Powell, Joel: *Satellites observe the mysteries of lightning*. Spaceflight, vol 44, no 8, August 2002.
39. O'Sullivan, John: *In the footsteps of Columbus – European missions to the International Space Station*. Praxis/Springer, 2016; Triplett, William: *Astronaut, cosmonaut, Euronaut?* Air & Space, August/September 2003.
40. Bernadini, Fabrizio: *Show and tell for Italian astronaut*. Spaceflight, vol 56, #12, December 2014.
41. Quine, Tony: *Austrian-German woman hopes to be next Soyuz space tourist in 2021*. Spacesleuth2 blogspot posting, 22 October 2019.
42. Powell, Joel: *Phantom heads and Matroshka*. Spaceflight, 52, no 12, December 2011.
43. Semkova, Jordanka *et al: Recent results for space radiation environment in the spherical tissue equivalent phantom on the ISS from Liulin 5 experiment*. Presentation, International Astronautical Congress, Naples, 2012.
44. Morfill, Gregor: *Plasma research from space – applications for Earth*. Max Planck Institute, ISS symposium, Berlin, 2–4 May 2012.
45. Demets, René; Weems, Jon; & McAvinia, Ruth: *From Eureca to EXPOSE and ExoMars – the evolving tools of astrobiology research*. ESA Bulletin, #172, Q4 2017; Demets, René; Weems, Jon & Walker, Carl: *How to live in space without a spacesuit – results of ESA's astrobiology research*. ESA Bulletin, #172, Q4 2017.
46. Dachev, Tsvetan *et al: Space radiation peculiarities in the extra vehicular environment of the International Space Station*. Aerospace Bulgaria, 25, 2013.
47. Burmeister, Kai & Hartmann, Jens: *20 years of fault tolerant computer operation on ISS – continuous sustaining maintenance and overcome of obsolescence issues*. Paper presented to International Astronautical Congress, Washington DC, October 2019.
48. Van Baarsen, Bernadette *et al: The effects of extreme isolation on loneliness and cognitive control processes – analyses of the LODGEAD data obtained during the Mars 105 and Mars 520 studies*. Presentation, International Astronautical Congress, Naples, 2011.
49. Urbina, Diego & Charles, Romain: *Enduring the isolation of interplanetary travel – a personal account of the Mars 500 mission*. Presentation, International Astronautical Congress, Naples, 2011; see also Šolcová, Iva Poláčkova; Šolcová, Iva; Stuchlikova, Iva; & Mazehóová: *The story of 520 days on a simulated flight to Mars*. Acta Astronautica, vol 126, 2016.
50. Salmon, Andy: *Science on board the Mir space station 1986–1994*. JBIS, vol 50, #8, August 1997.

4

Industrial cooperation

Cooperation in space science (see Chapter 2) and human spaceflight (see Chapter 3) were the most high-profile areas of European-Russian collaboration. However, what became known from the 1960s as the 'space industry' was at least as important, because of its practical impact on the economies, trade, workforce and technological know-how of both parties. As with the topics discussed in previous chapters, the politics of the cold war were never far away. They were first illustrated by the case of the fuel for Ariane, which leads us on to the general issue of the sanctions regime.

Fuelling Ariane

Chapter 1 noted how the American offer to fly European satellites combined both goodwill and a foreign policy objective of keeping Europe and its space industry within the sphere of influence and commercial interests of the United States. The tetchy nature of the relationship between Presidents Ronald Reagan and François Mitterrand indicated how the contours of that relationship continued to be tested.

The sphere of influence and commercial interests faced an early test with Symphonie, a French-German communications satellite project of 1963. It was scheduled for launch on Europe's Europa II rocket in 1972, but the Europa failed in November 1971 and was then abandoned, leaving Symphonie still on the ground. Despite German reservations, CNES proposed the use of a Soviet rocket, with the USSR agreeing a modest fee (under €1 m). The deal fell apart because of a problem that had become well known to the space scientists (see Chapter 2),

The original version of this chapter was revised. The correction to this chapter is available at https://doi.org/10.1007/978-3-030-67686-5_7

whereby the satellite would be handed over at the border and that would be the last sight of it until it was launched. This was not acceptable to the French or Germans, who now turned to the Americans for a launcher. They refused, on the basis that the United States had a global monopoly on the provision of satellite television services and were not in the business of launching rival providers. A compromise was agreed whereby the Americans would launch Symphonie, on the condition that it would only be used for experimental and not commercial purposes. The price of the Delta rocket at that time was US$11m. Two Symphonie communications satellites were duly launched in 1974 and 1975 on the Delta.

The Symphonie experience cast a long shadow and is not forgotten in French histories, even to this day. The French learned the lesson that Europe must never be put in such a humiliating position again and therefore needed its own launcher. In 1972, CNES proposed a rocket, the L3S, later renamed Ariane, which became the flagship project of the European Space Agency (ESA) that was formed in 1973 but officially founded in 1975. CNES pushed Ariane through both the French government and the early ESA with the help of the former prime minister who founded it, Michel Debré, now strategically located as defence minister and his colleague Maurice Lévy, CNES president and the man who put ESA together.

Michel Debré. CC Nationaal Archief.

Europe was not yet out of the woods with the Americans though, because the early Ariane launches would require 2,000 tonnes of Unsymmetrical Di Methyl

Hydrazine (UDMH) fuel for its first and second stages, something Europe did not have. The only known commercial source of UDMH was the United States. A proposal to buy the fuel from the leading UDMII company in Baltimore failed, officially on the grounds that UDMH was carcinogenic (though this did not prevent its domestic supply). This was regarded as a flimsy, fraudulent reason and there was suspicion that the company came under government pressure, because Ariane was regarded – correctly – as a serious threat to America's launcher monopoly.

CNES then returned to the USSR for UDMH, going through Interkosmos administrator, Vladlen Vereschetin, which led to political approval at a summit in Crimea between Georges Pompidou and Leonid Brezhnev. Thus Ariane got to fly in 1979 with Russian fuel – although most Europeans who saw Ariane's end-of-year take-off would have been blissfully unaware of the fuel's origins. The Americans were not and duly retaliated. The French company Aérospatiale, which built Ariane, was refused a consultancy contract with Martin Marietta to advise on the construction of the Atlas, Centaur and Titan rockets, on the basis of the risk of technology and know-how transfer to Europe. The US then issued a prohibition against the Swiss company Contraves, which was and still remains Europe's leading manufacturer of rocket fairings, from manufacturing the fairing separation system on the American Delta rocket [1]. In response, President Giscard d'Estaing's prime minister Jacques Chirac spoke of how France must not be a vassal of the United States [2].

The obstacle: sanctions

The Ariane experience showed how the European space industry had run up against a blockade originally designed to prevent technology or industrial transfer to the Soviet Union. Such sanctions had impinged little upon scientific cooperation or piloted flight, but the situation would be different in the industrial area.

Sanctions dated back a long time and would inevitably play a role in the nature of Russian-European-American cooperation. At the start of the cold war and the formation of the North Atlantic Treaty Organization (NATO) in 1949, western states established the Coordinating Committee for Multilateral Export Controls, (CoCom, also written COCOM) with a secretariat in Paris, in practice located in an annex to the US embassy. The legislative basis in the United States was the Export control Act, 1940, originally directed against Japan but later extended to other countries identified as hostile. Under the Battle Act, 1951, the United States could refuse assistance to any states that did not cooperate in the enforcement of sanctions against Russia and China. The legislative basis was updated by the International Traffic in Arms Regulations (ITAR) system introduced under the Arms Export Control Act (ASCA) in 1976. ITAR is the contemporary regulation. Sixteen countries were embargoed by CoCom.[1]

---

[1] Those countries were the USSR, the People's Republic of China, Poland, GDR, Hungary, Romania, Bulgaria, Czechoslovakia, Cuba, Vietnam, Laos, DPRK, Cambodia, Afghanistan, Mongolia and Albania.

CoCom itself had no legislative basis and published neither decisions nor documentation, so because it did not have a formal existence it was exempt from parliamentary control. Its weekly meetings were real enough and these considered applications for exemptions to trade (licences) with the embargoed countries, having typically been referred there by *national* trade or export control departments. The US was the dominant country, making the most amendments to the list, applying for and granting the most exemptions and, as the country with the most stringent legislation, setting the bar high for everyone else [3]. CoCom normally operated by unanimity, so for one country to say 'no' was normally fatal for exemption. Even to this day, sanctions are an instrument of choice in American foreign policy. In the period 1993–7 alone, 35 countries were sanctioned by 61 new laws and executive decisions [4]. The Russian prime minister described sanctions as a 'strategy of controlled technological inferiority', one that could be applied to friends and foes alike [5]. The Americans shamelessly used CoCom, directly or indirectly, to prevent Britain from selling aircraft to socialist countries, ostensibly because of the security risks involved, but in reality to win the battle for post-war civil aviation supremacy [6].

Although CoCom was largely American by inspiration, operation and application, it applied to all western European NATO countries and was, to say the least, a defined bar against European-Russian cooperation in the space industry (equally, it prevented cooperation with China). Under CoCom, member states proposing sales (exports) to the USSR and its allies had to notify all CoCom members, any one of whom could object [7]. CoCom had three categories of sanctions: military; atomic; and industrial and commercial, including dual-use technologies, civil and military. Defining 'dual-use' is inherently challenging, but CoCom's list included such items as jet engines, air traffic control equipment and computers. There were several embargoed categories, which included metalwork, petroleum, electrical and power-generating equipment, general industrial, transport, electronic and precision instruments, metals and minerals, chemicals and rubber. Some of these categories inevitably impinged on the space industry. Although France agreed with the principle of CoCom, it was also the country that had the most problem with the system for being too stringent and too dominated by the Americans [8]. All Soviet-French space cooperation had to go through the inter-ministerial commission for the examination of the export of war materials (Commission Interministériel d'Examen des Exportations du Matérials de Guerre), where the Secretariat General of national defence had the opportunity to contribute, before it went on to CoCom. American views of the potential of technology transfer to the USSR were, in the view of the French, 'harsh' [9].

An early outcome of the CoCom sanctions was that the USSR elected to set up its own trading group for the socialist block in eastern and central Europe, called the Council for Mutual Economic Assistance (CMEA) but better known as COMECON (1949). With regard to spaceflight, these countries formed Interkosmos

(1967) which led to instruments from these countries flying on Soviet spacecraft, a joint scientific programme (Interkosmos, from 1969, see Chapter 2) and the flight of guest cosmonauts to the Salyut and Mir orbital stations (see Chapter 3).

COMECON building in Moscow. CC Gennadiy Grachev.

As mentioned, the impact of CoCom was limited in the areas of space science and human spaceflight, but was always going to be greater in the area of industrial cooperation, where commercial interests came into play. CoCom was important, not so much for the cases that came before it for consideration and ruling, but for what did not. Both western and Soviet companies knew that there was no point in industrial cooperation in space research, because it would be prohibited and fail at CoCom. CoCom's disincentivizing effect was its greatest, but least visible, impact. An example was the Proton rocket, which the USSR offered for launching western satellites. The Russians wanted to take advantage of the opportunity created by the loss the Space Shuttle *Challenger* in 1986, which meant that the Americans no longer had their principal launching rocket. The Proton was the USSR's most powerful operational launcher, able to send four tonnes to 24-hour orbit – ideal for commercial communications satellites – or on to Mars or Venus. First launched in 1965, it had the highest performing rocket engines in the world at the time, Valentin Glushko's RD-253 engines. In 1979, the Soviet Union first offered the Proton to launch ESA's forthcoming maritime communications satellite Marecs, but, for reasons unknown, this did not go any further. The following year, the Proton was offered to launch satellites for INMARSAT, the London-based International Maritime Satellite Organization, initially at a cost of €20 m and then a mouth-watering, loss-leading €10 m. Initial discussions took place in Moscow, but they did not get very far either. The Russians assumed that the reason for this was a concern that they would inspect and strip the satellite to learn its secrets, so they offered round-the-clock security accompaniment of their precious cargoes – 'they can even be handcuffed to them', they said. The actual problem was that the satellite to be launched would be an 'export' and, accordingly, required CoCom approval. The Americans were actually unconcerned about security, the real problem for them was that the price on offer would undercut their launcher companies. Although 64 of the INMARSAT countries approved the deal, the Americans said no [10].

The USSR continued to push the Proton, making an in-your-face sales pitch at the Paris air show in 1987. Over the next two years, they won four contracts to fly satellites on the Proton: one for Eutelsat, another for Indonesia (with a piloted flight thrown in as a bonus) and two for American companies, including one for its leading maker, Hughes. Between the Department of State and CoCom, all were refused export licences. The unfortunate Indonesians tried to fly with the Chinese, but were vetoed there too.

The situation changed in 1991 with the dissolution of the Soviet Union. Political, institutional and economic chaos undermined a well-structured system, in particular destroying quality control. Unregulated capitalism was an invitation to corruption from which not even the space industry escaped. Even by the early 2020s, this process has not fully run its course. At one level, this introduced an uncertainty to having Russia as a partner, but at another it opened opportunities. Western companies, individuals and entrepreneurs were permitted and even encouraged to invest in Russia as part of the process of privatization and liberalization [11]. CoCom was replaced in 1996 by the current Wassenaar arrangement, named after the town in the Netherlands where it was agreed, with a secretariat in Vienna. Wassenaar was broader than CoCom (42 states, including Russia) and rather than economic warfare, its focus was on the non-proliferation of weapons and the nuclear industry to the 'axis of evil' countries (e.g. DPRK, Iran) and China.[2] The political environment

---

[2] The CoCom members were Australia, Belgium, Canada, Denmark, France, Federal Republic of Germany, Greece, Italy, Japan, Luxembourg, the Netherlands, Norway, Portugal Spain, Britain and the United States. Several self-proclaimed neutral countries also assisted in the application of CoCom: Austria, Finland, Sweden, New Zealand, Ireland and Switzerland. The original Wassenaar members were Australia, Austria, Belgium, Canada, the Czech Republic, Denmark, Finland, France, Germany, Greece, Hungary, Ireland, Italy, Japan, Luxembourg, the Netherlands, New Zealand, Norway, Poland, Portugal, Russia, the Slovak Republic, Spain, Sweden, Switzerland, Turkey, Britain and the United States.

also changed. The last General Secretary, Mikahil Gorbachev, had spoken of a 'common European home' of eastern and western European states. Russia disbanded the Warsaw pact military alliance and its space institutions, such as Interkosmos (1994) and joined the European telecommunications satellite organization, Eutelsat, becoming its 41st member. Russia also joined the Council of Europe in 1996, binding Russia to the European Convention on Human Rights.

The commercial environment changed as well. On 1 March 1991, price controls were lifted in Russia and an unregulated free market was introduced, referred to as 'shock capitalism'. Western companies, including those in the space industry, were free to trade in Russia. There was a fire sale of state industries – including parts of the space programme – of $3bn, 30 percent of an economy formally valued at $10bn, a sixth of the value of Wal-Mart. Because the economy was under-valued beyond belief, Russian and western investors made billions overnight. Western companies could take over Russian companies or set up joint enterprises with them (e.g. Lockheed Khrunichev in 1992). Whatever their past reservations about contact, western governments were not going to turn down ownership of parts of the Russian economy, opportunities for enrichment, or bringing Russia into its sphere of influence. Luckily for the space companics, most of them survived by remaining in state hands. Energiya had a close escape: part-privatized, it became quite unstable, but recovered.

Accordingly, the new Russia tried again. This time, approval was forthcoming in the form of a 1993 inter-governmental agreement signed by US Vice President Al Gore and Russian Prime Minister Viktor Chernomyrdin, with strings attached in the form of commercial conditions. Russia was permitted eight launches of western communications satellites into 24-hour orbit, with the condition of charging no more than $45m and no lower than 7.5% below American prices, as well as not making more than two such launches per year. In 1996, this was raised to 20 launches, with a -15% price cap. Although the licensing system was run by the Americans, it equally affected the Europeans, because the Americans defined a European-built satellite as American if it had any American components, which, given the global nature of technology, they all did.

The first satellite to be launched was the one that started this story, for INMARSAT (1996), now privatized as a British maritime satellite communications company (1988). The second was the Astra 1F satellite for Luxembourg. Although the terms were humiliating, the Russians were desperate and the cash provided fresh resources for Khrunichev and kept the Proton production line open. At the same time, Khrunichev argued that without those restrictions, it could have won 45 launches for western communications satellites on a truly open market and delays with building the International Space Station (ISS) might have been avoided. The west had the additional assurance of control, in that these launches went through the Lockheed Khrunichev joint enterprise. The change of American policy was due to the diminished perception of the Russian military threat, the impatience of their satellite-makers wanting to get their product off the ground, western economic benefits from the Russian economy and a fear that should the space programme collapse, Russian scientists might leave to work for rogue states.

Even so, industrial cooperation was far from easy. Satellites en route to the pad had 24/7 American supervision. Licences still had to be applied for and the process could be slow and bureaucratic. A licence for the Energomash rocket engine company was held up for 400 days while the Department of Commerce poked into all the company's corners, including its manufacture of pumps for fire engines. When the University of California, Berkeley, found a Russian launcher for its $13 m, 60 kg suitcase-sized CHIPS (Cosmic Hot Interstellar Plasma Satellite), it spent years struggling – and failing – to get an export licence. Nevertheless, despite these problems, Khrunichev's Proton opened the door for other Russian launcher companies to fly western satellites. Proton was the test case, as it addressed the most commercially valuable part of the market. Eventually, an informal price list came into operation for the price of a Russian launch (see Table 4.1).

Table 4.1:
Estimated prices of flights on Russian rockets

Rocket	Price
Proton M	€75 m
Zenit 3SL Sea Launch	€60 m
Proton	€50 m
Zenit 3 Land Launch	€35 m
Soyuz	€25 m to €30 m
Tsyklon	€20 m
Rockot	€15 m
Cosmos 3M	€12 m
Dnepr	€10 m
START	€8 m
Strela	€5 m
Shtil	€500,000

Russia launching European satellites

So what was Europe's experience? The first problem was that it did not know the Russian space industry, which was hardly a surprise since CoCom made it impossible to bring any commercial arrangements to a conclusion, or, arguably, made entering such discussions pointless. The Russian space industry was originally based around design bureaux (OKBs in Russian, later called NPOs) set up from the 1950s to do the work of designing and testing rockets and satellites and each with distinct fields of work. The original was OKB-1, now known as the Energiya design bureau. The others were Yuzhnoye in Dnepropetrovsk (Cosmos, Zenit, Tsyklon, Dnepr launchers); TsSKB Progress, Samara (the Soyuz launcher); ISS Reshetnev, Krasnoyarsk (communications satellites); Lavochkin, Moscow (lunar and interplanetary); and Khrunichev, Moscow (Proton, Rockot launchers).

The story of industrial cooperation and the commercial launching of western satellites by Russia runs through Britain. The person who started it was Gerry Webb, the originator of Commercial Space Technologies (CST), also well known for his contribution to the British Interplanetary Society (BIS). The BIS was an association of enthusiasts and engineers set up in the 1930s, its best-known early adherent being science fiction writer Arthur C. Clarke. In the 1980s, the BIS set up an annual forum on the Soviet space programme, so it had a base of knowledge of Soviet and Russian space activities, which included early knowledge of the design bureaux **[12]**. Gerry Webb worked in the Radio Research Station, then the Rutherford Appleton Laboratory. He attended international space conferences such as COSPAR and, from the early 1980s, the annual International Astronautical Federation Congress (IAF), which brought together the space industry from all around the world. CST was started as a promotional agency for the revolutionary British spaceplane HOTOL, along with its inventor Alan Bond.

Gerry Webb. Alistair Scott.

Arising from his meeting Soviet scientists and space programme administrators at the IAF, Webb received an invitation to attend an event called the *Space Future Forum* in Moscow in 1987. Although western space companies had to contend with CoCom and their own poor understanding of the design bureaux, for their part the Russians had little or no knowledge or experience of the western space industry in general, or Europe in particular. One country that did have trading experience, but not in this area, was neighbouring Finland, which dealt in timber, scrap iron, meat and gas and had the infrastructure in place, such as trading accounts. One of the leading companies was EKE Engineering, then trading £300 m a year, which had a big office building in Moscow, including connections to the Energiya design bureau and thereby the space industry.

From 1988, Webb began to travel to Moscow – six or seven times that year – facilitated by EKE, to see what could be done. The Russians said 'you give us a satellite, we'll launch it', suggesting an initial price of $8 m. From the Finns, Webb learned that the way to do business was not to set up an office with western staff, but instead to find Russian staff who could navigate their way around their own administrative, banking and bureaucratic systems. He duly converted a flat in Moscow into an office, now appropriately located on Cosmonaut Volkov Street and recruited skilled local staff. Webb began to meet the many different elements of the Soviet space programme, which had been portrayed in the west as a command-and-control monolith, but in practice was divided into independent-minded design bureaux. He was especially conscious of the need to work through the political officers as much as the engineers. Webb travelled to Dnepropetrovsk in Ukraine, home of the Yuzhnoye design bureau, the largest single designer and maker of Soviet missiles and rockets. Yuzhnoye had an important role in the civilian space programme, having built the original Cosmos launcher for the Cosmos programme of scientific satellites and even the lunar module in which a Soviet cosmonaut should have landed on the Moon. Some of its rockets, such as the Tsyklon, had both civilian and military uses and it made large numbers of military missiles.

One of these was a cold war silo-based military missile, the Dnepr, of which the bureau had a substantial stock and was now making them available for commercial purposes. With the end of the cold war, the dissolution of the Soviet Union and the subsequent Russian-American treaties reducing missile and warhead numbers, production had ceased and most of the missiles were decommissioned. By the 1990s, the Dneprs were typically 23 to 24 years old. Although their obvious destination was the scrapyard, it did not take a fine mind to realize that hitherto-secret military missiles could still launch satellites. According to Webb, the Dnepr was launched 'like a cartridge' out of a silo from two bases: Baikonour, which was Russian sovereign territory in Kazakhstan; and Dombarovska, a Soviet base near Yasny in southern Russia. In the 1990s, Russia was not prepared to turn any business down.

Webb's challenge was to match western satellite-makers with appropriate Russian launchers. Western institutes did not know how to meet Russian launching companies and for their part, Russian companies did not know how to meet and do business with western satellite companies. CST was the broker that bridged these two worlds and made things happen. Its task was to find western satellites looking for a launcher, match them against a suitable launcher at an affordable price and then handle the paperwork and logistics to get the satellite to Russia, integrated onto the launch vehicle and put into orbit. This included dealing with such issues as delays, customs, payment and currency issues, equipment transport, accommodation, launch site protocols, translation and insurance. Not only that, but CST would be able to offer earlier launching opportunities at much lower prices than were available from American launcher companies or Europe's Arianespace, both of whom had long manifests for American and European satellites. As noted earlier, this period coincided with the development of small satellites by the British company Surrey Satellite Technologies Ltd (SSTL). Most larger launchers had the capacity to carry these small satellites piggyback, but neither NASA or ESA, nor their launching companies, were interested in complicating things by doing so. At this time, SSTL was building small satellites on behalf of developing countries, for Earth observations, to aid economic development and for a consortium interested in forming a Disaster Management Constellation (DMC). For them, some of the Russian launchers would be ideal.

Webb was in a position to offer two options for the small satellite companies: dedicated launch or piggybacking. CST always warned that while piggybacking might be the cheapest option, it was always at the mercy of those responsible for the primary payload. A dedicated launch for a number of small satellites might be better and, in their tradition of simple engineering, the Russians made a new ring on the upper stage and simply bolted on the piggybacks. Later, the Briz stage fitted a 'customized aluminium structure'.

This would be a challenging enough environment in an orderly world, but it was far from that at this point. As Webb said, 'I had expected the Soviet Union to liberalize, but not collapse'. Suddenly there was shock capitalism, with economic chaos, a completely unregulated infrastructure, extensive privatization (including Energiya) and 10,000 banks touting for business. On the Russian side, there was little experience of handling commercial transactions, how to reduce costs to make price offers competitive, or even the concept of profit margins. Language difficulties were less of a problem than cultural ones: Russian companies were unused to western business methods, while westerners were unused to 'strange bureaucratic procedures, unusual taxes and odd travel arrangements'.

CST's first satellite was Fasat Alpha, a Chilean satellite built by SSTL that launched piggyback on a Tsyklon in August 1995. The first Dnepr payload was another Surrey satellite, Uosat, in April 1999. Many of the early payloads were SSTL satellites for companies in countries such as Algeria (Alsat), Germany (Rapideye), and Nigeria (Nigeriasat), but CST also picked up contracts from China (Tsinghua), though the satellite concerned was still built by SSTL. The payloads came from many countries, both in Europe and further afield and the traditional industrial players now took note. CNES commissioned CST to organize the launches of two of its small satellites, Demeter, launched on the Cosmos 3M and Picard, launched on the Dnepr from Dombarovska. Accompanying Picard was a pair of Swedish formation-flying satellites, Tango and Mango. In 2009, South Africa's Sumbandila flew on a Soyuz, the first of three Soyuz launches, which meant dealing with a different design bureau, TsSKB in Samara (see Table 4.2).

Gerry Webb was one of the first to travel to the military facilities from which these rockets were launched, such as Plesetsk. At Baikonour, he happened to see a Zenit rocket, one of the first westerners to see one at a time when it was not well known. Alsat was launched in temperatures of -30°C and 'it was snowing an awful lot', but in calm, clear conditions it was possible to follow the rocket right through its ascent. By 2012, CST had flown 33 cargoes on 17 launches, some on dedicated missions, others as multiple missions or piggybacks with larger payloads. Competition duly appeared over time, principally Eurockot, but CST was the pioneer [13]. As for Gerry Webb, he would later become President of the BIS.

Table 4.2:
CST launches

Date	Payload	Launcher
31 Aug 1995	Fasat Alpha (Chile, built by SSTL)	Tsyklon piggyback
10 Jul 1998	Fasat Bravo, TMsat	Zenit piggyback
21 Apr 1999	UoSat 12 (Britain)	Dnepr
28 Jun 2000	Tsinghua (China), SNAP 1 (Britain)	Cosmos 3M
26 Sep 2000	Megsat 1 (Italy), UNISAT (Italy), Saudisat 1A, 1B	Dnepr
	So-42 (Saudi Arabia), Tiungsat (Malaysia)	
28 Nov 2002	Alsat DMC (Algeria), Mozhayets (Russia)	Cosmos 3M
27 Sep 2003	Nigeriasat 1 (Nigera), Blisat (Turkey),	Cosmos 3M
	UK-DMC (Britain),Mozhayets 4 (Russia),	
	Kaistsat (Korea), Laretz (Russia)	
29 Jun 2004	Demeter (France)	Cosmos 3M
27 Oct 2005	Topsat (Britain), Chinasat DMC (China),	Cosmos 3M
	SSET Express, Sinah (Iran), Mozhayets 5 (Russia),	
	NCube (Norway), UWE (Germany)	
29 Aug 2008	Rapideye (5 satellites)	Dnepr
29 Jul 2009	UK-DMC, Deimos 1 DMC	Dnepr
17 Sep 2009	Sumbandila	Soyuz Fregat piggyback
15 Jun 2010	Picard	Dnepr
	Prisma (Tango and Mango)	
17 Aug 2011	Nigeriasat 2, Nigeriasat X	Dnepr
22 Jul 2012	ADS 1B	Soyuz Fregat piggyback
19 Jun 2014	KazEOSat 1 2	Dnepr
8 Jul 2014	TechDemoSat 1, UKube 1	Soyuz Fregat piggyback
28 Jan 2017	Hispasat 36W-1	Soyuz

Source: CST.

Eurockot

After Britain, Germany was next into the field, with the original approach made by Russia. In April 1991, the Salyut design bureau, part of Khrunichev, put its Rockot missile on display for commercial use at an international space exhibition in

Moscow. The Salyut bureau then visited the German company DASA (Daimler Benz Aerospacc) that December, leading to an agreement between DASA and the Khrunichev design bureau called Eurockot, with shares of 51 and 49 percent respectively.

Tests were carried out to look at how Rockot could be adapted as a commercial launcher. A linguistic note is that the word '*Rockot*' means 'roar of sound', not 'rocket' ('*raket*' in Russian). The Rockot was originally the UR-100 missile developed by Vladimir Chelomei's design bureau, a small, 30 m-tall missile using storable fuels, with 360 deployed in silos at four bases around Russia and Ukraine during the cold war. It was also an effective anti-satellite weapon. With the reductions in military forces in the 1990s, 105 were retained as missiles under the START II agreement, but 65 were ordered to be decommissioned, which meant scrapping them unless an alternative use could be found. Suborbital tests were made at Baikonour in 1990 and 1991 before a first orbital mission in 1994, carrying an amateur radio satellite. For additional capacity, a new third stage, called Briz K ('K' for *Kosmicheski*), was fitted to enable it to put two-tonne payloads into Earth orbit. This was a versatile stage, which could fly for seven hours and restart six times, dispensing different satellites into different orbits. With Briz K, Rockot could place up to 1.85 tonnes in low Earth orbit, or two tonnes for its subsequent version, the Briz KM ('M' for modified).

Rockot night time view Plesetsk. ESA.

One problem was that the Rockot launched from a silo, which was likely to impose intolerable acoustic burdens on a scientific payload, so a Cosmos 3M launcher site was adapted at Baikonour as an above-ground Rockot launch site. The Kazakh authorities got wind of the proposal and demanded a share of the profits of every Rockot mission out of Baikonour, but the Russians would have none of this and moved all Rockot launches to the military cosmodrome of Plesetsk in northern Russia, where they built an above-ground vertical support structure. Eurockot invested substantially (€34 m) in facilities in Russia, including a payload integration facility and mission control in the cosmodrome, with a hotel ('The Eurockot') and customer facilities in Mirny, the town near the Plesetsk cosmodrome. Payloads were flown to Archangelsk and then brought 200 km by rail to Plesetsk. Eurockot went to some efforts to reassure western clients that all their worries would be taken care of in the unfamiliar world of Russian launch sites, calling Plesetsk 'the European spaceport' and Rockot 'Europe's small launcher' [14]. Customers were assured of transport arrangements, hotels, clean room facilities, mission control and security. Plesetsk had changed a lot. When Czechoslovak scientists went there for a satellite launch in 1989, it was very controlled, 'with walls and military everywhere'. Not anymore.

Rockot payload loading in the snow. ESA.

In March 1995, the Eurockot consortium formally began to market Rockot, offering it at €7 m per launch. Within five years, it had built up an order book of 12 launches for €200 m. So confident was Eurockot that the company formally purchased 45 old Rockots to ensure its supply chain. In 2000, Eurockot was bought in turn by the German and French companies that became Airbus. Khrunichev still retained 49 percent.

Rockot's first operational launch for Eurockot in March 2002 was the American-German double satellite called GRACE, designed to measure small changes in the Earth's gravity. For Europe, Rockot was principally associated with the launch of Earth resources satellites. Eurockot launched GOCE (Gravity field and steady state Ocean Circulation Explorer), the first of ESA's *Living planet* programme, to provide an accurate map of Earth's gravity field. The Rockot subsequently launched ESA's Sentinel 3A, 2B and 5P missions, which became the principal instrument of Earth observations for the European Union. Sentinel was a programme well-publicized by ESA, so these Plesetsk launches became well-known and documented. There was also PROject on OnBoard Autonomy (PROBA, but written Proba) 2, a successful Belgian satellite orbiting the Earth to observe the Sun's corona and ultraviolet light from the dusk-dawn line, which generated 100 scientific papers in 17 years.

Rockot releases Sentinel. ESA.

Although it was generally a reliable launcher, Rockot was responsible for the loss of Europe's first ice monitoring satellite, Cryosat 1, on 8 October 2005. The command to shut down the second stage was accidentally omitted from the computer code, so it burned to depletion, preventing the second stage from igniting and dumping Cryostat in the very place that it was supposed to study, the Arctic Sea. Its replacement, Cryosat 2, used the Dnepr rocket, another cold war leftover.

Rockot's lifetime was always going to be limited and was squeezed from two directions. Firstly, there were only four flight-certified Rockots remaining by 2016 and further supply of their guidance systems was refused because of the breakdown of relationships between Russia and Ukraine. Second, Europe developed its own light launcher. In 1999, France killed off an Italian project for a small European launcher called Vega, on the basis that it would cost €16 m per launch compared to Rockot's €10 m. The Italians, though, could never take *Non!* for an answer and persisted with Vega, which entered service in 2012 as a successful European light launcher.

Vega was itself a cooperative project. It had three solid-fuel stages, but the fourth was a liquid-fuel engine, the RD-843. This was a Soviet-period design derived from the motor used on the Soviet lunar module that would have landed on the Moon, the LK, which had been successfully tested in Earth orbit. In Europe, it was known as the Attitude and Vernier Upper Module (AVUM). The exact circumstances in which its makers, the Yuzhnoye Design Bureau in Dnepropetrovsk, built their relationship with Vega are not known, except that the bureau was awarded the tender in December 2002. It was a small engine, 2 m tall, weighing 418 kg and using UDMH and nitrogen tetroxide, with a thrust of 2.45 kN, specific impulse of 315 seconds and a burn time of 317 seconds. Five firings were possible. Few European commentaries refer to Vega's use of Soviet-period engines; indeed, its origins are not mentioned in the principal booklet on the rocket **[15]**.

RD-843 on Vega launcher. ESA.

Sentinel 3B was the last European Rockot, while the last ever launch was a domestic one of a Gonets M satellite on 27 December 2019. In 2020, the Khrunichev design bureau announced that it would develop a new version for 2022, the Rockot M, which did not depend on Ukrainian parts, with the launch of foreign satellites in mind. A model of Rockot M was displayed by Khrunichev in August 2020, intending for it to have a future.

Over 1994–2019, Rockot made 31 orbital launches. Of these, 13 were for commercial users (e.g. Iridium), other space agencies (e.g. Japan) and Europe, where the Rockot launcher came to play an important part in the launch of ESA applications science, especially the *Living planet* programme (GOCE, SMOS) and the Copernicus Earth observation programme (Sentinel). At the institutional end, it was an example of a successful Russian-European consortium managing a launcher (see Table 4.3).

Table 4.3:
Commercial European Rockot launches

Date	Payload
16 May 2000	Simsat 1,2 (Iridium test)
17 Mar 2002	GRACE 1,2 (NASA/Germany)
19 June 2002	Iridium 97, 98
30 June 2003	Small satellites MIMOSA (Czech Rep), DYU, AAU (Denmark)
30 Oct 2003	SERVIS 1 (Japan)
8 Oct 2005	Cryosat (ESA, fail)
28 July 2006	Kompsat 2 (Rep Korea)
17 Mar 2009	Gravity Field and Steady-State Ocean Circulation Explorer (GOCE) (ESA)
2 Nov 2009	Soil Moisture and Ocean Salinity (SMOS) (ESA)
	Proba 2 (Belgian scientific satellite)
22 Nov 2013	SWARM (three satellites) to study magnetic field (ESA)
15 Feb 2016	Sentinel 3A (ESA)
13 Oct 2017	Sentinel 5P (ESA)
25 Apr 2018	Sentinel 3B (ESA)

All from Plesetsk

Bilateral launches

Most Russian launches of European and other satellites went through CST or Eurockot. There was no law to say they had to and a number of launches went through neither, instead being marketed directly by their parent company (see Table 4.4). The two main examples were the Cosmos 3M launcher (originally designed in Ukraine but later made by OKB Polyot in Omsk) and the Dnepr cold war missile (Kosmotras, a Russian-Ukrainian company formed in 1997). But which European companies and institutes would take up these opportunities?

Considering their low profile in European-Russian cooperation up to this point, it is perhaps surprising that German institutes and companies were the dominant users of these launch services. A precedent was set in 1989. One German company that did manage to fly payloads with Russia was Kayser-Threde, which flew small payloads on Resurs Earth observation satellites in 1989 and 1990. As for satellites, the Technical University of Berlin (TUB) was the principal developer of micro-satellites. A long-established body dating to the Königliche Bergakademie zu Berlin in 1770 and re-established as TUB in 1946, it was at the forefront of innovation. TUB developed an early series of micro-satellites, the first being a 35 kg store-and-forward electronic mail satellite flown piggyback with ERS 1 (Earth Resources Satellite 1) on Ariane in July 1991. TUB then persuaded the Russians to carry one on the Meteor 3-6 weather satellite in January 1994.

The next satellites were the most remarkable, as they had the distinction of being the first to be orbited from a submarine launch. The Russians were testing out the possibility of using the old cold war RSM-50 missile, launched from submarines, as a means of putting small satellites into orbit. TUB proposed two micro-satellites, 8.5 kg and 3 kg respectively, on a single rocket. The submarine *Novomoskovsk* duly dived under the Barents Sea. From the shore, observers could see the *Shtil* rocket break the surface, shake off the cold sea water and curve upward into orbit, placing the two little TUBSAT N and TUBSAT N1 satellites into an orbit of 401−777 km, 78.9°.

The Earth Research Centre in Berlin, the GeoForschungsZentrum (GFZ) of Potsdam, persuaded the Energiya design bureau to launch a small satellite from the Mir space station. Called GFZ after its makers, this €500,000 micro-satellite was 21.5 cm in diameter, weighed 20 kg and carried 60 reflectors for geodetic experiments, using stations in Potsdam itself and the Cuban National Centre for Seismological Investigations in Santiago de Cuba. It was brought up to Mir by the Progress M-27 freighter spacecraft. Although the United States and Japan later built elaborate systems for launching small satellites from space stations, the Russian system was always simpler: Vladimir Dezhurov and Gennadiy Strekhalov spring-launched GFZ through the Mir airlock.

Otto Hydraulic Bremen, popularly known as OHB, became Germany's principal satellite maker, having started life as a five-person company undertaking electrical work in 1981. OHB refocussed around satellites in 1985 under Christa Fuchs and soon became the third largest European aerospace company after Airbus and Thales. OHB formed COSMOS International Satellitenstart and went directly to Russian launcher companies, leading to the launches of Temisat (Tsyklon 3), the SAR-Lupe series (Cosmos 3M), Rubin small satellites (Cosmos 3M), ABRIXAS (Cosmos 3M) and later the Galileo system (Soyuz).

ABRIXAS (A BRoadband Imaging X-ray All-sky Survey) was launched for the Max Plank Institute in Germany on the Cosmos 3M on 28 April 1999. The launching marked the reopening after 12 years of the Kapustin Yar cosmodrome, the original Russian launch site dating to 1947. Sadly, although placed in the right orbit, the spacecraft's battery failed and the mission was lost, though this was not the launcher's fault. The idea was later revived in the form of Spektr RG (see Chapter 2).

Rubin satellites were testbeds for micro- and nano-technologies developed by OHB in Bremen with the Angstrom Aerospace Corporation in Uppsala, Sweden. They were either 300 or 600 mm^3, weighed 5 kg and could be easily attached piggyback to the orbital payload on a Cosmos 3M rocket. The first flew in 2000 with CHAMP and MITA; the second on a Dnepr in 2002; and then Rubin 3, 4 and 5 on Cosmos 3M rockets in 2002, 2003 and 2005. They tested communications and tracking systems, with Rubin 5 devoted to detecting meteoroids and space debris. Between them, they gave both companies low-cost opportunities to test out technologies that became widely used by the ever-growing numbers of micro-satellites and cubesats [16].

Jena Optronik and GFZ of Potsdam commissioned a Cosmos 3M launcher to orbit their CHAMP (CHAllenging Minisatellite Payload) geophysics research satellite on 15 July 2000, together with a Rubin micro payload (an Italian microsatellite also flew). The aim of the 522 kg CHAMP was to measure the Earth's magnetic field most precisely, using the Global Positioning System (GPS), laser reflectors, an ion-drift meter, a star sensor, an accelerometer and a magnetometer. An American instrument to measure the choppiness of the seas was also carried. MITA, launched on the same rocket, was a 170 kg Italian Space Agency experimental satellite, with a detector to observe energy particles and the nuclei of light elements in Earth's atmosphere.

Sweden managed to obtain two Cosmos 3M piggyback opportunities for its micro-satellite Astrid, named after the children's writer Astrid Lindgren and with the instruments named after her characters. Astrid 1, 27 kg, flew with a Russian Tsikada navigation satellite, while Astrid 2, 35 kg, flew with another navigation satellite, Nadezhda 5. The launch cost must have been low in the case of Astrid 1, as the entire mission, including launch, cost €1 m. Later, Sweden flew the 250 kg astrophysics and aeronomy satellite Odin, this time on the START 1 decommissioned missile from Svobodny missile base in the far east (the new cosmodrome of Vostochny was later built nearby). Odin had a 20 percent French participation of star sensor electronics and an acoustic-optic spectrometer.

The German space agency, DLR, used Russian rockets for a number of launches, starting on 15 June 2007 with Germany's first radar imaging satellite, TerraSAR. The German government used the Cosmos 3M from Plesetsk for its SARLupe military radar imaging satellites with a 3 m dish: SARLupe 1 (19 December 2006), 2 (2 July 2007), 3 (1 November 2007), 4 (27 March 2008) and 5 (22 July 2008), completing the constellation in a period of less than two years. DLR also developed a small, 120 kg test satellite to try out new techniques and technologies in orbit (e.g. solar cells, batteries, data storage, cameras and drives), the first being launched piggyback on a Soyuz carrying a Kanopus Earth observation satellite from Baikonour in 2012.

Table 4.4:
European satellites launched in bilateral cooperation with Russia

Date	Payload	Launcher
25 Jan 1994	TUBSAT 2	Tsyklon 3
24 Jan 1995	Astrid 1	Cosmos 3M
19 Apr 1995	GFZ	Mir
7 July 1998	TUBSAT N, N1	Shtil, Barents Sea
10 Dec 1998	Astrid 2	Cosmos 3M
28 Apr 1999	ABRIXAS	Cosmos 3M
15 Jul 2000	CHAMP	Cosmos 3M
	MITA	
	Rubin 1	
	Megsat O	
26 Sep 2000	Megsat	Dnepr
	Unisat	Dnepr
20 Feb 2001	Odin	START 1
20 Dec 2002	Rubin 2	Dnepr
	Unisat 1	
29 Jun 2004	Unisat 3	Dnepr
19 Dec 2006	SARLupe 1	Cosmos 3M
15 Jun 2007	TerraSAR	Dnepr
2 Jul 2007	SARLupe 2	Cosmos 3M
1 Nov 2007	SARLupe 3	Cosmos 3M
27 Mar 2008	SARLupe 4	Cosmos 3M
22 Jul 2008	SARLupe 5	Cosmos 3M
8 Apr 2010	Cryosat 2	Dnepr
21 Jun 2010	TanDEM	Dnepr
22 Jul 2012	TET-1	Soyuz
19 Jun 2014	Deimos 2	Dnepr

Exolaunch

With the exhaustion of the Cosmos 3M, Dnepr and Rockot fleet, the medium-lift Soyuz was used more frequently for European and other small satellites. Russian cargoes that had hitherto used smaller launchers now moved to the Soyuz which, being over-powered for the task in hand, meant there was now room for small cargoes. This acquired the name 'rideshare'. The classic case was the Gonets small communications satellite, which had used Tsyklon 3, Cosmos 3M and Rockot but from 2020 moved to the Soyuz 2. The versatility of the Fregat M upper stage meant that once Gonets had been delivered into orbit, smaller cargoes could then be delivered into quite different orbits. The European company most associated with this development was Exolaunch, set up in 2010 as a spinoff of the TUB, already known as a small satellite builder itself (TUBSAT) and now the most recent addition to the European-Russian launcher companies (see Table 4.5).

Exolaunch not only arranged missions on Russian launchers (it also had contracts with the American SpaceX, Electron and India), it also provided satellite deployment systems (EXOpod for cubesats and CarboNIX for micro-satellites) and deployment sequencers (EXObox) which ensured their safe and correct deployment on their intended paths. Exolaunch started small, with two BEESATS (Berlin Experimental and Education Satellites) and SOMP (Student Oxygen Measurement Project) from the Technical University of Dresden, which accompanied Bion M-1, but by the new decade had brought over 100 payloads into orbit. On the 2020 Gonets M launch, for example, the 15 Exolaunch payloads came from Europe (Germany, Lithuania, Finland), North America and the United Arab Emirates. It included four cubesats developed by the Würzburg Centre for Telematics. Illustrating how the launch served both Europe and Russia, Roskosmos provided nano-satellites developed by the Novosibirsk State University and Bauman Technical University in Moscow for studying the Sun. The Fregat M first delivered the three Gonets into 1,400 km high orbit at 82.5°, before making impressive plane and altitude manoeuvres to deliver the small payloads to orbits at 97.67°, 575 km.

Exolaunch small satellite deployment from Fregat. Exolaunch.

Preparing the rocket, with logos on the side. Roskosmos via Exolaunch.

Table 4.5:
Exolaunch missions

Date	Launcher	Launch site	Primary payload
19 Apr 2013	Soyuz 2.1a	Baikonour	Bion M-1
14 Jul 2017	Soyuz 2.1a	Baikonour	Kanopus V-IK
28 Nov 2017	Soyuz 2.1b	Vostochny	Meteor M2-1
1 Feb 2018	Soyuz 2.1a	Vostochny	Kanopus V-3, 4
5 Jul 2019	Soyuz 2.1b	Vostochny	Meteor M2-2
18 Dec 2019	Soyuz 2.1a	Kourou	CHEOPS
28 Sep 2020	Soyuz 2.1b	Plesetsk	Gonets M

Soyuz

Perhaps the most outstanding example of European-Russian launcher cooperation was that of the Soyuz rocket, which saw an entire part of Europe's launch base in French Guyana made over to the launcher. A rocket hitherto known for its ascents from the deserts, snows and forests of Russia could now be seen ascending from the jungle of South America. This project developed in two stages: first, a joint European-Russian venture using the Soyuz rocket from Russian soil to launch European and other payloads (Starsem); and second, the development of a Soyuz rocket base in the European Union (in French, *Soyouz à Kourou*).

The Soyuz rocket was the original Soviet Inter Continental Ballistic Missile (ICBM), the R-7 (R=*Raket*), first sketched by designer Mikhail Tikonravov in 1948, initially to carry a nuclear warhead over the Arctic to drop it on the United States. It was adapted as the rocket for the first satellite (Sputnik) and first piloted flight (Vostok), with the version known as Soyuz introduced in 1966. It has launched more satellites than any other rocket in history and has made almost 2,000 launches. Soyuz has a distinctive outline, with a big launcher cradle that releases the rocket when thrust reaches a certain level and four 'packet' stages that drop off and spin away two minutes into flight. Although technically obsolete, the Soyuz is simple, reliable, capable and profitable. The first time people outside Russia ever saw the rocket was in Paris in 1967. Until then, the nature of the rocket that launched Sputnik and Vostok was a mystery because of its military background. All was finally revealed when a 1:1 working model was crated and sent to Rouen in a cargo ship, its parts to the Paris air show at Le Bourget, where it was re-assembled and displayed, causing a sensation and giving Albert Ducrocq, France's original space writer, a worldwide exclusive.

Following on from CST working with Yuzhnoye to launch satellites from its family of launchers (Dnepr, Cosmos 3M) and DASA with Khrunichev (Eurockot), Europe's leading rocket company, Aérospatiale, set up a partnership with TsSKB Progress for the Soyuz launcher. Aérospatiale had been the principal company involved in development of the Ariane rocket since the 1970s, while TsSKB had been set up in Samara in the 1960s. TsSKB was the #3 branch of Korolev's original design bureau, OKB-1, now Energiya, which was established in an old aircraft plant, formerly a bicycle factory, on the banks of the Volga. In 1974, the #3 branch separated from OKB-1 and became TsSKB, or the *Tsentralnoye Spetsializorovannoye Konstruktorskoye Buro*, a sonorous title meaning Central Specialized Design Bureau, though it was also called 'the Progress works'. Its location was later renamed too, from Kyubishev to Samara. Led by one of the Soviet Union's great designers, Dmitri Kozlov, TsSKB became a sprawling design bureau and factory (*zavod*), with 30,000 people working there. After the crash of the Russian economy, the Progress plant diversified into everything from chocolates to vodka.

The French link-up with Soyuz arose from the difficult launcher situation in which ESA found itself in the 1990s. The original Ariane series (1, 2, 3 and 4) was reaching the end of its potential, with Ariane 4 making its last flight in 2003 and leaving Europe without a medium lift rocket and the prospect of a low launch rate. Europe had poured its launcher resources into the powerful Ariane 5, able to lift over 20 tonnes. Ariane 5 was perfect for launching large, commercially lucrative communications satellites – typically five tonnes in weight each and typically two at a time – as well as the big ATV, but also suited some of ESA's high-end science missions, like Rosetta. Europe, though, had no small or medium size launcher, until the Italian Vega launcher eventually filled the small launcher gap in 2012. During the 1990s, studies were made of what were called 'light Ariane derivatives'. François Calaque, the head of launchers at Aérospatiale, ran out of patience with the indecision and proposed that Europe adapt the Soyuz rocket as Europe's medium-lift launcher in February 1994. By the mid-1990s, France was the only country not to have a joint industrial project with Russia, having been beaten there by the Americans (Proton), the Germans (Rockot) and the British (CST). In the short term, the French proposal was to win commercial contracts, but in the long-term it was to establish Soyuz as an integral part of Europe's launcher fleet.

François Calaque (1940–98) is probably the key personality of the Russian-European Soyuz project. Born in 1940 in Homécourt in Lorraine, he qualified as an engineer and was responsible for the launch site for France's nuclear deterrent (Vaucluse). He joined Aérospatiale, becoming head of launcher facilities and then Ariane 5 production. He set up a joint company, Starsem, on 17 July 1996, comprising TsSKB (25%), the Russian Space Agency (25%), Arianespace (15%) and Aérospatiale (35%), with the blessing of the new French minister for postal services, information technologies and space affairs, François Fillon. Starsem made an opening pitch for the lucrative world communications satellite market, especially the Globalstar communications satellite constellation, but its initial problem was that Soyuz was not used to launching satellites to high Earth orbit, where most of these opportunities arose. Aérospatiale proposed a new upper stage, Irène (named after their interpreter), based on the Ariane upper stage and costed at FFR450m. This was well outside the CNES budget of mid-1995, but the Russians came riding to the rescue, with Ikar and Fregat.

Ikar was originally the propulsion module of the Yantar photo-reconnaissance satellite and was developed by TsSKB Progress. It was 2.9 m long, 2.72 m in diameter and weighed 3.29 tonnes, with 900 kg of UDMH and nitrogen tetroxide fuel. It was able to carry 3,300 kg into high Earth orbit. The motor was a 17D61, able to generate 2,943 kNs and equipped with 16 steering thrusters. As a civilianized version, Ikar offered a perfect method for getting groups of communications satellites into high, 1,400 km orbit.

Fregat was the same idea. Made by the Lavochkin design bureau, it was the lower stage of the new generation of deep space probes which first flew on the 1988 Phobos mission. It was small and toroidal in shape and although its engine was not especially powerful, it could be restarted many times and use long burns to achieve quite precise high-altitude orbits around Earth, ideal for a number of European payloads. Fregat had 28 attitude control thrusters, was 1.5 m tall and 3.35 m in diameter and could burn with a thrust of 19.6 kN for up to 877 seconds. Fregat had 5,350 kg of fuel and could be restarted up to 20 times. It now became the upper stage of the Soyuz to lift payloads to even higher orbit. So Irène led to Ikar and Fregat. Later, there was an improved version, Fregat M.

Starsem won its first order to launch American Globalstar communications satellites, with six launchings of four at a time. Earlier, Globalstar had suffered the disastrous experience of putting 12 satellites on a Zenit rocket on 9 September

1998, which crashed and burned. Ikar was adapted to include a 390 kg dispenser to spring each of the four satellites into their appropriate orbits. In 1999, the Soyuz Ikar made Globalstar launchings on 9 February, 15 March, 15 April, 22 September, 8 October and 22 November.

The Russians took advantage of the introduction of Ikar and Fregat to modernize the Soyuz, calling it Soyuz 2. This had many small improvements, but the principal one was the replacement of analogue controls by digital systems and 300 kg more lifting capacity. Gone was the old-fashioned countdown, supervised by 40 people at a dozen workstations in the bunker near the rocket and the old-fashioned turning of the launch key: instead, it was all done by computer.

Whether part of the original deal or not, an early priority was the crumbling facilities at Baikonour. When the Russian space programme contracted, cyclical maintenance was the first to go. Baikonour had become an embarrassment, with buildings falling down, debris dumped beside railway tracks, facilities well short of health and safety regulations and hospitality services that would attract notoriety in a tourist guide. Starsem invested an initial €30 m in three new white rooms of world standard – a payload preparation facility, a hazardous processing facility and an integration facility – plus a hotel, called Sputnik, for staff on site. Equipment was transported by rail all the way from Aquitaine to Baikonour. The assembly hall was brought up to standard. Later, additional resources were invested in improving the two Soyuz launch pads, 1 and 31. A repainting was completed.

The next opportunity for Soyuz to fly satellites emerged literally by accident. Over ten years in the making, Europe's flagship rocket programme for the end of the 20th century, the Ariane 5, was ready for take-off on 4 June 1996. Earlier that year, the Chinese had learnt the perils of installing valuable cargoes on new rockets on their maiden flights, with their Long March 3B having exploded on 14th February, taking with it a valuable Intelsat 708 communications satellite. For the maiden flight of Ariane 5, Europe had installed a four-satellite constellation called Cluster to observe the Sun, a project 20 years in preparation. In the biggest fireball ever seen over the Kourou launch centre, Ariane 5 and its precious cargo were destroyed 41 seconds into the maiden flight because of a software programming error.

Trying to salvage something from the disaster, ESA decided that the best option was to rebuild Cluster from the ground spares and launch them on the Soyuz. The charge was €210 m. Cluster required a more powerful upper stage than Ikar, with quite precise targeting into unusual orbits, so Fregat was chosen. Even so, Soyuz Fregat was much less powerful than Ariane so two launchings would be needed. The shroud on top of Soyuz was smaller as well, so some of the booms on Cluster had to be shortened.

There could be no more mistakes, so two demonstration missions were organized. On 9 February 2000, Fregat made a series of high-altitude burns, before concluding with an unrelated experiment to send rubber inflatable cones through re-entry for a series of tests of new ways of returning to the Earth (these were made by Lavochkin of Moscow and Daimler Chrysler of Bremen). On 20 March, Fregat was launched again for a second set of engine burns and mock separation tests. These missions paved the way for the real thing.

On 16 July and 9 August 2000 respectively, the Soyuz Fregat launched the two sets of Clusters, putting them into their final orbits of 19,000–119,000 km. They were initially put into a parking orbit and then the Fregat reignited, with six further engine burns required to settle them into their final orbit. They then used their own propellants to trim their position. By 24 August, all four satellites had reached their assigned points in the sky. Following a competition open to the public, they were then renamed Rumba, Salsa, Samba and Tango, to reflect the way in which they would dance in formation in their orbits facing the Sun, sometimes only 100 km apart and then moving to 20,000 km away from one another. Scientific results came in before the year was out. The four satellites measured the big solar storm of 9 November 2000, with the spacecraft drifting in and out of the magnetosphere as the storm reached the Earth. Not only did Europe get its Cluster satellites into orbit, it also witnessed an impressive display of Russia's oldest, most versatile and reliable rocket.

With this positive experience and the Cluster project saved, ESA subsequently commissioned the Russian Space Agency to launch its new generation of meteorological satellites, MetOp (also written Metop), under an agreement signed in May 2005. MetOp was Europe's flagship weather satellite programme, with three satellites to be launched four to five years apart. MetOp A was also the first European use of the new Soyuz 2, which had a new external fairing 4.1 m in diameter and 11.4 m long.

The first launch attempt of the new Soyuz 2.1a on 17 July 2006 was cancelled due to a programming error. On the second, the following day, there was a problem with the fuel gauges. On 19 July, there was a failure in the ground support system 185 seconds before lift-off and everything had to be sent back to the hangar until the autumn (there must have been some catty comments about how this would never have happened with the older versions). After a fourth and fifth delay, it was sixth time lucky for MetOp on 19 October (later, MetOp B was launched on the Soyuz 2.1a Fregat on 17 September 2012). Within six months, in addition to its forecasting duties, MetOp A had compiled a global map of nitrogen dioxide pollution and of ozone levels.

Europe used the new Soyuz a second time on 27 December 2006, when its sister, the Soyuz 2.1b, with a new, higher performance, longer-burning third stage (RD-0124), put into orbit the French planet-hunting COROT (COnvection, ROtation des étoiles et Transits des planètes extrasolaries). It found its first exoplanet in May 2007. The Starsem Soyuz was also used for the Israeli Amos 2 in 2003, its first mission to geostationary orbit. Table 4.6 lists European Starsem launches from Baikonour.

Table 4.6:
European Starsem launches from Baikonour 1999–2011

No	Date	Payload	Launcher
1	9 Feb 1999	Globalstar	Soyuz U Ikar
2	15 Mar 1999	Globalstar	Soyuz U Ikar
3	15 Apr 1999	Globalstar	Soyuz U Ikar
4	22 Sep 1999	Globalstar	Soyuz U Ikar
5	18 Oct 1999	Globalstar	Soyuz U Ikar
6	22 Nov 1999	Globalstar	Soyuz U Ikar
7	8 Feb 2000	IRDT	Soyuz U Fregat
8	20 Mar 2000	Dumsat	Soyuz U Fregat
9	16 July 2000	Cluster 1	Soyuz U Fregat
10	9 August 2000	Cluster 2	Soyuz U Fregat
11	2 June 2003	Mars Express	Soyuz FG Fregat
12	27 Dec 2003	Amos 2	Soyuz FG Fregat
13	13 Aug 2005	Galaxy 14	Soyuz FG Fregat
14	9 Nov 2005	Venus Express	Soyuz FG Fregat
15	28 Dec 2005	Giove A	Soyuz FG Fregat
16	19 Oct 2006	Metop A	Soyuz 2.1a Fregat
17	27 Dec 2006	COROT	Soyuz 2.1b Fregat
18	29 May 2007	Globalstar	Soyuz FG Fregat
19	21 Oct 2007	Globalstar	Soyuxz FG Fregat
20	14 Dec 2007	Radarsat 2	Soyuz FG Fregat
21	26 Apr 2009	Giove B	Soyuz FG Fregat
22	18 Oct 2010	Globalstar	Soyuz 2.1a Fregat
23	13 July 2011	Globalstar	Soyuz 2.1a Fregat
24	28 Dec 2011	Globalstar	Soyuz 2.1a Fregat

Soyouz à Kourou

Next came the second stage of the Soyuz relationship, one which built on the Starsem experience. The improbable idea of developing a Russian launch base in the European Union and French territory in the South American jungle arose from a confluence of factors. The original idea of launching Russian rockets from Kourou went back to 1993, when a senior Aérospatiale engineer, Patrick Eynar, wrote a memorandum to CNES proposing the use there of decommissioned missiles. The following year, Aérospatiale identified the potential for Soyuz to fill the gap in Europe's medium-lift launcher capability. Not only that, Soyuz would be able to deliver three tonnes to high Earth orbit or geosynchronous orbit from Kourou's near-equatorial location of 5.14°N, giving satellites a huge velocity advantage in reaching equatorial orbit. By contrast, Soyuz could send only 1.7 tonnes there from Baikonour's location at 51°N, so far north as to be uneconomical.

French Guyana is known to many people through the Dustin Hoffman film *Papillon*. The site was settled by France in the 1760s. Most perished in the jungles of the mainland and the survivors fled to the three islands off the coast, the Îles de Salut, or islands of salvation. These were Devil's Island, Île Royale and Île

Saint Joseph and in the 1850s they were turned into penal colonies known by 70,000 prisoners. The prisons closed in 1951 and when the rocketeers arrived in 1964 they found nothing but ruins and jungle. Guyana has a land area of 91,000 km², making it almost the size of Portugal, with 290,000 people mainly living along the coastal strip of a country that is 95 percent equatorial forest. It is a hot, steamy, wet climate, with a temperature of a steady 26°C and rainy from December to July (2.9 m a year). The first advice to visitors is still to 'bring your mosquito spray'. Tourists nowadays can visit Devil's Island, some of whose cells have been rehabilitated and where you can now stay in tourist accommodation. Kourou had the additional advantage of being geologically stable and out of the hurricane zone [17].

The launch site is near Cayenne, a coastal plain 29 km wide and 60 km long, served by a port where the rockets arrive by sea. The site covers 1,000 km², about one percent of the land area of the country and has about 1,400 residents. The interior is dense jungle and is used for training by the French foreign legion, which still has a base there. Piranhas and alligators infest its rivers and the native Amazonian people still hunt there in traditional ways, but entry into the interior by visitors is discouraged.

France began construction of the space centre there in 1964 and it was formally designated Europe's launching base two years later. The first launch, a Véronique sounding rocket, took place on 9 April 1968 and the first Ariane flew from there in 1979. Pads were built for sounding rockets, the small French Diamant launcher, Ariane 1 (later reconstructed for Vega), Ariane 4, Ariane 5 and a new pad for the Ariane 6. Around them are production and preparation facilities, integration halls, clean rooms and a launch control centre called Jupiter.

Following the initial ideas of 1993–4, the head of Arianespace, Jean-Marie Luton, announced in 1999 that a feasibility study was underway to look at the possibility of launching Soyuz from Guyana. One consideration was that Europe had a number of military observation satellites to launch, which they did not entrust to the security systems at Baikonour but could keep a closer eye on at Kourou. From the point of view of the Progress plant, the priority was to keep its production line open at a time when the Russian space programme was going through a difficult phase.

ESA hesitated and its June 2000 council meeting did not approve the project, considering Soyuz a threat to Ariane 5. There were still security concerns, as about 200 Russians would be stationed in Kourou for each launch campaign. The Russians asked the Europeans to reconsider, while senior French aerospace leaders warned of American dominance of the medium-launcher market. To stiffen European resolve, the Russians – who had by now learned a thing or two about commercial negotiations – threatened to take the project to Australia. President Chirac, visiting the Progress plant in July 2000, expressed his support for the project.

Soyouz à Kourou **previewed.**

In 2002, Claudie Haigneré became the French minister responsible for the space industry and she at once backed the proposal. More pre-studies were undertaken and a delegation of Russian engineers visited Guyana in December. At this stage, there was a growing acceptance that this was now a viable project, with only the funding package to be settled. The project was finally agreed at the ESA Council in Paris in May 2003 with a budget of €314 m, of which €193 m came from ESA and €121 m from Arianespace.

Texts were agreed by the Russian Space Agency on the one hand and the French government on the other, signed on 7 November 2003, the French side being represented by prime minister Jean-Pierre Raffarin and minister Claudie Haigneré, the Russian by President Vladimir Putin. It was supported by a later long-term cooperation agreement dated 19 January 2005. In effect, it was a three-sided governance arrangement between Russia, France and ESA. ESA was responsible for overall management, CNES was the system architect and Arianespace the operator, with Roskosmos responsible for Russian management and Lavochkin the upper stage. Although it was a European project, management of Kourou had always been delegated to the French, CNES and Arianespace, reflecting the substantial French financial investment.

Site plans were drafted in February 2004 and on 10 December, the French national assembly voted funding for the construction of the Soyuz pad. The global satellite market soon indicated its interest by booking a first launch in 2008, the Australian Optus D2 comsat, followed by three more orders. The financial agreement, though, did not seem as firm as it originally appeared. When Arianespace agreed to provide €121 m, it assumed that it would get a loan from the European Investment Bank (EIB), especially for a project supporting economic development in its periphery. The EIB only did so when guaranteed by the French government, to be paid back from the subsequent profits. This re-negotiation was lengthy. In the end, other countries joined the programme and France upped its contribution, so the final balance was France, 63.13%; Italy 8.71%; Belgium, 6.53%; Spain, 3.26%; Switzerland 2.72%; and Germany 5.65%.

The original plan was to erect the Soyuz site on the dormant Ariane 4 site, but this would have involved substantial demolition and a newer site to the north offered a granite subsoil, much more suitable for the flame trench. Bulldozers began clearing the 120 ha site in November 2004, with the official ground-breaking being on 25 October 2005, incorporating a piece of rock brought in from pad 1 at Baikonour, the *Gagarinsky Start*. Construction was carried out by Vinci Grands Projets and it involved 20,000 m^2 of building, 96,000 m^2 of roads and the excavation of 200,000 m^3 of rock. Construction of structures began on 26 February 2007 and this involved a Russian workforce from their cosmodrome builders, KBOM. A first Russian team of 50 arrived in July 2008, rising to 600 at its peak and contributing about two million hours of labour. The KBOM workers were given accommodation in Sinnamary 18 km away. The subsoil proved to be quite difficult and inconsistent to excavate and the flame trench took over a year. The deeper the builders dug, the more granite they found and they had to resort to blasting to clear the area. The earthworks were enormous and the outdoor work was not helped by tropical rain. The Russians, used to the arid atmosphere of Baikonour, greatly underestimated the corrosive effects of humid jungle temperatures, which could quickly cause rust. To complicate things, an archaeological survey found artifacts dating to 1200 BCE and they had to be dug out carefully. The pad was complete in early 2009, but the new mobile gantry took longer. The *Flinterland* arrived from St Petersburg in July 2008 with a shipload of initial equipment to install. It unnerved the Europeans to learn that the gantry had been sub-contracted to a company specializing in theme parks, but the Russians countered that theme park construction skills were ideal for building rocket gantries. ESA sent Italian engineers off to Sergeev Possad, close to the ancient, famous monastery, to supervise completion of the gantry there before it was loaded up and shipped to South America. It had to be disassembled, transported and reassembled and in the middle of all this the subcontractor experienced financial and record-keeping difficulties, so the gantry was not completed until March 2011.

Soyouz à Kourou **site. ESA.**

Soyuz had its own dedicated area at Kourou, 12 km north of the Ariane 5 site, called the Soyuz zone, with its pad called the ELS, *Ensemble de Lancement Soyuz.* Some, but not all of the facilities replicated the Soyuz pads at Baikonour and Plesetsk. Construction first involved a vehicle assembly building, called MIK (integration and test building hall in Russian). This was a low structure, only 20 m high but 56 m long, in which up to two Soyuz at a time could be assembled horizontally, having been shipped from Samara via St Petersburg. Beside the MIK were technical rooms, a 200-person reinforced control centre and storage tanks for the kerosene fuel. Each Soyuz would then be transported 600 m to the pad. Kerosene and hydrogen peroxide were brought in from Russia, while liquid oxygen was made locally.

The launch pad itself was built according to the blueprints used for the Gagarin pad in Baikonour and involved the similar construction of a launch platform and 26 m-deep flame trench. The Soyuz rocket was clamped by the same type of four-arm tower as at Baikonour, with the familiar system of weights and pulleys: when thrust builds up to 480 tonnes, the restraining arms are automatically released.

The principal difference between Kourou and the other Soyuz pads was the enclosed 42-tonne, 56.5 m-tall mobile tower for the vertical installation of the Fregat upper stage and payload. Vertical installation was not the Russian tradition, but it was the European procedure and ESA insisted on it, the reason being the need to protect the payload from humidity and rain (many technicians would add 'mosquitoes'). It also permitted easier access. In the event, the mobile tower worked so well that the Russians adopted it for their new cosmodrome at Vostochny several years later. *Soyouz à Kourou* was one of the biggest construction projects in the world at the time. It was the world's seventh Soyuz pad (two in Baikonour and four in Plesetsk, with an eighth to follow at Vostochny in 2015). In a health and safety triumph, there were no construction fatalities, just one broken leg.

Soyouz à Kourou **tower. ESA.**

Operating *Soyouz à Kourou* required a considerable logistical operation: fuel from Russia, the Soyuz from Samara, the Fregat from Moscow and the payload from Europe (Le Havre). The tracking network was extended. Soyuz used the existing tracking network of Montagne des Pères at Kourou, Natal (Brazil), Bermuda, Ascension Island, Libreville (Gabon) and Malindi (Kenya), but for the Fregat manoeuvres a range of other stations came into play: Cheju Island (Korea), Perth (Australia), Lucknow (India), Saskatoon (Canada), Santa Maria (Azores) and Aussaugel (Toulouse).

The first Soyuz launchers arrived on a cargo vessel in November 2009. A historic day for European-Russian space cooperation came on 21 October 2011, with the first launch of the Soyuz ST from Guyana, carrying two Galileo navigation satellites (see Table 4.7). There was a big crowd of Russian and French space officials, Commissioner Antonio Tajani of the European Union and 50 press. It was the first 'VS' (Vol Soyuz) mission, with other Kourou launches being 'VA' (Vol Ariane) and 'VV' (Vol Vega). In the Jupiter control room, the champagne and vodka flowed, because *Soyouz à Kourou* had been a tough test for Franco-Russian cooperation.

Soyouz à Kourou **on the move. ESA.**

Galileo provided most of the early payloads for *Soyouz à Kourou*. This was Europe's satellite navigation system, one co-funded by the European Union and designed to provide Europe with independence from the United States GPS. The project ran over time and budget, but Soyuz Fregat was ideal for lifting Galileo to its 19,000 km orbits. Galileo was divided into two stages: a predecessor series of two called IOV (In Orbit Verification) or Giove; and 26 operational satellites called FOC (Full Operational Capability). 'Giove' is the Italian for the planet Jupiter and it was the astronomer Galileo who devised the parameters of the planet's four main satellites in the early 17th century.

The second Soyuz launch was Pleiades 1A on 16 December 2011, a 970 kg dual-use Earth observation satellite for CNES, with five companions – a quartet of Elisa electronic intelligence satellites for France's Direction Général des Armaments (DGA) and a Chilean Earth observation satellite. Pleiades soon returned extraordinary images of land and sea, maintaining France's expertise that went back to the SPOT satellites. Pleiades 1B followed on 2 December 2012. The early flights went smoothly and the first problem launch was that of the first two operational (FOC) Galileo navigation satellites on 22 August 2014. The final burn of the Fregat went wrong (either it cut out early or else fired downward), leaving the two navigation satellites in 13,000 km orbits instead of 19,000 km. Controllers eventually found a way of using the spacecraft's thrusters to achieve an orbit close to that intended, albeit with the likelihood that their duration would be reduced.

Soyouz à Kourou **preparations. ESA.**

The Galileo launches were covered by ESA live broadcasting and viewers could follow the countdown from the glassy Jupiter mission control room, directed by the most important person in the room, the DDO (Director Des Opérations). It was quite like the shuttle launches from Cape Canaveral, except that the language was French, of course. Soyuz could be seen lifting off against the palm trees and the green jungle background. Sometimes, with clear skies, its engines could be seen making the famous 'cross of Korolev' and, two minutes into the mission, the four bottom stages and then its shroud spinning away. Kourou, though, was often cloudy and the Soyuz would frequently disappear into light cloud. ESA broadcasting would invite its viewers back several times in the hours that followed, for the first long Fregat burn, then coasting, then a second long burn, then deployment and confirmation of signal. Only then could the control room empty out and people go home.

An example of the versatility offered by the Russian system was VS23 on 18 December 2019. The Soyuz Fregat M launched no less than five payloads: the planet hunting CHEOPS (Characterizing Exoplanet satellite); an Italian military radar satellite called Cosmos-SkyMed 5, which would undoubtedly be used to spy on Russia; and three small satellites, the French ANGELS and EyeSat and the European OPS-Sat. The Fregat M made seven burns to leave each in the correct orbit. *Soyouz à Kourou* also made its first launch of the 03b satellites ('the other 3 billion') to bring the internet to developing countries.

***Soyouz à Kourou* Galileo launch. ESA.**

Table 4.7:
European launches on Soyuz from Guyana

Mission	Date	Payload
VS-01	21 Oct 2011	Galileo IOV 1, 2
VS-02	16 Dec 2011	Pléiades 1A
VS-03	12 Oct 2012	Galileo IOV 3, 4
VS-04	1 Dec 2012	Pléiades 1B
VS-05	25 Jun 2013	O3b F1
VS-06	19 Dec 2013	Gaia
VS-07	3 Apr 2014	Sentinel 1A
VS-08	10 Jul 2014	O3b F2
VS-09	22 Aug 2014	Galileo FOC FM1, FM2
VS-10	18 Dec 2014	O2b F3
VS-11	27 Mar 2015	Galileo FOC FM3, FM4
VS-12	11 Sep 2015	Galileo FOC FM5, FM6
VS-13	17 Dec 2015	Galileo FOC FM8, FM9
VS-14	25 Apr 2016	Sentinel 1b
VS-15	23 May 2016	Galileo FOC FM10, FM11
VS-16	27 Jan 2017	Hispasat
VS-17	18 May 2017	SES 15
VS-18	9 Mar 2018	O3b F4
VS-19	7 Nov 2018	MetOp C
VS-20	18 Dec 2018	CSO 1
VS-21	27 Feb 2019	OneWeb F6
VS-22	4 Apr 2019	O3b, FM 17-20 ST-B
VS-23	18 Dec 2019	CHEOPS, Cosmo-SkyMed 5, ANGELS, OPS-Sat, Eye-Sat ST-A

Some consideration was given to making piloted flights on Soyuz from Kourou in 2003. The principal problem appears to be that, like most launch sites – but not Baikonour – rockets head out over the sea. Soyuz is designed to be recoverable on land and although cosmonauts are trained for emergency sea recoveries, they are not ideal. There seems to be no great imperative to progress the idea.

Ural: Russian-European future launcher research

Soyuz met a key need in the European launcher market, that of a medium-lift launcher. A feature of launchers and their engines, though, is that they take many years to come to maturity. Barely was Ariane 5 off the pad before ESA began research on future launcher needs, the €40 m Future Launchers Technology Programme (FLTP) and Future Launcher Preparatory Programme (FLPP). For France, such early planning was imperative. In Aérospatiale (now Airbus) and the French engine companies (e.g. Snecma, Safran), France held the core of Europe's rocket design and rocket engine expertise, one which it wished to retain and keep in business. This required a long-term future pipeline of projects. When Britain cancelled Blue Streak, its rocket engineers all left, mainly for Canada and some for the United States, a great, permanent loss to its engineering workforce.

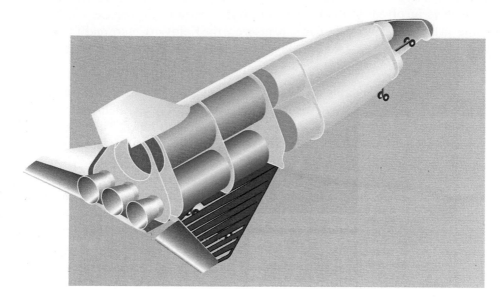

FLTP. ESA.

A strand of this research involved cooperation with Russia, hardly surprising given that Russian rocket engines were the best in the world. Russia had opened the first rocket engine laboratory, the Gas Dynamics Laboratory (GDL), in Leningrad in 1928, under its great designer, Valentin Glushko and this had evolved into the Energomash company. Even into the 2020s, as GDL approach its centenary, Russian engines – some of them of quite old design – powered the American Atlas V and Antares rockets with performance and reliability that the American could not match.

France provided €200 m to FLPP for cooperation with Russia, taking the form of accords dated 19 January and 15 March 2005 between the French space agency CNES and the Russian Roskosmos. The programme, called Ural (or, in French, Oural), was designed both to make studies and also ground test new systems and materials. The best description of Ural was that it was to develop a 'New Generation Launcher' for 2020 to take over from the Ariane 5, implying new technologies, capacities (e.g. flyback) and fuels. Demonstrator airframes and engines were envisaged [18]. CNES had already been making studies of next-generation launchers, such as EVolved European REuseable Space Transport (EVEREST), a two-stage-to-orbit flyback rocket which involved TsNIIMash, so this was the next logical step. Oural was deliberately so named to reflect General de Gaulle's phrase of 'Europe from the Atlantic to the Urals' [19]. Under Ural, there were five fields of work: launch vehicle concept studies, methane propulsion, advanced cryogenic tanks, first stage demonstrators and re-entry vehicle demonstrators. The main companies involved on the French side were Astrium, Cryospace and Snecma, while on the Russian side the participants were the Keldysh Centre and TsNIIMash. Ural was very much a forward-looking programme, aimed at the 2020s rather than the 2010s. The key elements were:

- Volga sub-programme: Snecma working with the Keldysh centre, Energomash and KhimAutomatiki on the oxygen-methane RD-0162 engine, with 200 tonnes thrust and relightable 50 times. This also drew in the large European aerospace company EADS in Germany, Techspace Aero (Belgium) and Volvo (Sweden).
- Barzugin sub-programme: flyback winged recoverable launchers.
- EADS with KhimAutomatiki, to test a reusable next-generation engine.
- EADS with Khrunichev and the Keldysh centre on ion engines.
- Pré-X, a lifting body which could be tested by the European Vega or the Russian Dnepr launcher, involving CNES, the French defence development agency ONERA, EADS, Snecma, Dassault and, in Germany, MAN Technologie.
- Flex, a first stage reusable demonstrator involving France, Germany and Russia (Khrunichev, Energiya and Molniya).
- Structure X, ground tests of the materials needed for hydrogen engines, with a programme of work being undertaken by France (Air Liquide, Cryospace), Germany, Russia and Switzerland [20].

There was no financial exchange with Ural. CNES funded the work done on the European side and Roskosmos on the Russian side. Good progress was made on Ural in the first year. Exotic-powered engines had long been a theme of Russian rocket engine development – a flourine ammonia engine, the RD-301 had even been built – and the Ural programme adapted the hydrogen-powered KVD 1 engine, originally intended for the Moon landing in the 1970s, as a methane-fuelled engine. Called the KVD 1.2, it made a 17-second test run in December 2005. Ural set up a working group in August 2008 to design and build a demonstrator reusable first stage, able to launch micro-satellites and using methane. At a seminar held in Miass in October 2008, attended not only by CNES but by DLR and the Spanish space agency, it was agreed to proceed with a methane-powered demonstrator. The importance of the Ural programme was underlined at a meeting of the Russian and French prime ministers in Sochi in September 2008. Ural and Barzugin were described in the press as 'at the heart of French-Russian cooperation in launch systems for the next years'.

RD-301 - the original exotic fuel engine.

Barzugin was a concept study for a future cost-reducing 'deep modernization' of the Ariane 5, replacing its pair of solid rocket boosters with two liquid-propelled, hydrogen-fuelled flyback rockets with air-breathing turbojet engines and canards. The principal participants were EADS, Dassault and Senecma on the French side and TsNIImash and the Molniya Design Bureau on the Russian side. Using existing technologies, like the Vulcain 2 engine, this offered cost savings through the use of re-useable boosters, with a target of 50 times re-use. The technique of recovering side boosters had already been used by the American Space Shuttle, its solid rocket boosters parachuting into the sea for recovery by two NASA ships operating out of Port Canaveral. Here, though, Barzugin envisaged the boosters gliding back to the runway at Kourou, saving them a drenching in corrosive, salty seawater. Designs were completed by autumn 2006 [21].

The Ural, Volga and Barzugin programmes were talked up frequently at international space industry events in the late 2000s. The last recorded mention was a roundtable on Ural during the *2010 Anneée France-Russie*. In the honoured tradition of cancelled projects (Roseau, Éos, Hermes) they seem to have evaporated without formal announcement or obituary and people no longer talked about them.

At around this time, 2008–11, ESA, CNES and the other key industrial players in France and Germany settled on their approach to post-Ariane 5, with the financial environment not one that encouraged launcher innovation. This was confirmed by the November 2014 ESA ministerial summit, which halted post-Ariane 5 development and went straight on to Ariane 6, a conservative choice that emphasized keeping costs down, reliability, meeting market need and the use of existing technology rather than the heady, earlier ideas of new ones.

In the event, Ural's ideas were brought to fruition by the Chinese, where Landspace developed methane-fuelled engines. The Raptor engines on Elon Musk's extraordinary Starship design, unveiled in 2019, also used methane engines. SpaceX also dramatically developed fly-back rockets, which tail-land their Falcon 9 first stage either on drone ships in the Atlantic or at Cape Canaveral itself. In 2020, Russia announced a reusable, methane-fuelled rocket called Amur. Ural and its related programmes are now seen as a 'French-Russian story' that disappeared into the mist.

Other industrial cooperation

Cooperation in launching services, such as *Soyouz à Kourou*, Starsem, Eurockot and CST, represented the most visible, high-profile part of European-Russian industrial partnership. They were matched by a range of cooperative and commercial arrangements at mezza and micro-level. In 1999, a study by the Centre for the Analysis of European Security compiled an inventory of European-Russian cooperation and itemized 87 joint projects and enterprises. Because of the commercial and contractual sensitivities involved, some have been poorly publicized and documented.

Cooperation in the area of satellite-making was a key factor. The principal maker of communications satellites in Russia was in the closed city of Krasnoyarsk, also called Zhelenogrosk, near Novosibirsk. It was originally called design bureau OKB-10, then NPO PM and is now ISS Reshetnev. The Soviet Union developed communications satellites from the 1960s, first in high-altitude orbits (Molniya) and then 24-hour orbit (Raduga). Western commentaries took the view that they were inferior to western satellites and this may have been true of electrical power, the number of channels and design lifetimes. It is quite possible that communications satellite technology fell behind in the 1990s as a consequence of the collapse of funding for investment in research.

On 1 March 1992, CNES set up DERSI to promote Russian-French industrial space cooperation (DÉveloppement des Relations Spatiales Industriales internationales), managed by Hélène Bourlakff, which prompted more than FFR4m of business in less than a year. Alcatel was quick to spot an opening in the area of communications satellites. It first met with NPO PM in 1992 and this led to a 'design partnership' the following year. This principally took the form of the Ekspress series of satellites starting in 1999, which were NPO PM designs with Alcatel updates and equipment. Alcatel subsequently contributed to the Geo-1K and Luch series. From 2007, NPO PM developed the Ekspress 4000 design, based on Alcatel's Spacebus 4000, 5.9-tonne satellites using 12 kW of power. This continues to the present, with a number of name changes on the European side as well (consolidated as Thales Alenia Space and Airbus).

Thales Alenia logo on Proton. Roskosmos.

In 1992, Alcatel also formed a partnership with the Energiya design bureau and Gazprom for the development of the Yamal communications satellite, the most recent being a Thales contract to build the Yamal 601 comsat launched on 30 May 2019. The Ekspress 80 and 103 communication satellites, launched together by Proton on 31 July 2020, were largely built by Thales Alenia in Italy. Airbus and Energiya made an agreement in December 2016 for the construction of an advanced medium-class satellite communications platform.

Although this relationship has been portrayed as Europe supplying Russia with the high-tech instruments that it lacked for communications, some of the traffic went the other way, with the prime example being the Siberian European Satellite, called Sesat, built by ISS Reshetnev for Europe. Also called Eutelsat 16C, it was launched by Proton on 17 April 2000 and was not retired until it was de-orbited nearly 18 years later. Sesat 2, also known as Ekspress AM22, launched in 2003.

The second area of cooperation is electric engines, which have been used increasingly on communications satellites. SEP, later Snecma and then Safran, sent a first mission to the Fakel design bureau in Russia in February 1991, the outcome of which was the development with Fakel of electric engines that were subsequently used on communication satellites. Fakel SPT 100 plasma thrusters then found their way onto Thales- (now TAS) and Astrium- (now Airbus) made communication satellites, such as Intelsat 10 and two Inmarsats (4F1 and 4F2). A version of this thruster, the PPS 1350, was used for Europe's first Moon mission, SMART, though its origin was not much advertised.

A final aspect of cooperation in danger of being overlooked is that of the media. France has possibly the biggest concentration of spaceflight writers and journalists in Europe and they were involved in reporting the Russian axis from the start. Albert Ducrocq, founder of *Air & Cosmos*, visited Moscow as early as 1957 and he gave a letter of recommendation to Christian Lardier to visit 20 years later. Albert Ducrocq, the premier early French space journalist, worked with Serge Berg, who was the Agence France Presse representative in Moscow until he returned to Paris in 1961. Together, they doorstepped the Russian delegations at international conferences and anywhere else they could to get their stories. CNES organized a first press visit to Star Town in June 1974 and for the PVH, the journalists got to see all the legendary places of the Soviet space programme: TsUP, Baikonour and the recovery zone near Arkalyk. There were restrictions on what they could see but not photograph, such as the Energiya pads. In 1995, they were able to visit Plesetsk for the first time for Foton 10 and then attended the landing site. In 1996, they visited the Reshetnev design bureau in Krasnoyarsk. As a result, the press played an important role in making the Soviet and Russian programme known to the west, an important foundation for cooperation.

Industrial cooperation: conclusions

Industrial cooperation was the most difficult, challenging and controversial area of European-Russian partnership. Regardless of the level of cooperation in scientific research and human spaceflight, industrial cooperation tested the commercial relationships of the USSR, Russia, Europe and the United States, set in ever starker relief by the cold war, or here the economic cold war. This was often a three-way battle, with the Americans making it clear as early as 1963 that Europe should not enter the communications launcher market which it dominated – and then making life difficult for Europe's independent launcher, Ariane, by refusing to sell fuel for it.

In reality, barriers against industrial cooperation were well established and dated back to 1949, making such cooperation difficult. They forced the USSR to set up its own socialist block channels of economic cooperation and specialization in the form of the CMEA (Comecon) and its distinctive spaceflight institutions, like Interkosmos. The end of the cold war and the privatization of the Russian economy suddenly lifted most of those barriers, but the test case was again the commercial communications launcher market – specifically Russia's Proton launcher – with the Americans eventually permitting the launch of western satellites under tight conditions. The lower end of the launch market was less problematical: quite the reverse in fact, with the market flooded with decommissioned cold war launchers. This was where the entrepreneurs entered, people like Gerry Webb (Commercial Space Technologies) in Britain and companies like Daimler Benz in Germany (Eurockot), who set up the structures of industrial space cooperation that made possible the launch of European and other satellites on Russian rockets. They were the people who had the vision, skills and persistence to bridge the worlds between western satellite makers and the Russian design bureaux and did so in a country going through the throes of shock capitalism. German companies like OHB were to the fore in taking advantage of these opportunities, illustrating the country's quiet, late emergence as a major country of cooperation with Russia. Exolaunch in Berlin is the latest of these successful companies.

It was Russia's oldest rocket, the Soyuz, that was to be the occasion of the highest level of integrated cooperation between Russia and Europe. This was the area where France made its move, with Arianespace setting up a joint project with the company responsible for Soyuz, TsSKB Progress in Samara, in 1996, to launch European satellites from Baikonour. Although short of development resources, Russia had just enough available to develop two new upper stages, Ikar and Fregat, to make Soyuz an increasingly versatile launcher for the type of satellite payloads Europe now needed. The 2020 Gonets M launch was a classic illustration of Fregat's capabilities.

Soyuz's successful demonstration of its abilities by launching Europe's Cluster scientific satellites came at a key and fortuitous moment, as Europe had a big gap in its launchers in the area of medium lift, especially for its forthcoming satellite navigation series, Galileo. What became the *Soyouz à Kourou* project was a case study in lateral thinking. Building a Soyuz launch site in the South American jungle to take advantage of its equatorial location was an inspired move.

Although a European project, it was French-driven and largely French-financed. Developing the *Soyouz à Kourou* site took time (ten years from proposal to launch), planning and much effort to integrate the Russian and European operations. The design, financing, engineering and management challenges were substantial and it was a cooperative construction project on a very large scale, one of the largest in the world at the time.

In the subsequent decade, Soyuz launches from Kourou averaged two a year. Since the first launch on 21 October 2011, Soyuz has carried into orbit a long manifest of European cargoes, such as Galileo, Pleiades and more. Its most important role was to put into orbit the European navigational satellites, from the Initial Operational Version (IOV) to the Fully Operational Constellation (FOC). The launcher proved reliable, with an underburn on only one launch, although the satellites subsequently reached their intended destination. One irony was Composante Spatiale Optique (CSO 1), a 3,500 kg spacecraft for the French military, with agreements for data sharing with Germany, Sweden and Belgium, that was launched by Soyuz on 19 December 2018 – and would most likely be spying on Russian military activities. Television broadcasts of the DDO calling out, in French, the various stages of the Soyuz launch and subsequent Fregat burns have become routine. Who would have imagined, in 1979 when the first Ariane flew from Kourou, that a Soyuz launcher would later make its home there?

It is not clear how long Soyuz will continue to fly from Guyana, but it still has a manifest of up to 20 possible future launches at Kourou. The end of the road for the Soyuz launcher has been predicted many times, but it has always outlived these Cassandras. The Ariane 6 rocket, especially the lighter 62 version, due to enter service in the 2020s, will certainly challenge some of its business (CSO 3 has already moved from Soyuz to Ariane 6). Ariane 6 has been seen – indeed touted as such by its builders – as a replacement for the Soyuz. At the other, lighter end of the market, the Vega rocket will challenge what might have been smaller payloads for Soyuz. In theory, Ariane 6 and Vega between them might squeeze Soyuz out of business in Kourou, but Arianespace and Starsem are likely to hold on to the advantages of Soyuz as long as they can. Officially, the policy is reserved: the director of launchers at ESA, Antonio Fabrizi has stated, accurately enough, that Ariane 6 occupies *some* of the market position of Soyuz 'but we do not yet have a position on the launcher's future' **[22]**. Vega's use of a Soviet-period engine is a little-known part of the cooperative story.

Starsem and *Soyouz à Kourou* were the most visible side of Russian-European cooperation, but it has also been present at mezza level in the form of communications satellites (Alcatel, Thales Alenia, Airbus, ISS Reshetnev, Energiya, Khrunichev) and engines (Fakel); and at micro-level (TUBSATs, Rubin). Much of this behind-the-scenes work is not well-publicized nor documented, but has led to a substantial level of cooperation between enterprises.

In the meantime, France undertook a launcher and engine exploratory programme called Ural with Russia, with a number of sub-programmes such as Volga and Barzugin. There was a confluence of factors here, with France having the biggest concentration of rocket and rocket engine making expertise in Europe and looking to obtain new business for this workforce, while Russia is the world's leader in rocket engines. It is hard to avoid the conclusion that in dropping Ural, moreover without even publishing or disseminating what outcomes were achieved, Europe lost a major opportunity to move into advanced launcher technologies, with Russian help, ground that is now being taken up by SpaceX and China.

There have been gains on both sides. For Russia, facing an economic crisis, cooperation with Europe was a means of keeping its design bureaux, *zavods* and enterprises in business, keeping production lines of rockets, engines, satellites and equipment open and in the case of Kourou, retaining its cosmodrome builders. Western funding enabled the physical upgrading of facilities (e.g. Baikonour, Plestesk) and of technologies – notably for communications satellites – and should have enabled some fresh funding for research and development. Russia was able to add impressive, high-performing upper stages to its inventory (Ikar, Fregat, Briz) and modernize the analogue Soyuz with the digital Soyuz 2 version. Income-generating uses were found for obsolete but entirely functional cold war missiles.

For Europe, the gains were principally in much-reduced launch costs and more flexible launch opportunities. Europe gained access to Russia's rocket fleet, which ranged from heavy launchers like Proton to submarines capable of getting micro-satellites into orbit at the other extreme. Thanks to CST and others, the launch of small satellites became affordable. Soyuz managed to fill a broad range of Europe's medium lift requirements, including sending fleets of satellites to previously underused medium-height orbits (e.g. Galileo). Europe was able to get an entire fleet of Earth observation satellites into orbit on Rockot from Plesetsk, as did Germany with its SARLupe military radar fleet on the Cosmos 3M from Plesetsk. Without Russia, this would have cost much more and taken much longer. Although seasoned observers pointed to the obsolete nature of the Soyuz (1948 design), it was reliable and profitable for its makers and users. The Progress plant in Samara must hold a record for producing the same rocket continuously since 1956.

A final comment is that industrial cooperation is largely a Franco-German story. France took the most visible, institutional role, most evident through *Soyouz à Kourou*, Starsem and communications satellites and its role in ESA, but Germany was heavily invested through Eurockot, Sentinel, SARLupe, OHB and Galileo. CST stands out as a pioneering British venture, but there is a remarkable lack of visibility of other European countries, enterprises and projects. Even if one makes allowances for subsidiaries of the main European aerospace companies in other countries (e.g. Italy, Spain), they are difficult to identify.

References

1   Krige, J; Russo, A & Sebesta, L; *A history of the European Space Agency*, vol II, 1973–1987. Nordwijk, ESA, 2000.
2.   CNES Magazine, #14, January 2002.
3.   Cupitt, Richard & Grillot, Suzette: *CoCom is dead, long live CoCom – persistence and change in multilateral security institutions*. British Journal of Political Science, vol 27, #3, July 1997.
4.   Freeman, Marsha & Scanlon, Leo: *Fight intensified over US sanctions policy*. EIR, vol 25, #25, 19 June 1998.
5.   Joao, Alexandre: *A matter of honour? Russia's reaction to western sanctions*. Thesis presented for the degree of Master of Arts in European Interdisciplinary Studies, College of Europe, Warsaw, 2016.

6. Engel, Jeffrey: *Cold war at 30,000ft – the Anglo-American fight for aviation supremacy.* Harvard, Harvard University Press, 2007.

7. Lipson, Michael: *The reincarnation of CoCom – explaining post-cold war export controls.* Nonproliferation review, winter 1999.

8. *Multilateral export control policy – the coordinating committee,* from Technology and east-west trade, Princeton University, 1978.

9. Office of Technology Assessment: *US-Soviet cooperation in space.* US Congress, OTA, Washington DC, 1985.

10. For background on this period, see Baker, David: *Behind the curtain – a view from inside the Soviet Union.* Presentation, British Interplanetary Society, 3 June 2017.

11. Browder, Bill: *Red notice.* London, Corgi Books, 2015; Klein, Naomi: *The shock doctrine – the rise of disaster capitalism.* Canada, Knopf, 2007.

12. Phelan, Dominic: *Cold war space sleuths – the untold secrets of the Soviet space programme.* Praxis & Springer, 2013.

13. Webb, GM: *The use of former Soviet Union launchers to launch small satellites.* Paper presented to the BIS Chinese/Soviet forum, 5 June 2004.

14. Eurockot: *Stepping into Plesetsk – the European spaceport.* and *The way up – the European launch provider for small satellites.* Eurockot, Bremen; for more on Rockot, see Hendrickx, Bart: *Naryad V and the Soviet anti-satellite fleet.* Space Chronicle, vol 69, 2016.

15. Vega – the small European launcher. ESA, Noordwijk, 2007; see Zak, Anatoly: *Europe's Vega launch vehicle,* from www.russianspaceweb.com, accessed 17 August 2020.

16. Ziegler, Bent; Kalnins, Indulis; Bruhn, Frederik; & Stenmark, Lars: *Rubin – a frequent flier testbed for micro- and nano-technologies.* Paper presented to International Astronautical Congress, Valencia, October 2006.

17. For the development of this project, see Lardier, Christian & Barensky, Stefan: T*he Soyuz launch vehicle – the two lives of an engineering triumph.* Praxis/Springer, 2010; and de Angelis, Laurent: *Soyuz in the jungle,* in Brian Harvey (ed): Space exploration 2007. Praxis/Springer, 2007.

18. Bonhomme, C; Theron, M; Louaas; Beaurain, A; & Seleznev: *French/Russian activities on LAX/LCH4 area.* Paper presented to International Astronautical Congress, Valencia, 2–6 October 2006.

19. Speech given at University of Strasbourg, 22 November 1959.

20. Astorg, Jean-Marc; Louaas, Eric; Yakutchin, Nikolai: *Oural – cooperation between Europe and Russia to prepare future launchers.* Paper presented to International Astronautical Congress, Valencia, 2–6 October 2006; Gogdet, Olivier; Arnoud, Emilie; Prampolini, Marco; Prel, Yves; Talbot, Christophe; Kolozezny, Anton; & Sumin, Yuri: *Next generation launcher studies – preparing long-term access to space.* do.

21. Sumin, Yuri; Kostromin, Sergei; Panichkin, Nikolai; Prevl, Yves; Soin, Mikhail; Iranzo-Greus, David; Prampolini, Marco: *Development of RFBB Barzugin concept for Ariane 5 evolution.* Paper presented to International Astronautical Congress, Valencia, 2-6 October 2006.

22. Ministerielles. CNESMag 57, April 2013.

5

ExoMars

ExoMars is the climax of European-Russian cooperation, a joint double mission and the second largest area of cooperation after the International Space Station (ISS). This chapter looks at the origins of the mission in an early century European rover project, the development of the two-stage project and then the first mission in 2016, which saw the Trace Gas Orbiter enter Mars orbit but the Schiaparelli lander fail. The mission took place against a background of rising political tensions between Europe and Russia over Ukraine and Crimea, with the reinstatement of the sanctions regime once again affecting their relationship.

VEX and MEX

Since 1971, France had flown instruments on Soviet missions to Mars and it was joined by other European countries for the Phobos and Mars 8 missions. Inevitably, Europe began to think of its own mission to Mars and the project had even acquired the name 'Kepler'. Within weeks of Mars 8 going down, France and Germany had begun a study on how to use the type of instruments developed for Mars 8 on a smaller mission. Because it was normal practice to build spare instruments, the practical legacy of Mars 8 was quite a number of spares and parts looking for a home. The new mission was put together quite quickly using backup parts and hence attracted the name Mars Express, sometimes called MEX. This was approved by ESA in late 1998, allocated €180 m and set for a June 2003 launch. ESA commissioned the

The original version of this chapter was revised. The correction to this chapter is available at https://doi.org/10.1007/978-3-030-67686-5_7

Soyuz Fregat rocket as the launcher, with a target of reaching Mars orbit in December 2003. Regardless of the problems with Mars 8, the Soyuz had proven itself reliable.

The mission was successfully launched from Russia on 2 June 2003 by Soyuz Fregat, arriving in Mars orbit on Christmas Day. Despite its modest size, Mars Express was hugely successful, barring the loss of a small British lander, Beagle 2 (that was located in January 2015 on photos sent back from the American Mars Reconnaissance Orbiter, which had pinpointed the lander, parachute and cover). It was a European mission, with Russia providing the lifting power and contributing to some experiments.

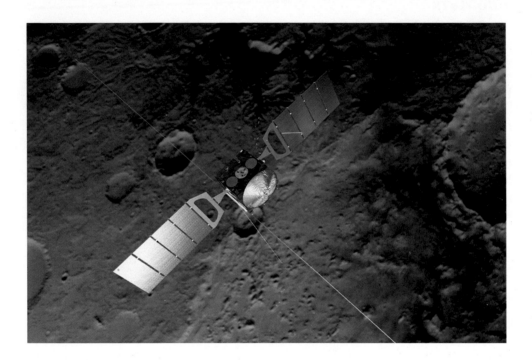

Mars Express over Mars. ESA.

Mars Express had six instruments weighing a total of 115 kg, with some derived from Mars 8: a high resolution stereo camera; a subsurface radar sounder; a Fourier spectrometer; a plasma analyser; a radio science experiment; and a spectrometer. The cameras were based on those originally developed in the GDR and provided extraordinary images of the poles, hazy clouds, dunes, basins, past flooding, mammoth volcanoes, old river valleys, glaciers and traces of catastrophic floods. The announced discovery of methane by the Fourier spectrometer at up to 30 ppb (parts per billion by volume) was considered one of its most important findings, one which sent headlines around the world because methane was a possible indicator of microbial life. By 2009, ESA was able to publish a 294-page summary of mission outcomes, with the results from the different instruments [1]. It included a chapter co-authored by Oleg Korablev concerning the SPICAM instrument (SPectroscopy for the Investigation of the Characteristics of Mars) derived from Mars 8, intended to profile the atmosphere. The original was 46 kg, but this had to be slimmed down to 4.7 kg for the much smaller MEX. It enabled them to remodel the Martian atmosphere. The idea of the 'warmer, wetter' Martian history originated with Vasili Moroz from his studies of data from the 1970s, but it now became the consensus [2]. Mars Express was still returning data and photographs more than 17 years later, leading to 1,300 papers by 2020.

Even before Mars Express had departed, ESA took the view that a nearly identical spacecraft could be adapted economically for a Venus mission. Going to Venus was less demanding on solar power (the solar wings could be half the size, but it required additional thermal protection instead – 23 layers). Venus Express used three instruments from Mars Express (including the camera), two from the Rosetta comet probe and two other instruments. It launched on a Soyuz FG Fregat on 9 November 2005 and arrived in Venus orbit following a 50-minute braking manoeuvre on 11 April 2006. By 2014 it had transmitted 5.5TB of data, more than all previous Venus missions put together: There were 400 scientific papers published in the first ten years. The mission lasted until 28 November 2014, when contact was lost.

Like Mars Express, Venus Express had an important effect in revising theories on the origin and development of the planet. Venus Express tracked the weather patterns of the planet: hurricanes, lightning and its super-rotating cloud system which went round the planet every four days, leaking atmosphere [3]. Scientists painted a picture of a world that had once been comparable to Earth: similar in size, density and solar energy absorbed and with an abundance of water and carbon dioxide, but which had evolved quite differently. The main breakthrough was in vulcanism and here the outcome was very much an international one. In summer 2015, Venus Express scientists announced the best evidence yet for present-day vulcanism on the planet. The international team was led by James Head of Brown University on the American side, Mikhail Ivanov on the Russian side and Dima Titov for ESA, with Eugene Sholygin and Wojciech Markiewicz of the Max Plank Institute of Germany, as well as Alexander Basilevsky (Vernadsky Institute) and NI Ignatiev (Institute of Space Research, IKI). They found that there had been a sharp rise in sulphur dioxide over 2006–7, which had fallen subsequently, consistent with volcanic eruption. Then, in 2014, four hot spots between 1 km^2 and 200 km^2 across, spiking several hundred degrees, were located in the Ganiki Chasma rift zone near the volcanoes Ozza Mons and Maat Mons. This suggested outflowing magma, but they would cool again, like Earthlike magma, over a few days. Between them, these observations offered compelling evidence of a currently volcanically active Venus.

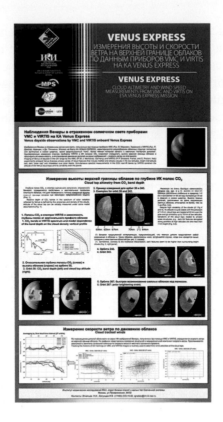

Venus Express outcomes.

Mars Express and Venus Express were both quickly assembled, low-cost missions with an unexpectedly high scientific return that salvaged some of the instrumentation opportunities lost by Mars 8. Although essentially European missions facilitated by Russia, they confirmed the value of working with Russia on Europe's planetary programme and instrumentation and became the firm base for a new, ambitious mission. The Beagle 2 failure was a particular learning point. Beagle was very much a one-person project by British scientist and enthusiast Colin Pillinger, who persuaded ESA to fly his project. The damning report released subsequently was a textbook example of how not to plan a Mars landing – and a warning of just how difficult it could be.

ExoMars: origins

With Ariane 5 in service, the ISS flying and budgetary pressures easing, Europe felt that with the arrival of the new century it should set medium and long-term goals. The agency issued a 'call for ideas' in 2001, which led it to a consensus on completing the ISS, developing a strategy for robotic exploration to Mars and beyond and stimulating new technological approaches in a civilian programme with wide public appeal. This became Aurora, which was approved at the ESA council in Edinburgh in 2001 as 'ESA's master plan to explore the universe', with Mars as a primary focus. Officially, Aurora was a long-term programme of robotic exploration of the solar system, no less, but while the brochures issued were colourful, they were also vague about how or when this would be done, or by or with whom. There was even talk of a human eventually landing on Mars. ESA may have also intended to draw in the European Union for funding, but the EU limited itself to the Galileo navigation system and Earth observations (the Copernicus programme).

Aurora went through a preparatory phase from 2002–4, attracting an initial budget of €48 m (Italy being the biggest supporter) and was approved by the ESA council in Berlin in December 2005. The name ExoMars appeared for the first time, an initial flagship mission focussed on finding life (hence the 'exo-'), with a Mars sample return flight as the follow-up. At that stage, ExoMars was a single mission, with the rover as the centrepiece and a static experimental package on the lander. The launch was set for a Soyuz from Kourou in 2011, with a slow, two-year trajectory to Mars and an airbag landing. Alternatively, an Ariane 5 might be used to carry an orbiter for relay purposes. There was a nice promotional brochure, *Aurora and ExoMars*.

ExoMars early rover.

In reality, a big Mars rover was the only constant in the programme, its key instrument being a drill to penetrate deep into the soil to find water or indicators of life. Originally, the rover's starting weight was around 700 kg, just less than the Soviet Lunokhod Moon rovers from the 1970s, but it was downsized over time. Models of a rover were displayed at international exhibitions from this point, while various mission profiles were studied over 2002–5. There was much discussion on the landing method and on an instrument package for the static lander. Like the successful small American rovers, *Spirit* and *Opportunity,* it would land on airbags, but they would deflate on landing rather than bounce. At the Bremen International Astronautical Congress in 2003, it was mentioned that Russian help would be sought with the rover because of their lunar experience. Following these preliminary studies, ExoMars was approved by the ESA council 2005 as a €650 m project, with Italy designated as lead nation. Britain would build the rover, with Astrium in Stevenage identified as the prime contractor [4]. The details had been better defined, with an ambitious instrument suite. The rover was designed with six wheels and was intended to drive 125 m a day. Its principal instrument was a 2 m drill to lift samples to an analytical laboratory. Other instruments were a panoramic camera (pancam), an infrared spectrometer (to identify mineral composition), a close-up imager, a Mossbauer spectrometer (to search for iron), a Raman and Laser Induced Breakdown Spectrometer (to characterize rocks), an x-ray diffractometer, an infrared microscope and the Urey biological analyser (Mars Organic Detector and Mars Oxidant Instrument) designed to resolve the ambiguities of the Viking analysis of Martian soil in 1976.

Although the project had a well-established place in Europe's list of future missions, concluding the funding package for such an ambitious and costly enterprise remained embarrassingly elusive, to the point that the model was put away and no longer exhibited. The principal difficulty was the launcher. In 2005, following the success of MEX and VEX, there was discussion of it flying to Mars on a Soyuz Fregat, though that would mean reducing the instrumentation, or on the Proton, which would require even more funds beyond the €1.2bn already committed. The 2008 ESA council in the Netherlands decided on a change of course – described as a 'project sharing approach' – to bring in the Americans, renaming the mission the Mars Exploration Joint Initiative (MEJI). This offered lower costs (€1bn for Europe and €1.5bn for NASA), but spread the mission over a two-part, 2016 and 2018, timeframe.

Under MEJI, the Americans would provide an Atlas V rocket for both missions. The first, in 2016, would carry a European orbiter and a surface meteorological station, while the Americans contributed two instruments. Europe's rover would have to wait till 2018 and would be scaled back to 310 kg to make way for a second, American rover, the Mars Astrobiology Explorer Cacher, MAX-C. The pair would land alongside one another to explore in a coordinated programme, with MAX-C to cache samples for their later collection by another mission. The ultimate intention was to progress jointly toward a soil sample return mission in the

2020s. For the Americans, there was a growing conviction, expressed in its 2011 decadal survey *Vision and voyages for planetary science in the decade 2013–2023* by the National Science Council, that a sample return was the top priority, with an acceptance that its high costs should also be shared.

For Europe, the mission would be later and smaller than it had intended, but at least the funding problem was solved. Except, perhaps, that they had forgotten the unfortunate experience of Claude Allègre and Netlander, as the American side soon got into financial trouble. NASA was already building its own big rover, the Mars Science Laboratory (MSL), later known as *Curiosity*, intended for a 2009 launch, but this ambitious project got so behind schedule that it was delayed to 2011, the postponement costing $400 m [5]. At the same time, there was also a big overrun on the cost of the James Webb Space Telescope, a project begun in 1996 and scheduled for launch in 2007 (its latest launch date is 2021) and increased demands for funding for the Space Launch System rocket to return Americans to the Moon. The Office for Management and Budget (OMB), a critical point in the financial decision-making chain, always opposed Mars sample return as too expensive and viewed killing off MAX-C as the best way to stop it. NASA's Mars budget for 2012 was cut by 35 percent and that was the end of it. It left the Europeans high and dry, just like the French ten years earlier. Ironically, NASA did get its caching mission in the end, as this task was added several years later to the mission of *Perseverance*, launched to Mars in July 2020.

What about Aurora in the meantime? People do not speak much about Aurora nowadays, even though ESA's Aurora webpage, dated 2004, is still there, describing Aurora as the roadmap that could lead to a human mission to Mars by 2025.

ExoMars: a European-Russian joint project

Within days of it becoming known that the Americans had pulled out of ExoMars, officials from Roskosmos and IKI in Moscow were on the plane to Paris, offering the Proton rocket (in the event, two) that would enable an ambitious ExoMars to reach the planet. Officially, the European record is that the Russians were 'invited' to join, but in practice it was Russia that made the opening. It was the chance to get back into the business of planetary exploration.

The offer of the Proton showed how desperate the Russians were to do so, since its scientists had been told in the late 1990s that the Proton was too expensive for scientific missions and it was assigned only to money-making commercial launches and military missions (navigation and communications satellites) out of the military side of the space budget. Its production line at Khrunichev had greatly slowed down. This was true, with scientific missions delayed for years as missions were tediously reconstructed so that they could fit on other rockets (e.g. Phobos Grunt, Spektr R on Zenit). Apparently, the reason why two Protons magically appeared out of nowhere for the ExoMars mission was that IKI had had the foresight to insure the Phobos Grunt mission, from whose failure it received a R1.2bn pay-out.

The Proton could lift a large payload to Mars – over four tonnes – so it was an offer the Europeans could not refuse. But what were the Russians looking for, in return? The big problem for Russian space science was the lack of scientific instrumentation on its own space probes. The lifeblood of any scientific community is its own original research from its own instruments. Starved of a budget for scientific missions, the Russians were reduced to begging to include the occasional instrument on someone else's mission and were even scanning American datasets and going back through the tapes of their own missions from the Soviet period to see if new science could be extracted. So Russia wanted new instruments and full access to the scientific data.

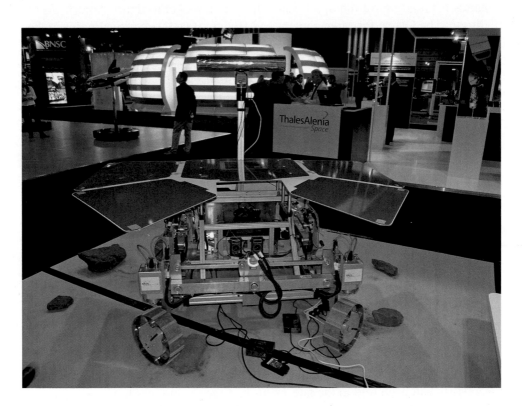

ExoMars rover re-design.

The Russians were determined to get their own science from any joint mission. At one stage there were some awkward exchanges between Moscow and Paris, with Russian scientists annoyed at what they perceived to be a European approach of 'just launch it for us, but we don't otherwise want you involved'. Before this got out of hand, IKI and ESA organized a discussion of the possible joint science in November 2011. This led to an agreement that the Russian instruments would comprise the Atmospheric Chemistry Suite (ACS) and a spectrometer, with Russia also supplying the lander's radioactive heat generator. It paved the way for ESA and Roskosmos to sign a declaration of intent on 6 April 2012. The first industrial meeting took place in Turin, Italy, on 2–4 May 2012, bringing together Italian prime contractor Thales Alenia with the rover builder Astrium from Britain, as well as the builder of interplanetary spacecraft Lavochkin and the Proton-maker Khrunichev from Moscow. Conferences on the instrumentation were convened immediately, the sixth being held in Moscow in February 2014.

ExoMars signing. ESA.

Despite the American withdrawal, Europe still wanted to keep to the 2016 and 2018 target dates – delaying them would cost more money – but this was now quite a tight schedule. Many key contracts had still to be signed and the winter of 2012–13 saw continued negotiations over the respective roles of the two sides. The original planned signing of the all-up formal agreement set for 27 November 2012 did not take place until 14 March 2013. This included all the arrangements for construction, the mission itself and the sharing of scientific information (the Russians got what they were looking for). Once agreed, construction of the orbiter and lander picked up sharply at various construction and testing sites across Europe. Clarity quickly emerged on the shape of the joint project:

- In 2016, a European orbiter, the Trace Gas Orbiter (TGO), to operate until at least 2023; with a European EDL (Entry and Landing Demonstrator) to test landing techniques, later called Schiaparelli.
- In 2018, the European ExoMars rover; with Russia to provide the landing platform.

Europe tried to hold the cost at the original €1.2bn (ESA), but with all the delays and redesigns, this was never going to be enough. Financial shortages were a recurrent theme. At the November 2014 ministerial meeting, Britain stepped forward with a welcome addition of £47.7 m, a vote of confidence more important than its actual value. ExoMars was funded by 14 member states, the largest contributor being Italy, followed by Britain.

At an industrial level, ExoMars was a big project. Although the most prominent contributions came from Britain, Germany, France and Italy, the project also involved Switzerland, Spain, Portugal, Belgium, the Netherlands, Austria, Poland, Denmark, Sweden, Norway, Greece, Finland and Ireland. Put another way, very few member states were uninvolved. Italy, for example, had responsibility for the TGO, Schiaparelli and the AMELIA and DREAMS instruments (2016 mission); as well as the carrier module, rover module, drill and MaMiSS instruments (2018 mission).

The ExoMars mission came during an intense period of Mars exploration. ExoMars was straddled by the American Mars Science Laboratory (MSL) rover *Curiosity* (2012); the Mars Atmosphere and Volatile EvolutioN (MAVEN) orbiter (2013); India's Mars Orbiter (2014); the lander InSIGHT (2018); and, in 2020, Curiosity's successor, *Perseverance*; plus Hope, of the United Arab Emirates, launched by Japan; and China's first Mars probe (orbiter, lander and rover), Tianwen.

First ExoMars: Trace Gas Orbiter (TGO)

Like MAVEN, the TGO was focussed on the atmosphere of Mars. Scientists of Mars, worldwide, had taken the view that the understandable follow-the-water approach to investigating Mars had meant that the planet's atmosphere had been overlooked. Indeed, understanding the dynamics of atmosphere was the key to learning the planet's history, which in turn was essential for the study of water and life. Here, the role of trace gases – those which constituted less than one percent of the atmosphere – was critical: methane, water vapour, nitrogen dioxide and acetylene. The main target was methane, a gas which had previously but sporadically been detected on Mars but whose role was frustratingly difficult to pin down. Its irregularity and source were uncertain, the possibilities being geological or metabolical. The American *Curiosity* rover had found methane at up to 0.7 ppb, with a spike of 6 ppb, while MEX once found a spike of 15 ppb. TGO aimed to provide a threefold improvement in its measurement, doing so across the seasons for a Martian year, to construct a map of methane by location and density set against a profile of temperature, aerosols and ozone. As well as determining trace gases, the TGO would map evaporation levels and its instruments would also be able to search for subsurface hydrogen down to 100 cm, to identify water ice below the surface.

ExoMars TGO team. ESA.

TGO was a large spacecraft of over four tonnes (4,332 kg), measuring 3.2 m by 2 m by 2 m, including 112 kg of scientific instruments and 600 kg for the lander. Its solar wings were 17.5 m across and able to generate 2 kW. For propulsion, TGO had a motor of 424 N. Communications were through a 2.2 m, 65 W high-gain x-band transmitter receiver and three low-gain antennae. TGO was designed to work for two years and act as a relay for the lander two years later. Its orbit was set for 400 km at 74°. TGO carried a suite of instruments, developed on the European side by the Belgian Institute of Aeronomy and on the Russian side by IKI (see Table 5.1).

Table 5.1:
TGO instruments

NOMAD	Nadir Occultation for Mars Discovery (two infrared, one ultraviolet spectrometer) (Belgium)
ACS	Atmospheric Chemistry Suite (spectrometer) (IKI) (three infrared instruments) (Russia)
ACS-TIR	Temperature and aerosols
ACS-NIR	Atmosphere profile – CO, water, 02, night glow
ACS-MIR	Methane and aerosols
CaSSIS	Colour and Stereo Surface Imaging System, resolution 5 m (Switzerland)
FREND	Fine Resolution Epithermal Neutron Detector (IKI) to locate hydrogen and ice to 1 m (Russia)
	With Liulin MO, radiation meter (Bulgaria)

NOMAD was intended to detect methane and other hydrocarbons and give sufficient information to determine whether their origin was biological or geological. CaSSIS was designed to characterize candidate sources of trace gases, investigate dynamic surface processes that may contribute gas (e.g. sublimation, erosion, vulcanism) and certify potential future landing sites for slopes, rocks and other potential hazards. Because it was a stereo camera, it would be possible to build digital elevation maps as well as explore geological features of interest. CaSSIS was developed by the University of Berne, Switzerland, with the assistance of scientists from Windisch, Warsaw and Padua, Italy. The Principal Investigator (PI) for CaSSIS was Nicolas Thomas of the University of Bern, Switzerland, while Ann Carine Vandaele of the Belgian Institute for Space Aeronomy was the PI for NOMAD.

The ACS and FREND were the Russian contribution. The PI for ACS was Oleg Korablev, a scientist who had been prominent in a number of recent space missions, such as Spektr R and was deputy director of planetary exploration at IKI. He held two degrees in the physics and mathematics of the solar system. ACS comprised three infrared spectrometers covering the range 0.7 to 17 μm (micrometres). They were subdivided into:

- Near Infra Red (NIR) channel 0.7 μm to 1.6 μm, to look for water vapour, aerosols, and molecular oxygen, by looking down on the planet, by solar occultation of the light passing through the atmosphere and through limb measurements.
- Mid Infra Red (MIR), 2.2 μm to 2.4 μm, looking for trace gas content, methane, aerosols, hydrogen and deuterium.
- TIRVIM, Thermal Infra Red Fourier Spectrometer, 1.7 μm to 17 μm range, to make atmosphere temperature, dust content and surface temperature profiles up to 60 km, searching for ozone and hydrogen peroxide. The full title came from from TIR, to which were added the initials of VIM, for Vasily Ivanovich Moroz, Russia's great Martian scientist.

According to Korablev, the purpose of ACS was to achieve high accuracy, high resolving power and a wide spectral range that would give unambiguous detection of trace gases of geophysical or biological interest. Its purpose was to achieve breakthrough science in determining the limit of trace gases in the atmosphere, as well as novel science in the form of an understanding of the atmospheric cycle set in their climatological background. The key question was how Mars had lost its earlier, thicker atmosphere. Dmitry Shaposhnikov had already developed a complex model of how seasonal weather changes, combined with dust storms, had pumped water from the lower atmosphere (up to 30 km altitude) to the upper (60 km−160 km) where it turned into ice clouds, some of which may contribute to the northern polar cap, but part of which may then be lost under the effects of ultraviolet rays. A specific purpose was to clear up the meaning of the earlier discovery of methane, so the ACS instruments were a hundred times more sensitive [6].

FREND was directed by Igor Mitrofanov, one of Russia's most experienced and well known space scientists. It would detect hydrogen down to a depth of 1 m, either associated with water or hydrated minerals and improve on the first water map compiled by FREND on Mars Odyssey in 2001. FREND brought in the Bulgarians, the great experts in space radiation, in particular the Liulin MO. This was an experiment developed by IKI and the Space and Technology Institute in Sofia, a successor to an instrument carried on Phobos in 1988. The Liulin dosemeter spectrometer was designed to measure the effect of Galactic Cosmic Rays (GCRs) from outside the solar system and Solar Energetic Particles (SEP) on the Martian environment, both during cruise and Mars orbit, especially with a view to their danger to astronauts travelling there. Cosmonauts were currently limited to a maximum exposure of 0.5Sv a year, or 1Sv a lifetime [7].

Igor Mitrofanov.

The TGO's structure, thermal control and propulsion systems were completed by the principal builder, OHB Systems in Bremen and handed over to Thales Alenia in Cannes, France on 4 February 2014 for assembly, integration and testing. Ground stations in Russia and Europe (Darmstadt, Germany) tested the communications systems in simulations.

First ExoMars: Schiaparelli lander

By 8 November 2013, ESA was sufficiently confident to name the lander 'Schiaparelli', after the 19th century Italian astronomer who first mapped what he imagined to be the channels of Mars, the *canali*, mistranslated into English as 'canals'. The chosen landing method was parachutes with rockets, the most traditional form of landing on Mars (exceptions were Pathfinder (1997) and the rovers *Spirit* and *Opportunity,* which had used landing bags, while *Curiosity* pioneered the sky crane). It would use a combination of a heat shield, parachute, pulsed engines (turned off at 1 m) and crushable structures to absorb the touchdown.

Europe had no prior experience of landing on Mars, nor the Moon for that matter, only the sad experience of the British Beagle 2. Rather than risk an all-up payload on its first Mars landing mission, Europe first decided to test out the technologies involved (e.g. heat shield, parachute, engines, altimeter) and, as a secondary objective, to install some limited scientific instruments. Schiaparelli had a diameter of 1.65 m, or 2.4 m with the heatshield, a mass of 600 kg and was encapsulated in an aeroshell. The total landing mass would be 310 kg. Key moments in the descent were: the separation three days out; atmospheric entry at 122.5 km at 21,000 km/hr, temperature on the heat shield of 1,500°C; 12 m parachute opening at 11 km, speed 1,650 km/hr; base heat shield jettison and radar activation at 7 km; upper heat shield and parachute jettison at 1,300 m, speed 270 km/hr, followed by ignition of the three 400N pulse mode engines; and shutdown at 2 m, speed 2 km/hr, with freefall to touchdown. Unlike the Chinese Chang e landings on the Moon, the descent was not guided for obstacle avoidance. Schiaparelli was designed to survive an impact at 11 km/hr, rocks up to 40 cm across and a slope of 12.5°. Sensors in AMELIA (Atmospheric Mars Entry and Landing Investigation and Analysis) included an accelerometer, a system to record atmospheric parameters (e.g. density) and a camera to film the descent.

ExoMars Schiaparelli first design

The EDL was built by Airbus at Saint-Medard-en-Jalles, near Bordeaux, France. The 80 kg front heat shield was made of 180 cork tiles between 0.5 mm and 14 mm thick, designed to withstand the re-entry of 21,000 km/hr, surface pressure of 8,000 tonnes and a temperature of 1,850°C. The 20 kg rear shield comprised 93 tiles made of 12 different elements settled on a carbon structure. There were five instruments (see Table 5.2).

Table 5.2:
Schiaparelli instruments

Instrument	Country of PI
DREAMS	Italy
COMARS+	Italy
AMELIA	Germany
INRRI	Italy
DECA	ESA

The descent instruments were AMELIA (to measure the descent profile, density and wind); COMARS+ (to measure heat); and DECA (a camera to film the descent every 1.5 seconds and build a topographic map of the landing zone). Once landed, the spacecraft was expected to operate for eight Martian days (sols). The surface package comprised DREAMS (Dust characterization, Risk assessment, Environmental Analyzer on the Martian Surface), with sensors to measure wind (Metwind), humidity (MetHumi), pressure (MetBaro), temperature (Mars Tem), atmospheric transparency (ODS) and electric fields (MicroARES). It was intended to pinpoint the lander from orbit using the reflection from the four laser reflectors on top. As construction went ahead, airborne tests were made of the parachute, radar and engines. Vibration, acoustic and vacuum tests followed.

Schiaparelli's landing site was on Meridiani Planum, not far from that of the NASA rover *Opportunity*, which had roved a record 45.1 km over 2004−18. *Opportunity* had landed at 1.9462°S, 354.4734°E, while Schiaparelli's target was at 2.0524°S, 353.7924°E.

ExoMars Schiaparelli landed. ESA.

First ExoMars: launch campaign

In 2015, the launch was set for January 2016. May 2015 saw another failure in the Proton rocket intended to launch the mission, with Proton continuing to suffer from inadequate quality control in the still financially-disrupted Russian space industry. The cause was identified reasonably quickly (vibration shutting down a third stage turbo pump) and Proton returned to flight that August, with a second successful mission a month later.

The ExoMars spacecraft was completed and integrated in November 2015, but at a late stage there were delays from the European end. Cracks appeared in the seals of the welds for the pressure sensors of the lander, which had to be removed and re-set, leading Europe to ask Russia to delay the launch to 16−30 March. The last tiles were bonded into place on the lander.

The two spacecraft were shipped to Baikonour in early 2016 and were mated in mid-February. For transit, they had to be split in two and then re-integrated and they arrived in two Antonov 124 transports from the final assembly point in Turin, followed by 65 mainly Italian and French engineers. Another Antonov brought in all the equipment and consoles needed for testing, including those required to certify that there was no biological contamination, especially of the lander. There was a biological clean room at both Turin and Baikonour to ensure 'super-sterility'. Schiaparelli was then re-attached to the TGO, connected mechanically, then electrically and fuelled, with the final tiles sealed and bonded. Each day, a bus would bring the engineers 3 km from two hotels to the integration site, with most working a long day shift and some through the night as well.

The dangerous procedure of fuelling Schiaparelli with hydrazine for its three 17.5-litre tanks for nine thrusters and 15.6-litre pressurizing helium at 170 bar was done by night, as it was less hazardous and fewer people would be there. Every morning there was a briefing for ESA staff and a meeting with the Russian personnel from Khrunichev, the makers of the Proton. Alignment, fit, leak, pressure, seal, fuel, signal, battery and electrical checks were made, with sniffers to detect any leaks. It was a tense, nervous stage, with plenty of scope for something to go wrong. While this process was underway, over in America engineers checking the InSIGHT (Interior exploration using Seismic Investigations, Geodesy and Heat Transport) mission, which had been scheduled to fly to Mars during the same launch window, found a leak in a key instrument. Despite their best efforts to repair or replace it in time, they were not satisfied with the fix, so the mission had to be postponed to the next launch window, an exasperating two years later. At this stage at Baikonour, the worst that happened was that one instrument (FREND) had to be replaced with a spare.

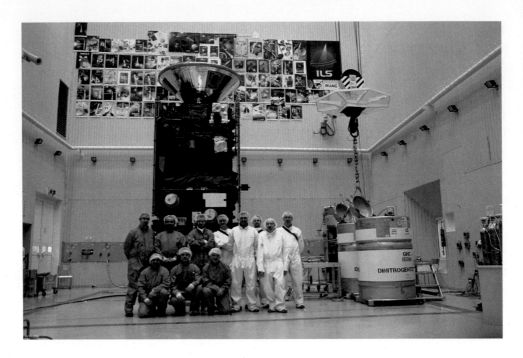

ExoMars fuelling TGO. ESA.

On 21 February, the TGO itself was fuelled with 2.5 tonnes of fuel brought in from Germany, benefitting from an exemption to the embargo of European fuels reaching Russia imposed after the Crimean crisis. On 28 February, ExoMars was installed horizontally on the Proton rocket, with encapsulation completed on 2 March and the fairings attached on 5 March, when the team posed beside the Proton in the hangar. The Proton was loaded horizontally in the high hangar, hung from an orange and yellow crane and placed on a flat blue trailer, with 'ESA', 'ExoMars' and a red illustration of the spacecraft painted on the top, as staff in white coats and light blue overalls swarmed over the rocket.

It was a busy time at Baikonour, which was also preparing to launch an Earth resources satellite (Resurs P3) that flew on 13 March and the piloted Soyuz TMA-20M which launched six days later. ExoMars rolled down to the pad early in the morning of Friday 11 March, silhouetted against the early rising sun. Two men in high-viz red jackets walked ahead of the train, comprising engine, service wagon and flatcar, as it was pulled past the brown sage brush and occasional thorn trees. When it reached the pad, the erector crane raised it to the vertical against a clear blue sky. It was fuelled and made ready to go.

ExoMars Proton team in hangar. ESA.

The launch trajectory for ExoMars was an unusual one, which Russia had never used before. Typically, the Proton flew a single revolution around the Earth in a parking orbit, at the end of which the block D or block DM (in Russian, 'blok D' or 'blok DM') upper stage gave a powerful, single blast to send the payload into deep space. This time, the Briz M was used (see chapter 4). Briz was a versatile upper stage capable of multiple re-starts, but was low thrust, so the only way to carry out the manoeuvre was through a series of lengthy firings, thereby increasing the danger of error. The Briz first had to fire to get the payload into Earth orbit, then make two firings to reach increasingly extended orbits out from Earth and finally a fourth time to escape Earth orbit.

By the time of the launch, at 9.31 am GMT on 14 March (lunchtime locally), the clear sky had clouded over. Back in Europe, ESA assembled a large crowd of media to watch the launch on television and put on a show of interviews with scientists and other experts associated with the mission. Just before 9.31 am the live feed transferred to Baikonour. Russia had never introduced a countdown clock, a leftover from the early days when they did not want the Americans to know when a launch was due, so watchers just had to wait for things to happen. No pre-launch commentary was provided either, but viewers could hear barely-audible loudspeaker announcements in Russian. Eventually, the Proton lit, spewing out smoke on either side of the pad as it quickly began to rise. A close-up camera also caught sight of the launch, including debris kicked up by the engines, which also sent a startled crow flying. The Proton launch is distinct, because its use of storable fuels means that there are no wisps of cold liquid oxygen flowing off the side of the rocket, as is the case with the Soyuz. Instead, the nitrate fuels gives them an orange-brown colour. In no time, though, the Proton was lost in the clouds.

ExoMars 2016 liftoff. Roskosmos.

The first stage fell away after 2 minutes at 43 km (normally visible but not on this occasion), the second stage after 5 minutes 30 seconds at 129 km and then the third stage after 9 minutes 41 seconds at 153 km. With the Proton now above the atmosphere, the fairing fell off at 5 minutes 46 seconds, leaving ExoMars exposed. The Briz M began its first burn over Siberia at 11 minutes 16 seconds – for 4 minutes 29 seconds – to achieve an initial 175 km parking orbit around the Earth. Over Russia an hour later, the Briz fired for a second time, for 18 minutes 3 seconds, sending the ExoMars-Briz combination into a high orbit over the Pacific. Returning to the far side of the Earth over Portugal, Briz then fired again, this time for 14 minutes 31 seconds, to send them to an ever-higher and more elliptical orbit as far out as 21,000 km. Then, 10 hours after launch, the Briz fired for a fourth and final time for 12 minutes 29 seconds, with the payload separating 14 minutes later some 5,000 km above the Earth. Briz then made a deflection manoeuvre so that it would not later collide with the payload.

The danger was not over yet, as ExoMars still had to make contact with ground control, deploy its solar wings and confirm that all was in order. This moment had to wait until a full 12 hours after launch, the moment of confirmation being due when the spacecraft was over the tracking station at Malindi, Kenya (a station which also plays a key role in tracking Chinese spacecraft). Back at ESA, the media had returned for the critical moment, all eyes on a circular cathode ray display that was showing a flat signal, to see if it would detect the arriving signal. After some anxious moments, the flat line suddenly peaked, confirming the inbound signal from ExoMars. Solar panel deployment was confirmed less than two minutes later. Remarkably, visual images were made by astronomers in Brazil of ExoMars heading away from Earth, with the Briz stage close by.

Schiaparelli crashes

ExoMars – TGO and Schiaparelli together – cruised toward Mars over the summer of 2016. Arrival was set for 19 October. In advance of arriving, Schiaparelli separated from its mother ship for a three-day independent coast before beginning its descent. As it began its six-minute descent, the lander relayed data to the mother ship for onward relay to Earth.

ExoMars Schiaparelli final descent. ESA.

Landing on Mars is the moment most dreaded by mission controllers. Entry, descent and landing, the very phases under test, generally take six to ten minutes and there is nothing that ground controllers can do because it takes 15 minutes for a signal to travel there. All they can do is wait and hope for the best; that being the long-waited signal that the probe has reached, or is about to reach the surface. During the descent, good telemetry was picked up by TGO, MEX and the Indian radio telescope in Pune. Schiaparelli was barrelling in at 21,000 km/hr from an altitude of 122 km, warming up its heatshield to 1,850°C. Its cameras were set to turn outward at 3 km altitude to film the final descent, while 600 MB of signals were received through the descent, identifying all the key milestones right down to the final moments.

And then silence. As the time passed, it quickly became apparent that the probe had not made it to the surface. The following day, the wreckage was spotted from orbit by the high-resolution cameras of America's Mars Reconnaissance Orbiter. This identified the crash site, a darkened area with a 50 cm deep crater, 2.4 m across − about what one might expect from a 300 kg impact at 100 m/sec − and, at 1.5 km distant, the white spot of the parachute and the back shell. It was little consolation that Schiaparelli had landed right in the middle of its intended site.

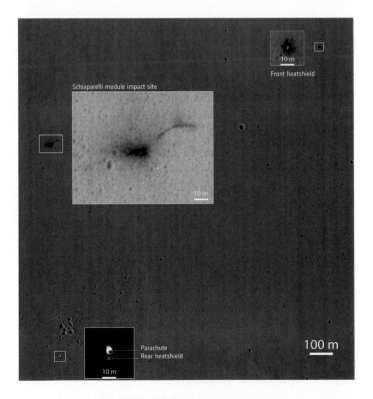

ExoMars Schiaparelli debris. NASA.

An investigation was ordered, which comprised a ten-person panel chaired by ESA's Inspector General. The 28-page outcome was later published. It first set about establishing the timeline (see Table 5.3) and then a fault-tree analysis. Its findings were that, at three minutes into the six-minute descent, the 12 m-diameter parachute had deployed and the lander underneath rotated rapidly as the strings of the parachute straightened; so far, nothing untoward. At this stage, Schiaparelli was 11 km high and its speed down to 1,650 km/hr. Such rotation was not unusual, because parachute deployment can be quite a violent event, as cosmonauts returning to the Earth will tell you given that the Soyuz swings wildly as the parachute first deploys and straightens out. Schiaparelli's rotation rate, as measured by the lander's Inertial Measurement Unit (IMU) exceeded the pre-set limits, thus saturating the computer, but its software design permitted it to stay saturated. This saturation in turn persuaded the computer that the lander was below ground level, prompting the release of the parachute and back shell. Although the landing engines turned on for three seconds, they were then shut down because of the incorrect belief that it was already on the surface. The lander crashed about 35 seconds before its intended set down.

Table 5.3:
Schiaparelli timeline

Time	Event
13.29.48	Exit hibernation
14.42.22	Enter Mars atmosphere
14.45.23	Parachute deployment
14.46.03	Jettison shield
14.46.19	Activate Radar Doppler Altimeter
14.46.49	Jettison back shield
14.46.51	Activate Reaction Control System (switch off 3 seconds later)
14.47.28	Impact at 150 m/sec.
14.48.05	Intended touchdown

Ironically, the initial swinging motion damped out quite quickly and fell to only 3°. There had been nothing wrong with the deployment and the swinging motion was well within normal limits, but the persistent saturation of the computer led to incorrect estimates of the attitude of the lander (165° when it was actually normal) and that the spacecraft had actually landed, thereby turning off the landing engines. The premature cut-off meant that the lander crashed from 3.7 km at a speed of 540 km/hr, as evidenced by the debris pattern.

In effect, the design permitted the rotation to induce permanent saturation of the computer and did not anticipate that such saturation could lead to the computer fatally believing that the spacecraft had already landed. Not anticipating the consequences of saturation was a basic design error and there had been insufficient worst-case *What if?* planning. The board report warned that parachute deployment is one of the most difficult aspects of this type of mission ('complex, dynamic, affected by several uncertainties and difficult to model and predict'). No 'sanity checks' had been built into the computer (it determined the lander to be *below* ground level, which was impossible; and that it was upside-down at an angle of 165° while at the same time receiving radar echoes from the ground. Altitude was 3.7 km one second, then below zero the next, another impossibility). There was no recovery strategy for the computer. Even with the saturation problem, it should still have been possible to recover the computer. These were expensive errors, costing €230 m.

The inspection board issued warnings for the second mission, pointing out those aspects that used the same equipment. The board made 15 recommendations, covering the interpretation of parachute-induced oscillations, saturation, peer review (especially of the computers) and an integrated engineering team around Lavochkin. It also warned that there could still be problems lurking in the final descent phase, which remained to be tested. Although this side of the failure was not publicized, the flight software was European and the IMU was American. It was realized later that there was an incompatibility between them, so next time both would be European **[8]**.

This was not the first time that an error like this had happened. The Americans had lost their Mars polar lander when the jolt of deployment of the landing legs was mistaken by the computer for the jolt of the actual landing, similarly leading to it turning the engines off so that the lander then crashed from a great height. Beagle's loss was partly due to the lack of an immediately-deployed omnidirectional antenna, a problem which could also have been avoided had the designers studied the Russian experience at Venus. There, Venera 7's faint single-directional signal had almost been missed, so an omnidirectional antenna was installed thereafter. This was Europe's second failure to land on Mars and get a signal from its surface. On the positive side, it was better and less expensive to learn this lesson on the EDL than the planned rover. At least they had sufficient and appropriate telemetry to tell them what had happened.

The failure of Schiaparelli cast a shadow over the biennial ESA ministerial council six weeks later. ExoMars was once again running out of money, but to the relief of ESA's directors, the ministers agreed to a further injection of €430 m to keep ExoMars going. The second launch had originally been planned for 2018, but on 2 May 2016 it was put back to 2020 for 'technical and funding issues'. Stretching a project is always costly and with €1.5bn already spent on ExoMars, the total now rose to €1.93bn.

TGO science mission

The good news for ExoMars was that the TGO had successfully entered orbit around Mars on 19 October 2016. Its original orbit was 300 km by 98,000 km, taking four days to orbit the planet on a near-equatorial path. This orbit was a function of the trajectory to Mars and the release of EDL, but the intention was to manoeuvre TGO into a near-circular, polar orbit of 400 km in 18 months, from which it could carry out its science mission. Most of the orbit alteration would be undertaken by aerobraking, descending to the point where the thin Martian atmosphere could slow, lower and reshape its orbit. Aerobraking had been used before, both by American spacecraft orbiting Mars (Mars Global Surveyor, Mars Odyssey and Mars Reconnaissance Orbiter) and a European one over Venus (Venus Express). Using nature to reshape the orbit would save 600 kg of fuel.

Accordingly, TGO made engine burns on 19, 23 and 27 January 2017 to alter its 7° equatorial orbit to achieve a near-polar orbit at 74°, from where it could survey almost all the planet. A final trim was made on 5 February to lower the perigee from 250 km to 210 km to begin the aerobraking, the purpose of which was to reduce the high point of the orbit from 33,000 km to 400 km and the low point from 200 km to 113 km. Orbit-lowering burns took place in March and April 2017, commanded by the European Space Operations Centre in Darmstadt, Germany.

By the end of February 2018, TGO had reached its intended near-polar science observation orbit of 400 km, circling the planet every two hours. March was spent running calibration tests of the instruments.

ExoMars TGO aerobraking. ESA.

The four science instruments had been tested for the first time over 20–28 November 2016. The spacecraft's infrared camera was able to track clouds, especially the thicker ones from 45 km to 60 km, more effectively than previous spacecraft. Photography then became impossible because a thick dust storm engulfed the planet, described as 'thick soup', which led mission controllers to switch off the Colour and Stereo Surface Imaging System (CaSSIS). One of the casualties of the storm was NASA's rover *Opportunity*, which had so much dust dumped on it for so long that it was unable to recover its battery and was formally declared lost. CaSSIS was switched on when the storm began to abate on 20 August and was clear enough by 2 September to show dark streaks, possibly dust devils left on the surface by the storm, in Ariadne Collis in the southern hemisphere.

The main science observations had to wait until the final, lower orbit was achieved. The first results came in from FREND in September 2018, when Jordanka Semkov of the Bulgarian Academy of Sciences told the European Planetary Science Congress in Berlin that had any astronauts been flying on TGO, they would have accumulated 60 percent of a lifetime's radiation exposure in a year.

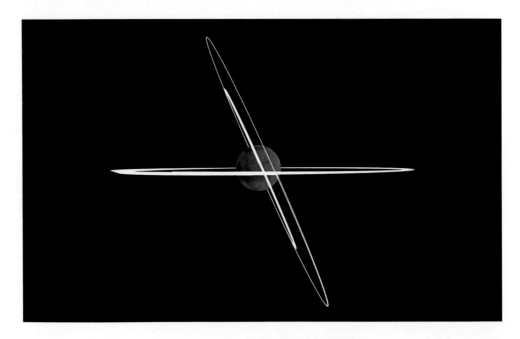

ExoMars TGO changing orbit. ESA.

On 10 April 2019, ESA published its initial results from TGO. First, using FREND, it was possible to compile a map of subsurface water distribution, showing water-rich permafrost at the poles and localized wet and dry regions in between. Igor Mitrofanov published more detailed maps of individual regions. Second, using the NOMAD and ACS instruments, Oleg Korablev and Ann Carine Vandaele presented a paper on the dust storm. They found that before the storm, there was no detectable dust or water in the atmosphere below 40 km. Within a few days of the storm starting, significant levels of dust and water had been funnelled up to 40−80 km altitude, more so in the northern hemisphere [9].

Third, in 2020, Oleg Korablev of IKI and Frank Montmessin (University of Paris Saclay) and their colleagues tackled some of the fundamental problems of the Martian atmosphere that TGO had been sent to answer. Using the ACS, they found that during the warmest, stormiest part of the Martian year, large amounts of water ascended into the upper atmosphere, meaning that the mechanism for Mars to lose its water was much more effective than previously thought. They estimated that Mars had now lost 90 percent of its water through this mechanism, whereby water decomposes in the lower atmosphere in a complex chain reaction, rises, gains thermal energy and escapes the planet's gravity. In the meantime, NOMAD had detected glowing green oxygen in Mars' atmosphere, being most evident 80 km up. This phenomenon was well known for colouring polar aurorae on Earth and the ISS had captured many entrancing photographs of the green glow around Earth's limb. Something similar had been predicted for Mars but never before found. According to the researchers, Jean-Claude Gérard of the University of Liège, Belgium and Ann Carine Vandaele of the Institut Royal d'Aéronomie Spatiale de Belgique, it was formed as carbon dioxide, CO_2 that subsequently broke down into carbon and oxygen.

The 11th Moscow Solar System Symposium, held on October 2020, had the benefit of the collection and analysis of data from TGO since the formal start of the science mission in May 2018. The FREND instrument, searching for water resources through hydrogen content, indicated small, water-rich oases close to the equator in the upper layers of the regolith − wet subsurface areas where there might be remnants of past Martian life. The particle flux around Mars was measured, together with how it was affected by going in and out of the planet's shadow. There was a new picture of the role of dust and water ice in the thermal structure of the atmosphere up to 70 km. The humid mesosphere was measured at 110−120 km. Halogen gas was detected for the first time, widely distributed in the 1−4 ppb range, 20 times more than expected, which increased during the 2018 dust storm and then declined, but suggested that Martian photochemistry in general should be revised, in particular the origin of oxychlorines. During the storm, water rose to 70−100 km and saturated. A striking feature of the papers presented was their Russian-European authorship, with France being the European country most in evidence [10].

Anticipating the arrival of the second part of ExoMars, in June 2019, TGO made an inclination change manoeuvre of 30.9 m/sec (twice) and 1.5 m/sec (once) to be in the right orbit for the 2021 landing. TGO's orbit was a gradually wandering one, affected by the planet's uneven gravity field, hence the need for such advance planning.

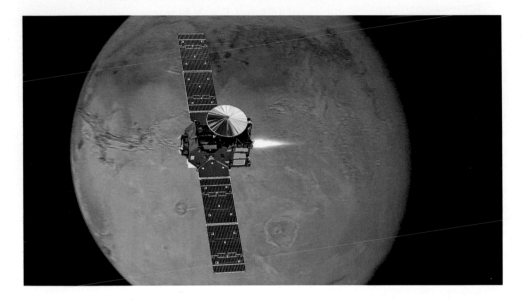

ExoMars TGO engine firing. ESA.

Interval: sanctions

Between the first phase of the ExoMars project (2016) and the second (2018), the mission came under strain from a radical deterioration in relations between Russia and Europe, with obstacles to cooperation reminiscent of the 1949–91 CoCom period. As noted, sanctions had directly affected the fuelling of the TGO with 2,300 kg hydrazine, but they also provided an increasingly problematical environment that potentially threatened cooperation. Furthermore, the prime target of the sanctions was none other than the head of the Russian space agency, Dmitry Rogozin, making them a direct threat to cooperation, one which might even jeopardize the ExoMars project.

Ultimately, the conflict could be traced to 19 February 1954, when a poorly attended – technically inquorate – Praesidium of the Supreme Soviet transferred Crimea to Ukraine to mark the 300th anniversary of the Treaty of Pereyaslav uniting Cossack Ukraine with Muscovy. Crimea had been part of the Ottoman empire from 1475 until the Turks were ousted by a Russian army during the time of Catherine the Great and her military advisor Prince Grigori Potemkin, who formally proclaimed it part of the Russian empire in 1784. France protested but Russia offered to overlook France's recent annexation of Corsica if the matter were not pursued any further [11]. The transfer was done at the suggestion of the General Secretary at the time, Nikita Khrushchev, as a gesture toward that part of the country where he had previously been party leader. In the Soviet period, given the unitary nature of government-and-party decision-making, such a gift had little more than symbolic significance – until 22 February 2014, when President Viktor Yanukovich, the elected leader of a now-independent Ukraine, was ousted in a street coup. Russia responded by retaking the Crimea, holding a referendum which approved the decision and then formally annexing the territory into the Russian Federation.

Relationships between the European Union and Russia had been on a positive path ever since the signing of a Partnership and Cooperation Agreement (PCA) in 1994. In the area of spaceflight, formal cooperation with ESA dated to a first Framework Agreement on Cooperation (1991); a permanent ESA mission to Moscow (1995); and a three-sided Memorandum of Understanding between the European Commission, ESA and Roskosmos – *New opportunities for a European-Russian space partnership* – (2001) which encompassed navigation, Earth observations and launchers. A European-Russian summit in St Petersburg in May 2003 saw agreement on cooperation in four 'common spaces' (in French, *espaces communs*): freedom, security and justice; external security; education and culture; and economic, with spaceflight being included in the last. In 2006, the European Union and Roskosmos opened a 'space dialogue', signed by Anatoli Perminov (Russian space agency), Jean-Jacques Dordain (ESA) and Commissioner Günter Verheugen (EU) respectively. This set down seven areas of interest: launchers; exploration; ISS; technology; and three grouped under applications (navigation, communications, Earth observations), underpinned by potential funding from the Seventh European Framework Programme for Research and Development. Seven working groups were established. There was even fresh discussion about Russia joining ESA [12].

Although the subsequent unravelling of the relationship between the European Union and Russia is attributed to the events in Ukraine in 2013–14, it is also explicable in terms of the Eastern Partnership (EaP) launched by the European Union in May 2009, which Russia saw as a direct attempt to draw the rest of eastern Europe and even the Caucasus into its sphere of interest, with the danger that this would be followed by NATO membership and NATO armies on its doorstep [13]. In the view of the Kremlin, this could mean American tanks in Kharkov (Kharkiv in Ukrainian) only 650 km from Moscow, which would be no more welcome than the German panzers in 1942. Whatever the politics, cooperation in the space field, so promising back in 2006, came into play.

The Crimean annexation was declared illegal by the United States on 6 March 2014 and the European Union on 17 March (Council Decision 2014/145), with a travel ban and an asset freeze imposed on persons deemed responsible. The original list was compiled with the US decision and subsequently expanded to 149 persons and 38 organizations. The individuals were nationalist politicians, Donbass separatists and oligarchs who led Ukrainian companies turned into Russian companies, while the companies concerned were state-owned banks, oil and energy firms and arms manufacturers. High on the list was Dmitry Rogozin, then minister responsible for the Russian space industry, who was identified because he had 'publicly called for the annexation of Crimea'. Unlike most of the others, Rogozin seems to have had no direct connection to Crimea or the subsequent conflict in eastern Ukraine: his main offence appears to have been to have supported the annexation and to have been considered close to President Vladimir Putin. Indeed, one commentator referred to the selection as having been intended 'to hit the president's cronies', another that Rogozin was 'in Putin's inner circle' **[14]**. Sanctions were also applied against Alexander Karamian, with the reason given that he was a protégé of Rogozin and, in effect, guilty by association.

Dmitry Rogozin (right) meeting CNES in 2016. Roskosmos.

This was not universally agreed, since another analysis did *not* list Rogozin as being in the inner circle [15]. Although there was a high degree of overlap between the American and European approach to sanctions, the Americans went after companies and the Europeans after people. The theory behind the punishment of individuals was that they could no longer travel, use their financial assets abroad, nor send their children to study there. Sanctions would tarnish their reputation, impede their access to social networks and curtail their involvement in political or economic cooperation [16].

Further regulations (833/2014, 960/2014, 1290/2014) were issued from July, their combined effect being to restrict cooperation and exchanges in the areas of arms, dual-use technologies, finance, oil and sensitive sectors. For its part, Russia imposed counter-sanctions, primarily affecting European agricultural products and individual members of the European Parliament. Western sanctions were intensified in 2018 following the allegations of Russian interference in the American presidential elections and the Skripal spy case in Britain.

These events heralded a return to something comparable to the cold war of 1949–91. There had already been a significant change in the balance of power across Europe since this period. The Warsaw Pact had disbanded in 1991 and NATO had always pledged to do the same once its rival folded its tent. Instead, NATO doubled its size, taking in the former socialist block countries with their history of earlier subordination to the Soviet Union and who now sought the reassurance of NATO's protection. From the Russian point of view, it was now encircled by a NATO that was twice as large and much closer than ever. This mattered little when there were no points of contention between Russia and western Europe, but the coup in Ukraine changed all that. Ukraine joining NATO – a goal pressed for by western-inclined Ukrainian leaders – would bring NATO even closer to Moscow. By 2018, there was already a significant American military presence in Ukraine, with NATO exercises taking place there.

Most analyses of the sanctions regime focussed on its economic, industrial and political impacts, but little on the scientific. A general view is that they have been costly to both European and American companies trading with Russia and *vice versa*, such trade falling 25 percent over 2014–2017, with the most impact on Russian consumers but with significant costs to the imposing countries too, estimated at up to €600bn. In terms of intended outcomes, they did not cause a reversal in Russian policy, nor weaken Vladimir Putin, whose popularity soared.

There is no similar analysis readily available of their effects on the space programmes of Russia, Europe and the United States. The principal effect on the space programme may be indirect, insofar as reduced economic activity led to a fall in government revenues which in turn affected funding of the space programme. Although spaceflight was not identified as sanctions target, the inclusion of aviation, avionics and computer technology brought the sanctions close to companies supplying the space industry (e.g. Almaz Antei, Mashinostroeniya). At the same time, the Pentagon fought for and obtained an exclusion for three companies – Energomash, Motostroitel and Kuznetsov – because they made the RD-180 and NK-33 engines used by the American Atlas and Antares rockets [17]. Nevertheless, the selection of the head of the space industry for individual sanctions, not to mention his prominent place on the list, was a remarkable development. One could only imagine the controversy were the head of NASA to head a list of Russian targets.

The British experience was illustrative of the way sanctions worked. Chapter 2 showed how British scientists working on Kvant in the 1980s had to be careful not to supply electronics that might breach sanctions and how they had to keep their computer under 24/7 guard. In the 2020s, British companies were forced out of Spektr UV and M because their American suppliers adjudged them to be breaching rules that dated to the 1940s. Such an environment could only have an inhibiting effect on any possible cooperation that never even got under way.

So who is Dmitry Rogozin? His background was in politics in the nationalist Rodina ('motherland') party, for which he was elected to the Duma from 1993 to 2007. Rodina had been banned for racism in 2005, but he shifted to the political ground that supported Putin and was rewarded by being made Russia's ambassador to NATO in 2008. In 2011, Rogozin was appointed deputy prime minister, which included defence and space, taking considerable interest in the latter and trying to find ways of bringing its different industries together more efficiently. He acquired a reputation for outspoken remarks and he reacted badly during what became known as the Moldovan incident. According to one version, the United States persuaded Romanian air traffic control to refuse his aircraft landing permission in the nearby state of Moldova; another version is that he did land in Chisnau, having given false identity information when he transited through Bulgarian and Romanian airspace. When found out, he had to take a regular airline flight from Chisnau back to Moscow and was quoted as saying that the next time he would come in a Tupolev 160 (a bomber). He also said that tanks did not need visas either. The following year, Rogozin visited Svalbard on his way to a drifting Russian ice station. When Norway objected, Russia pointed to a 1920 treaty guaranteeing Russia unrestricted access there, so Norway amended the treaty to expel anyone on the sanctions list [18]. In 2018, he was appointed head of Roskosmos, the space agency.

Dmitry Rogozin with Vladimir Putin. Roskosmos.

Considering Rogozin's position as deputy prime minister responsible for the space industry – and subsequently responsible for Roskosmos – and given the importance of European-Russian cooperation in space, the decision to sanction him must not have been taken lightly. How were the potential repercussions and consequences weighed? What was the assessment of potential damage to such cooperation? Was the potential loss of such cooperation considered acceptable? What was discussed in the documents compiled to brief foreign ministers, prime ministers and their officials? The home of such a decision would be the sanctions division of the European Union Council's External Action Service.

The Council conclusion stated that persons would be sanctioned, but did not give a justification or explanation. Rogozin was not on the original 17 March list but was one of 12 new names added in subsequent days. Because of their sensitivity, these 12 names were not circulated by e-mail for security reasons during this period, but following their approval by the Council on 21 March the identities of those concerned were declassified shortly afterwards **[19]**. Reasons were given in the annex and in the case of Rogozin, this was that he had 'publicly called for the annexation of Crimea', though no reference source was given and it was potentially quite a long list.

No other reasons were given, known or circulated during the period when Rogozin was a government minister. Given that he became head of Roskosmos in May 2018 and allowing for the fact that the sanctions were put through a process of regular renewal, it is important to ascertain if the reasoning changed to reflect his altered status. Here, two papers (14081/18, 19 November 2018; and 11963/2019, 29 October 2019) are available which concern the continuation of these sanctions and they confirm that the European Council was indeed aware of Rogozin's new status as head of Roskosmos. The first, entitled *Rogozin – underlying/supporting evidence – territorial integrity*, comprised Russian text, taken from the Roskosmos website, of the directors of Roskosmos and a machine translation of their names from the Cyrillic text, correctly confirming his identity. Roskosmos is an open website, with its information and photographs widely used by space enthusiasts. The second document, with the same title, is an extract from *RBC news*, with the following machine-translated text from 28 August 2018:

> *Rogozin gave Putin photos of the Crimean Bridge from space. The head of the state-owned corporation Roskosmos, Dmitry Rogozin, presented President Vladimir Putin an album with photographs of the Crimean Bridge taken from satellite. This is stated on a message on the Kremlin website. 'Since 2014, we have been shooting the Crimean Bridge, how it was built', Rogozin said. Putin thanked the head of Roskosmos for the donated photo album. Rogozin noted that the surveys were carried out monthly by ten satellites. Since that time, they have shot 240 routes.*

This is based on the following original text:

Общество, 08 авг 2018, 16:44 10 545 Поделиться. Рогозин подарил Путину фотографии Крымского моста из космоса. Глава госкорпорации «Роскосмос» Дмитрий Рогозин подарил президенту Владимиру Путину альбом с фотографиями Крымского моста, сделанными со спутника. Об этом говорится в сообщении на сайте Кремля. «Мы с апреля 2014 года проводили съемку Крымского моста, как он строился», — рассказал Рогозин. Путин поблагодарил главу «Роскосмоса» за подаренный фотоальбом. Рогозин отметил, что съемки проводились ежемесячно десятью спутниками. С этого времени они отсняли 240 маршрутов.
Подробнее на РБК: https://www.rbc.ru/rbcfreenews/5b6aef2f9a794788b2 d7776b

RBC is a Russian business media conglomerate for internet, television and print. The news item in question could still be accessed some time afterwards and Google translate gives a word-for-word identical translation.

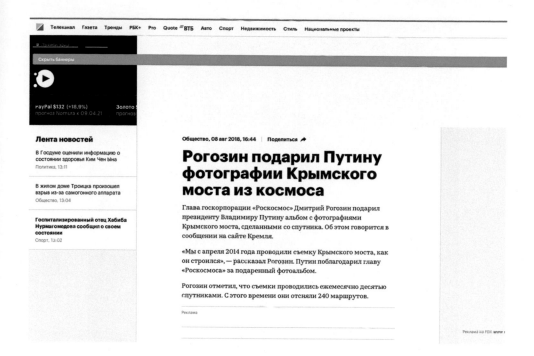

Original RBC text.

So how did the Crimean Bridge become such a determining factor in the sanctioning of Dmitry Rogozin? The Crimean Bridge, also called the Kerch Strait Bridge, was built between the Caucasian, eastern side of the Sea of Azov (Krasnodar) and the western Crimean side (Kerch), starting in 2016 and concluding in 2019. At 18 km, it is the longest bridge in Europe, even longer than the Øresund bridge linking Denmark and Sweden. The bridge was condemned by the European Union − its first known condemnation of a bridge − for connecting Russia to Crimea. The idea of such a bridge had been an on-off project since 1903 and was an agreed project between Russia and Ukraine as recently as 2010, so it was hardly a new idea devised to annoy the European Union. Russia made the point that the direct train line from St Petersburg to Crimea, which used to bring the Tsar and his families on holidays there, had been broken only twice before 2014 − during World War I and II − and was now restored.

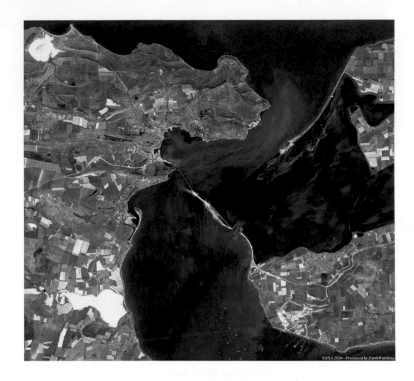

Kerch bridge. ESA.

As for the photograph album, such imaging was hardly a novel idea. Taking photographs of such a bridge would have been unchallenging for either military or civilian satellites. Roskosmos published a photograph of the Kerch Strait Bridge on 2 September 2017, taken by its civilian imaging satellite Resurs R. Indeed, using its Sentinel 2 satellite, ESA had published photographs of the Kerch Strait Bridge in its popular *Images of the week* series, the photographs having been taken on 13 July 2017 and 5 December 2019. Overall, it is striking that a photograph album of the bridge, pictures of which were widely available through open media, should be in effect the *casus belli* for the sanctioning of Dmitry Rogozin. It is also surprising that the justificatory text used an awkward machine translation, rather than getting a professional translation from the EU's own translation service (the European institutions have 4,300 professional translators). The use of the term 'shoot' twice in the machine translation may have conveyed a more belligerent impression of the bridge than might be strictly intended and it is likely that a professional Russian-English translator would have used the term 'photograph'. On the first occasion, съемку would normally be translated as 'surveyed' or 'filmed', while in the second, отсняли, the more normal translation would also be 'filmed'. Both are a long way from 'shooting'. It seems curious that such an important diplomatic decision should be based on a machine translation of doubtful fidelity.

Dmitry Rogozin proved to be the most dynamic head of Russian spaceflight for many years. If he had political views at this stage, he kept them to himself. As for his job, Rogozin appeared to be on a path of reversing the decline of the Russian space programme. The age profile had become quite elderly, with the French having referred to the *génération perdue* (lost generation) of middle-aged people (35–55) who had left and the difficulty of attracting the young [20]. The government had taken the extreme measure of offering an exemption from military service to work in the space industry, but space no longer attracted the imagination of young people when they could find more enticing and lucrative careers in finance, commerce and other parts of private industry. The Russians had to rebuild the technical capacity that they had lost since 1989 when the socialist block broke away. The French saw the disintegration of research and education under President Boris Yeltsin, as well as the demoralization of those who worked in the laboratories at first hand. At Lomonosov University, the toilets no longer worked, the tiles were falling off the walls and the copper cables were stolen for re-sale, all in a place with the most advanced robots in the world. In the view of one European visitor, conditions in the laboratories were 'deplorable, their labs dirty and poorly equipped'. Russia had fallen behind in electronics. One French engineer related an experience when he was testing the St Petersburg-made cameras for the rendezvous system for the ATV. Although it worked perfectly well, the test bench was something out of the 1950s, the electric testers did not work and they had to get replacements from another laboratory. There was growing obsolescence. Russia's technical failures were really management failures and one of their reasons for participation in ExoMars was to observe European management methods, especially between the space agency and their industrial companies.

Under Rogozin, this began to change. He brought a high level of energy to his new post and was seen daily visiting cosmodromes, welcoming foreign delegations, checking on production plants, meeting specialists, exhorting workers and speaking with students, young people and children. Programmes were introduced to bring young people into the industry. Rockets were sent all the way back to the plant in Moscow if they did not check out correctly on the pad. A new pad was built for the Angara rocket in Vostochny. Corrupt contractors were charged. New projects were started, like the Soyuz 5 rocket. This was not just about *grands projets*, because there was attention to detail, such as new equipment in the laboratories. There was a sense of purpose. One might have thought that Europe would regard him as an ideal person with whom to do business, rather than blacklist him.

Being on the sanctions list meant that Rogozin was forbidden entry to the European Union. This was a problem, given the need for cooperation between Roskosmos and ESA, not least around the ExoMars project. Sanctioned or not, he came to speak at the Unispace conference in Vienna in June 2018 but seems to have been there and back before he could be interdicted. Although invited to

Germany to speak at the heads of agency meeting at the International Astronautical Congress in Bremen three months later, he was then denied a visa under the European sanctions. When Rogozin was invited to visit the United States the following year, congressional leaders managed to get the visit cancelled. Personal relationships with Rogozin between NASA Administrator Jim Bridenstine and European leaders remained good, as the friendly letter to him from ESA Director General Jan Wörner on the death of Alexei Leonov in October 2019 illustrated.

Dmitry Rogozin in Vienna. Jacqueline Myrrhe, Go Taikonauts!

Sanctions and ExoMars

The sanctions most affected ExoMars with regard to hydrazine. Hydrazine is used for spacecraft engines, manoeuvring and station keeping. It is a derivative of H_4N_2 and has high thrust, but it is toxic and requires careful handling (mishandling of hydrogen peroxide fuels in a torpedo, for example, led to the tragic loss of the submarine *Kursk* in 2000). Russia produced its own hydrazine until 1993 at a factory in Novosibirsk, but when that went out of production, Russia bought hydrazine from western companies, principally the German manufacturer Kayser Threde **[21]**.

European supplies of hydrazine to Russia were banned under the 2014 sanctions. Without hydrazine, TGO would be without fuel and could not fly, so the European Union had to facilitate ESA by lifting its ban in October 2015 for ExoMars. The sanctions prompted Russia to resume the manufacture of hydrazine at its own factory, first in Nizhny Novgorod which was able to make 15 tonnes a year and then a second at the Nefketim Salavat plant in Bashkortstan, the latter also making sanctioned naphthalene fuels. Thinking ahead, the European Council formally adopted decision CFSP 2017/2214 on 30 November 2017, permitting the export of hydrazine to be exempt from the Common Military List for the fuelling of the descent module of the second mission. As the rover was made in Britain for export to Russia en route to Mars, Britain amended the Export Control Order 2014 on 9 September 2019 to include (this is the legal language), 'under §3, activities which required prior authorization under §4(2b) of the Russian sanctions regulation in §5 of the 2014 order to reflect amendments to allow certain activities within the ExoMars 2020 mission framework'.

Western satellites on Proton (see wall). Roskosmos.

Inconsistency has long been a feature of the sanctions regime, so ExoMars is not unusual. The German space company OHB politely described sanctions as having an 'uncertain effect' [22]. The persistence and consistency of the sanctions on Rogozin were in marked contrast to the inconsistency of their application in other areas, as some examples illustrate.

Although sanctions applied mainly to the defence, energy and financial sectors, no less than 125 public and oligarch Russian financial houses had located their assets offshore in the tax haven of Ireland's International Financial Services Centre (IFSC). Although the notoriously ineffective Central Bank of Ireland appeared to have made no enquiries as the to the purposes or use of this money, the more interesting point is that the European Central Bank – an arm of the European Union which applied the sanctions – also seemed to be unconcerned [23]. Member states seemed enthusiastic enough about enforcing sanctions, as long as it did not affect their own state. Also in Ireland, a huge aluminium plant on the Shannon estuary, Aughinish Alumina, employing 450 workers, had come under the control of Rusal, one of the companies of oligarch Oleg Deripaska. When it was affected by sanctions, the Irish government rushed to Europe and the United States to get an exemption [24]. The capricious nature of sanctions was evident at consumer level, where supplies of Russian *Baltika* beer became erratic. Eventually, the non-alcoholic version was allowed in, but not the real thing. Germany did not let sanctions get in the way of its Nordstream gas supply from Siberia [25]. Russia was suspended from, but then re-admitted to the Council of Europe [26].

By contrast, American sanctions became harsher, but they had a collateral effect in Europe. In May 2019, the Department of Defence issued a rule change to the National Defence Authorization Act to add Russia to the axis of evil countries (China, DPRK, Iran, Sudan, Syria) not permitted to fly American spacecraft or use commercial satellite services from 2022. Presumably, the definition 'American' would include any European satellite using American parts. This decision drew fire from Dmitry Rogozin, who rarely commented but described this as an 'extremely unpleasant' move. In the area of communications satellites, Airbus became involved in a joint project with Energiya in 2016. The Airbus director in Moscow, Vladimir Terekhov, argued that such a transaction was civilian and commercial and thereby outside the sanctions regime. It seems that the United States took a different view and, in spring 2020, Airbus was fined an extraordinary €3.6bn for breaches of the American ITAR rules and other offences, including sales to the Russian Satellite Communications Company [27]. Many Russian speakers were denied visas to attend the International Astronautical Conference in Toronto, Canada in October 2014, something that had never happened during the previous cold war.

Western machinery for the space programme, Voronezh. Roskosmos.

The charge of inconsistency could also be applied to the United States. Aside from their use of Russian rocket engines, Boeing's piloted spacecraft, the Starliner, used power converters built and supplied by ZAO Orbita in Voronezh, a city with a long record in the space industry. Orbita, originally Electrical Repair Plant #17, had provided power converters for the Mir space station and Energiya-Buran and was chosen by Boeing for its 'mission assurance, customer reliability' and the compactness of its components.

Students of sanctions believe that a key impact on the space industry is electronics. One calculation, in *Voprosy ekonomiki*, was that 65–79 percent of electronic components for the space industry were imported. Such dependence was confirmed by the esteemed Belgian researcher, Bart Hendrickx, who found that in one 2016 contract for the electronic parts of the Burevestnik satellite, 45 components were Russian and 457 were non-Russian. Ultimately, Russia realized that sanctions presented it with the need to replace imported electronics with home-made ones – import substitution (*importozameshenye*). In the meantime, it was reported that Russia subsequently bought in nearly €2bn worth of electronic components from the world leader, China [28].

Even relations between Russia and its most reliable European partner deteriorated. In September 2018, President Macron's defence minister, Florence Parly, accused Russia of spying on France's Athena Fidus military communications satellite. For some time, western analysts had been following a Russian satellite, Luch Olymp, in geostationary orbit. Whereas it was normal for all countries to move their satellites in geostationary orbit from one orbital position or slot to another, Luch Olymp seemed to be on an inspection tour of the orbit, moving in close to the satellites of other countries while doing so and 'scooping up' or listening in to their communications.

This 'cold war on again' atmosphere became the backdrop to the continuation of the ExoMars project. ESA was eventually prompted to come out and make a stout defence of its continuing participation in ExoMars, arguing that ESA was not part of the European Union that imposed the sanctions. In an interview with TASS, the head of the ESA office in Moscow, René Pischel, said that regardless of the political issues, it was important to maintain cooperation at a cultural and scientific level [29]. He had been the head of the ESA permanent representation in Moscow since 2007. Born in 1959, Pischel had studied mathematics in Kharkov, now Kharkiv, where there were also famous institutes (e.g. Physics and Technology (1928) and Aviation (1930)), before working in the Institut für Kosmosforschung (Institute for Space Research) in Berlin, the GDR's institute for spaceflight, where he was awarded a doctorate for his thesis on image processing. He worked on the Mars 8 and Mars Express missions and had a lifetime of experience in cooperation, so was likely to take the long view.

Several of those affected by the sanctions have challenged them under the common human rights law code shared by Russia and the European Union [30]. The Treaty of Lisbon, 2007, specified that any sanctions decisions of the Union must be based on clear, distinctive criteria and guarantee judicial review to those affected, be they within or outside of the Union. The European Court of Justice affirmed in 2014 that it had the right to review blacklisting where it was based on facts that were materially inaccurate, or a manifest error in assessment of the facts. The European Council circumvented this by renewing sanctions so regularly that it left insufficient time for challenges to be made to the previous round, which by then were out of date. Other challenges also failed. In the case of *Almaz Antei*, a space and missile company, the court ruled that the risk of its illegality was sufficient to justify sanctions. With *Rosneft*, the court ruled that the European Union had the broad right to take political decisions, making precise reasons immaterial. For *Kiselev*, the court ruled that the journalist concerned had the continued right to express his opinion, despite sanctions, so the sanctions were not relevant. In the opinion of the Jean Monnet chair of the Department of Integration and European Law at Moscow State Law University, these judgements were 'bitterly disappointing' and made future challenges futile. All this provided a difficult backdrop for the ExoMars project.

Rover *Rosalind Franklin*

Sanctions and renewed cold war or not, Europe and Russia proceeded to the second part of the ExoMars project. The second ExoMars mission comprised a small 800 kg carrier module, later called the cruise stage (Russia); a landing platform (Russia); and a rover (ESA). The latter remained the most constant part of a project whose total weight was 1,500 kg. The cruise module, the key to maintaining the spacecraft during the long coast from Earth to Mars, had 16 20N thrusters using 136 kg of hydrazine, while its solar panels could generate 2 kW power, supplying a 24.4 kg battery. On arrival at Mars, the landing would be by braking, parachute, rocket engines and shock absorbers, with the rover then driving off the landing platform. In 2019, the landing platform was called Kazachok ('little Cossack') while the rover was named the *Rosalind Franklin* (1920–58) after the scientist who determined the structure of DNA. Her name was chosen from 36,000 entries in a Europe-wide competition.

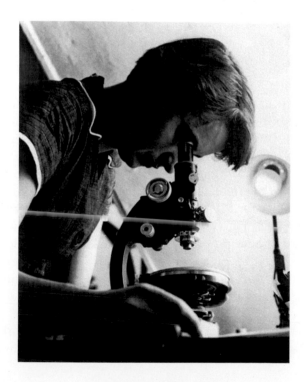

Rosalind Franklin. ESA

Rosalind Franklin naming announcement. ESA

The rover was now smaller than originally planned, at 310 kg, making it medium size by international comparison. It was much larger than the small Chinese and American rovers (e.g. *Sojourner*, 16 kg, Yutu, 120 kg), but much smaller than the big American and Russian rovers (Lunokhod, 840 kg; *Curiosity*, 889 kg). Built at Astrium's works in Stevenage, Hertfordshire, the rover made Britain the second European contributor to ExoMars **[31]**. Astrium had already built prototype Mars rovers called Bridget, Bruno and Bryan for use on a 30 m by 13 m simulated Martian surface made from 300 tonnes of sand – called the Mars Yard Test Area – to test motion and autonomous avoidance navigation on a realistic surface. The new rover, with six wheels on three bogie systems, would have programmed routes based on detailed orbital photography, but with hazard avoidance cameras to prevent a misstep. Power came from two 2.4 m^2 450 w solar panels, with three 8.5 w radio isotopes to maintain heat at night. Its mission control was the Rover Operation Control Centre (ROCC) in Turin, Italy.

The target was to travel 100 m each Martian day, or sol and work for a total of 218 sols (230 Earth days). To travel safely, the rover had two wide-angle panoramic cameras (pancams), 50 cm apart, able to survey the landscape from the height of a human but in different wavelengths. There was also a high-resolution camera, provided by Germany.

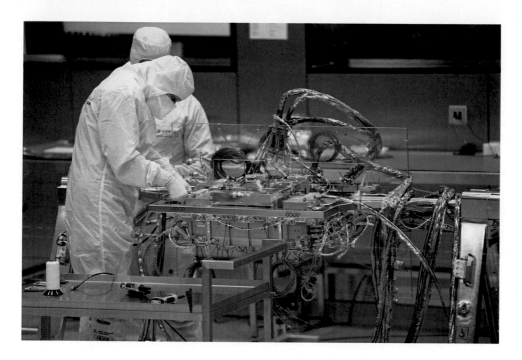

ExoMars rover under construction. ESA

The 'Exo' part of ExoMars would work like this. The rover had a ground-penetrating radar (3 m depth), a Raman spectrometer to identify the molecular structure of rocks and a micro-imager to search for biological markers. Once a suitable drilling site was identified, the drill would be guided by an imager, CLUPI. The further down it could drill the better, as evidence had mounted that radiation had killed off most signs of life close to or immediately under the surface. The drill had a power of 60 w to penetrate 2 m down, into a range from gypsum, to sedimentary rocks, to basalts. The drill would collect 17–22 samples of 1 cm diameter and 3 cm long. Once collected, each sample would be distributed on a carousel to the nine instruments of the 55 kg Analytical Laboratory Drawer (ALD).

ExoMars CLUPI imager up close. ESA.

At the heart of the sampling system was the Mars Organics Molecular Analyser (MOMA), developed by the Max Planck Institute, which would apply two tests: gas chromatograph mass spectrometry and laser desorption mass spectrometry. The laser was tiny, only 220 g and 20 cm long and was the first ultraviolet space laser, emitting pulses of 130 μJ at 266 nm. It held out the hope of detecting organic molecules that might indicate past life on Mars. MOMA had an especially complex route to the launch pad. Built in Laser Zentrum Hanover and developed by the Max Planck Institute for Solar System Research in Göttingen, it was delivered to the Goddard Space Flight Centre in Maryland for testing of the Laser Desorption Mass Spectrometer to ensure that it was shock-absorbing and able both to endure wide temperature ranges and withstand radiation. Then it travelled back to Europe where it was attached to the *Rosalind Franklin* in January 2020 for 18 days of temperature and vacuum tests. Russia contributed two instruments: the Infrared Spectrometer (ISEM) and the neutron spectrometer (ADRON; see Table 5.4). Oleg Korablev and Igor Mitrofanov, prominent on TGO, again featured as PIs, for ISEM and ADRON respectively.

Table 5.4:
ExoMars rover instruments

Instrument	Full name (country of PI)
Pancam	High resolution, wide-angle panoramic camera (Britain)
ISEM	Infrared Spectrometer for ExoMars (Russia)
WISDOM	Water Ice and Subsurface Deposit Observations on Mars (France)
ADRON	Neutron spectrometer (Russia)
CLUPI	CLose UP Imager (Switzerland)
MaMISS	Mars Multispectral Imager for Subsurface Studies (Italy)
MOMA	Mars Organics Molecular Analyzer (Germany)
RLS	Raman Laser Spectrometer (Spain)
MicrOmega	Imaging spectrometer (France)

As ever, one of the biggest challenges was finding the best landing site and the selection process began in 2013, with the first site identification workshop taking place in Spain in March 2014. The rover would be sent to those places where ancient rocks of Mars could be found, be they with water or biological markers, commensurate with the spacecraft trajectory and ease of landing. Scientists' choice was shaped by their reconstruction of the history of Mars. They divided its history into three: the Noachian epoch, 4.6bn years ago and earlier, a period of formation and bombardment; the Hesperian, from 4.6bn to 3bn years ago; and the Amazonian, from 3bn years ago to the present. The Hesperian was the time of volcanoes, valleys, oceans and lakes; the 'warmer, wetter Mars' when it was most habitable. During the Amazonian period, the lakes around the equator (30°N to 30°S) retreated, most of the atmosphere evaporated and impacts diminished, although vulcanism continued. Although glaciation persisted, it also retreated. There continued to be transportation of material – liquids and dust – even in a diminished atmosphere, as witnessed by occasional global dust storms and even dust devils roaring across the plains [32]. So ExoMars was reaching back to the Hesperian, especially the channels that flowed out of the northern uplands into the south.

The formal requirements were: suitability to the spacecraft entry angle; low-lying to maximize parachute time; low horizontal and vertical wind speeds; gentle slopes, boulders not higher than 35 cm; the absence of loose surface material; the best prospects for scientific sites of interest with 2 km, and in the range 5°N to 9°N, close to the equator. The four front-runners were Mawrth Vallis, Oxia Planum, Hypanis Vallis and Aram Dorsam. Of these, Mawrth Vallis and Oxia Planum were clay-rich and had exposed rocks, indicating past water, eroded only in the past few hundred million years. Mawrth Vallis also had an old outflow channel between the highlands and lowlands. Hypanis Vallis was a fluvial fan, possibly an ancient river valley, with fine-grained sedimentary rocks and estimated to be 3.45bn years old. Aram Dorsam was a curving channel of alluvial sediments like the river Nile, with indications of sustained water activity. Oxia Planum was selected in October 2015. It was considered to have the best combination of clay rocks likely to contain water, exposed rock and previous volcanic activity. Either Aram Dorsum or Mawrth Vallis could be the backup. Oxum Planum was on the boundary where many channels emptied into the lowland plans and exhibited layers of clay-rich minerals formed in wet conditions 3.9bn years ago. It was considered to be representative of Hesperian Mars. Oxia Planum was at 18.14°N, 335.7°E, east of the site of Pathfinder and *Sojourner* (1997).

For a reality check, the close-up imager CLUPI was tested on a real Martian rock. Exhibit #0102.226 of the Natural History Museum in Bern, Switzerland was one of 124 rocks recovered by geologists on Earth that could be identified as a meteorite from Mars, having been ejected by an asteroid impact and propelled toward Earth. Most such meteorites ended up in the preservative environment of Antarctica, but this fragment was discovered in Sayh al Uhaymir, Oman in 2001.

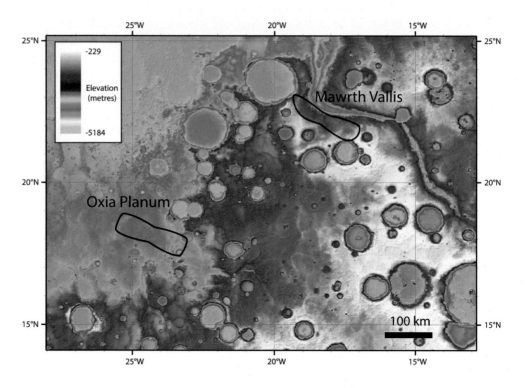

Oxia Planum elevation. ESA.

Carrier module, landing platform Kazachok

The primary function of the landing platform was to ensure the safe arrival of the rover. Lavochkin had overall responsibility for Kazachok, but in Europe, Thales Alenia was responsible for the computer, navigation control, the parachute, the

IMU and radar. Kazachok had a new propulsion system, one with which the Europeans were unfamiliar and necessitated their reassurance. The braking engine had four hydrazine chambers, three throttleable valves and 14kN thrust.

As the rover wandered, there was some scope for science on the lander (albeit more static by nature), concerned with the environment, the weather, chemistry, water, radiation and the atmosphere. Of its 828 kg weight, 45 kg was devoted to instruments, which were selected in December 2015 (see Table 5.5). The aim of the surface platform was to operate for a Martian year (1.8 Earth years), take photographs, monitor the climate, investigate the atmosphere, analyse radiation, look for subsurface water and make geophysical observations of the planet's internal structure.

Table 5.5:
Kazachok experiments and sensors

LaRa	Radio science (Belgium)
MAIGRET	Magnetometer (Russia)
WAM	Wave Analyzer Module (Czech Republic)
HABIT	Habitability, brine irradiation and temperature (Sweden)
METEO M	Meteorology (Russia)
METEO P, H	Pressure and humidity (Finland)
RDM	Radiation and dust sensors (Spain)
AMR	Anistropic Magneto Resistance sensor (Spain)
TsPP	Camera system (Russia)
BIP	Instrument interface and memory (Russia)
FAST	Infrared Fourier Spectrometer (Russia)
ADRON	Active neutron spectrometer and dosimeter (Russia, Bulgaria)
MDLS	Multichannel Diode Laser Spectrometer (Russia)
PAT-M	Radio thermometer for soil temperatures to 1 m (Russia)
Dust suite	Dust particle size, impact (Russia)
SEM	Seismometer (Russia)
MGAP	Gas Chromatograph Mass Spectrometer (Russia)

The purpose of the Lander Radioscience experiment (LaRa) was to investigate the internal structure of Mars, make precise measurements of the rotation of the planet and detect variations on angular momentum due to the migration of ice from the polar caps to the atmosphere. LaRa came from the Royal Observatory of Belgium. It is an x-band receiver and transmitter, which will receive signals from Earth – likely from Bear's Lake – and relay them back again, the idea being that tiny shifts in movement in the Martian surface will become detectable over time. Although a European experiment, LaRa was developed by a combined team and tested at both IKI in Moscow and in Europe. The purpose of the HABitability, Brine Irradiation and Temperature package (HABIT) was to investigate the amount of water vapour in the atmosphere, daily and seasonal variations in ground and air temperature and the ultraviolet radiation environment.

On the Russian side, prominent scientists Igor Mitrofanov and Oleg Korablev were again on the list, being responsible for ADRON and FAST respectively. Kazachok included instruments that dated back to the Interkosmos period in the socialist block (see *Interkosmos after Interkosmos*, chapter 3). The WAM wave analyser was proposed by Professor Santolík of the Czech Institute for Atmospheric Physics to examine electromagnetic emissions of atmospheric origin, magnetic anomalies on the surface of Mars, the planet's internal structure and the impact of space weather, with a particular view to detecting lightning discharges in dust storms and whirlwinds. It would also determine whether radio waves from outer space reach the surface of the planet. ADRON included the Bulgarian Liulin ML, designed to match, on the surface, the work of FREND on the TGO orbiter. Liulin ML was intended to capture the secondary particles of neutrons and gamma rays from galactic and solar cosmic rays when they strike the surface of the planet. MDLS would make a continuous study of the atmosphere for a Martian year. For the Kazachok instruments, there was a dedicated mission control, the Surface Platform Payload Operations Control Centre (SPOCC) in Moscow.

ExoMars LaRa. ESA.

ExoMars LaRa team. ESA.

Preparing the second ExoMars

With the 2018 departure for 2019 arrival date postponed, new launching dates were published in October 2017, not difficult given the implacable nature of celestial mechanics. The new date chosen was 24 July 2020, utilizing a launch window lasting until 13 August, with arrival at Mars in March 2021. Weather-wise, the new arrival date offered less danger of dust storms than the 2019 season. The new launch date required the lander and rover to be delivered to the cosmodrome in May 2020 at the latest. Preparations were complex, as they involved three spacecraft (rover, cruise stage and landing platform) and three test locations (Cannes, France; Turin, Italy; and Lavochkin, Moscow). Some individual instruments also had complex journeys. For example, the two European scientific instruments, LaRa and HABIT, had to go to Russia for connection to the electronics system and then return to Italy for assembly in 2018.

In January 2019, the cruise stage was completed at the Lavochkin design bureau, so that it could be integrated with the lander and rover for vibration and thermal tests. Kazachok was shipped from Lavochkin to OHB Bremen in Germany on 22 March and then to Thales Alenia in Turin on 2 April 2019 in a big Antonov Volga carrier aircraft. It was carefully wrapped in plastic to prevent contamination. Once there, it was fitted with avionics, the carrier, the rover and thermal protection. The instrumentation for the rover *Rosalind Franklin* was completed in Britain in August 2019 and was sent to Toulouse for environmental testing.

ExoMars Kazachok arriving. ESA.

ExoMars Kazachok unloading. ESA.

ExoMars Kazachok testing. ESA.

In April 2019, the first worries that all might not be ready in time began to emerge. Referring to concerns arising from vibration tests, the Director General of ESA, Jan Wörner, warned against moves to delay ExoMars 2020 beyond its scheduled lift-off date, though there was no sense that a show-stopper was lurking. The integration stage was bound to uncover some problems – which is why it is so thorough – and nerves were probably a bit on edge. Then a real problem did emerge: the parachute, a European responsibility. Wörner affirmed that 2020 was a 'must' and that the project would lose support among the member states were it to be delayed a further two years.

ExoMars problem parachutes. ESA.

The landing technique included a parachute system weighing 192 kg. It consisted of two parachutes, each with its own pilot chute, the first of 15 m diameter to open at a speed of 1,700 km/hr and then the second of 35 m diameter to open 20 seconds later at 400 km/hr. These were large parachutes as Mars landers went, with previous ones having been 12.7 m diameter (Pathfinder), 11.7 m (Phoenix and InSIGHT), and 15 m (*Spirit* and *Opportunity*). Indeed, the 35 m diameter parachute is the largest-ever on a Mars mission and folding it correctly took three days. The engines would bring in the lander for the last 1,000 m of the six-minute descent.

The first, low-altitude test of the 35 m second main parachute took place successfully in 2018 when it was dropped from 1,200 m at the European test range in Kiruna, Sweden. The following year, they went on to the critical high-altitude tests. The first, on 28 May 2019, involved the full deployment sequence of all four parachutes being dropped from a balloon at 29 km. Although the deployment sequence worked fine, the two main parachutes were torn, with multiple tears on the small one and one on the larger one. Modifications were made for the second test on 5 August 2019, but the main parachute tore again and the cabin landed under only the pilot parachute. At the MAKS air show in Moscow at the end of August 2019, Jan Wörner assured Roskosmos that there would be time to sort out the problem. The parachute failures were not a catastrophe, but a problem to be fixed in good time. Thankfully, the parachutes could be loaded at a late stage, which gave them a margin.

ExoMars parachute tests. NASA.

There was nothing wrong with the parachutes themselves, the problem was with their extraction. After the second failure, ESA put together what it called an 'aggressive plan of action' with the principal companies involved (Vorticity, Arescosmos, TASinF and TASinI) to solve the issue in time. NASA, which had considerable parachute extraction experience, was asked and agreed to help. These tests were set for January and March 2020 in Oregon, with a final test at Kiruna in May. The purpose of the fix was to make the parachutes open like petals without frictional damage and the revised plan involved two bungee tests, two extraction tests and two high-altitude drop tests. These additional tests had financial implications and the ESA Council meeting in Seville, Spain agreed to provide another €52 m which was disarmingly entitled 'precautionary risk mitigation'. There were some delays on the Russian side too, up to six months, such as verifying Kazachok's braking system. One of the electronic boxes had to be replaced and there were cracks in the descent module front shield, which had to be repaired.

The spring of 2020 was hectic, with the institutes and contractors working flat out three shifts a day, seven days a week, to get the spacecraft ready in time. In February, integration of the European part was completed in Cannes. The tests of the landing engine for Kazachok took place in Lavochkin on 27 February and were regarded as entirely successful, so Russia signed off on its side of the mission. The Proton rocket for the launch was ready.

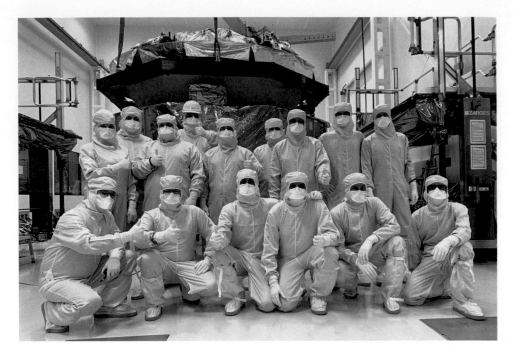

ExoMars: Russia signs off. Roskosmos.

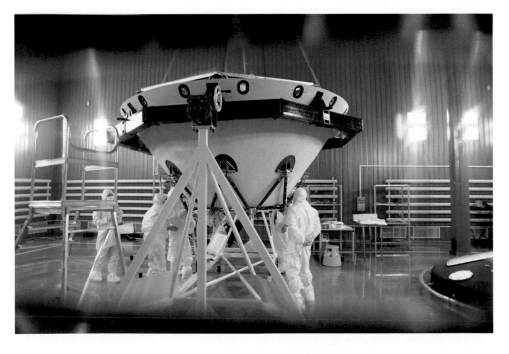

ExoMars final tests. Roskosmos.

The following week, ESA announced that it was convening a press conference in Paris on the status of the project on 12 March, but there was no hint of what was to follow. Generally, delays or cancellations leak in advance or are announced in the lowest-key way possible, rather than in a formal way. There was a teleconference between the Russian-European project teams and between the head of ESA, Jan Wörner and Dmitry Rogozin of Roskosmos (under sanctions, Rogozin could obviously not meet him in person), the outcome of which was to delay the mission by two years. It was quickly renamed ExoMars 2022.

The decision caused quite a shock, as only the previous week senior ESA officials were confident of a successful outcome to the tests and an on-time launch in July. The reason cited was the need for additional hardware and software tests. The coronavirus raging at the time in Europe cannot have helped, but it did not stop launches of other spacecraft to Mars that July from Japan (Hope), Cape Canaveral (Perseverance), nor Wenchang (Tianwen). It is possible that there was a subliminal fear that the final stage was being rushed, with the risk of an uncorrected fault being unidentified and unfixed. The few comments made on the decision were that it was 'the difficult, but the right decision'. The Russians cannot have been pleased, given that they had done their tests successfully on time and now had an expensive Proton rocket on their hands with a time-limited warranty. Interviewed in *Kosmsolskaya Pravda*, Dmitry Rogozin was clearly frustrated that they had kept their side of the project on schedule, despite the virus, but that the Europeans had run out of time.

Further explanations emerged later. Wörner later referred to 'competing priorities' at the Oregon test site, implying – but not explicitly saying – that the Americans could have helped them more by holding the test earlier. He also referred to four electronic boxes that had to be replaced, which meant that their software would have to be re-tested. The lesson from Schiaparelli, he said, was 'do all the tests. Don't cut corners on tests'. The ExoMars programme leader, François Spoto, attributed the delay to ESA's decentralized structure, the late delivery by Russia of avionics equipment, incomplete environmental tests and 18 outstanding issues sent on a list to the two agency heads. Another commentary faulted the Russians for the delay, with the weak propulsion system on its lander obliging the Europeans to build a bigger and more complex parachute system than would otherwise have been necessary [33]. If that were the case, the issue should have been resolved at the design stage.

ESA took advantage of the delay to make a number of fixes to ExoMars. The rover was kept at Thales Alena in Turin for battery, cleanliness and microbiological checks. Cracks to the brackets holding the solar panels were discovered and fixed. Some consideration was given to improving the instruments and their software, especially the Mars Organic Molecule Analyser (MOMA); the infrared spectrometer ISEM; and the Close Up Imager, CLUPI. New lines were made to reduce parachute tears during the split-second 200 km/hr extraction. Then the hardware was drained of anything that might degrade in storage and the spacecraft put into hibernation. As for the parachute tests, these were postponed to the autumn, principally because of the virus.

ExoMars 2022 integrated. Roskosmos.

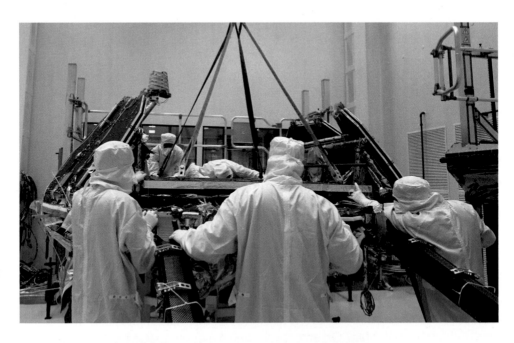

ExoMars 2022 going into storage. Roskosmos.

In September 2020, the ExoMars 2020 spacecraft was moved from Turin to Cannes for additional thermal, vacuum, acoustic and leak tests. ESA published new mission dates, with launch at 16:10 Central European Time on 20 September 2022 and landing on Mars at 17:32 on 10 June 2023.

Despite this hiatus, the politics of Mars missions moved ahead. In spite of being burned with Netlander and the first ExoMars, Europe returned to the Americans for the successor mission to Mars. President Trump had somehow overcome the principled opposition of the Office of Management and Budget to the Mars sample return mission and the *Perseverance* rover had been permitted to collect and cache samples. In April 2018, NASA and ESA entered an agreement to develop a Mars sample return mission, with first funding approved by the congress and ESA ministers the following year. This involved a NASA lander with an ascent stage and an ESA rover (called 'fetch') to gather the samples collected by *Perseverance* (2026), followed by an ESA orbiter and return craft (2028). This was also called the 26-26-28 plan. The pre-mission study of NASA and CNES, as outlined at the International Astronautical Congress in Washington DC in October 2019, bore uncanny similarities to the unmentioned mission planned 20 years earlier by Claude Allègre (see Chapter 2).

The ExoMars experience, therefore, remains incomplete, half-finished. This has been a largely technical description, but the human factors in the experience are worth recording too. There were always going to be problems for two such diverse sides to work together, but they were nothing like as difficult as in the 1980s when the French battled even to see a *zavod*. On the Russian side, there still remained not so much secrecy but rather issues which the Russians saw no need to disclose to the other side, though they were nonetheless desirable for cooperation. The Europeans sometimes lacked clarity as to whether they should be dealing with Roskosmos or individual companies (e.g. Energiya, Lavochkin). Paperwork on the Russian side could be challenging, something mainly put in place across Russian industry as an anti-corruption measure, which meant dealing with many financial controllers.

There were language difficulties, of course. Those educated in the USSR did not learn English, or came to learn it late. Scientists in IKI had good English, but this was not the case at the contractor companies. Understandably, the Russians insisted on their technical texts being in Russian, which the Europeans then had to translate, but getting good technical translators in Europe was difficult, with the main ones being in France, Switzerland and Romania. On the Russian side, they would not deal with documents sent to them in English, insisting on their translation into Russian first. Such translations could take a day, if top priority, or a couple of weeks if not. Over time, some of the Europeans came to learn Russian, with many coming to understand the language even if they were not confident enough to do business in it. In the Bion programme for example, some Europeans, like Italian Antonio Verga, could speak fluent Russian, but that was unusual. There were always a few polyglots who picked up the language quickly, while the rest began with a few words, then learned to make sense of text and understand what was being said at meetings. There was less incentive for the Russians to learn English, as most meetings took place in Russia (e.g. Samara) rather than in Europe. The Europeans found that Russian equipment had much less documentation, but had much more practical testing, while European equipment had more paperwork and procedures and took longer. One engineer concluded that 'at another level, the Russians are easier to work with than the Americans. The Russians value and invest in human factors and personal cooperation more. Their mentality is more like the Europeans. Because of that, we can talk openly together and criticize. They are really clever people'.

The close level of cooperation on ExoMars brought into sharp focus many of the practical problems when two quite different countries cooperate, or attempt to do so. Technical standards are different between Europe and Russia and this is a big challenge. Tests of equipment in Europe might happen earlier or later in Russia, while there were different standards and functions for qualification models and flight models. Systems for project management and review were different too, being much more documented and written down on the European side, with greater scope for asking questions early on. Tests relied more on written calculations, requirements and analysis on the European side, but on demonstration on the Russian side. Funding follows at each step in Russia once a new stage is reached, but Europe has a different system of payments. The systems are different, but that is 'not a matter of one better, one worse,' according to one experienced observer.

ExoMars team. ESA.

ExoMars: conclusions

ExoMars is the culmination of the story of Russian-European cooperation, not just because it is the most recent project, but because it is the space science project with the highest level of joint planning and integration. After 1996, Europe had come to be wary of cooperative deep space projects with Russia because of the country's financial difficulties, which had translated into poor quality control in general and the heart-breaking loss of Mars 8 specifically. Faith in Russia's inter-planetary capacity recovered when the Soyuz rocket launched the highly success-ful European Mars Express and Venus Express missions in 2003 and 2005.

The origins of ExoMars lie in the Aurora programme, in which Europe declared Mars to be its principal objective with a rover as the key instrument for explora-tion. At this stage, Europe re-learned the hard lesson already previously encoun-tered by Claude Allègre in 2000 when he decided that France should team up with the Americans to go to Mars. Three years were lost in developing the project with the Americans when the congress pulled the plug on it in 2011. One German observer sympathized: 'the Americans cancelled one contract after another. By contrast, the Russians never broke contracts'.

Barely had word reached Europe of the debacle in Washington DC before a Russian delegation was winging its way hotfoot to Paris bearing an insurance pay-out sufficient to buy two Protons. For Russia, this was a golden opportunity to get back into Mars exploration after the Phobos Grunt debacle. The trade-off involved Russia providing two Protons, a landing platform and much-cherished instru-ments, with Europe providing an orbiter, a demonstration lander and a rover. Making up for lost ground, events moved at a rapid pace, contractors were arranged, agreements signed and a complex two-part mission authorized.

The first ExoMars, 2016, was partly successful. The successful part, the Trace Gas Orbiter, has circled Mars since October 2016, sending back a stream of fresh data from its Russian-European instrument package and adding to our knowledge of the dynamics of the Martian atmosphere. Its science mission benefits from the best instrumentation that Europe and Russia together could devise. Their value was more than apparent at the Moscow solar system symposium in October 2020. The demonstration lander, Schiaparelli, crashed when a software fault commanded the engines to switch off 3.7 km up, with predicable results. The function of a demonstration is to isolate problems that might emerge during the real mission, so that was one gain from an otherwise unsatisfactory outcome. The inquest exposed shortcomings in Europe's engineering skills at several levels.

The second part of the ExoMars mission, originally scheduled for 2018, was postponed to 2020, the original schedule proving too tight. By 2019, the different elements of the project were coming together, albeit enduring the nerve-testing experiences bound to attend the integration of three different spacecraft in multiple locations. Ultimately, the problems were technical on the European side rather than an outcome of problems in cooperation, because the teams themselves worked successfully together on the basis of the experience and networks that had built up over many years. Europe's decision to delay to 2022 speaks well of its patience, caution and thoroughness, but the outcome does not compare well with Japan and the United Arab Emirates, the United States or China, which did get their 2020 missions under way.

At this point, ExoMars is still half-finished, but is the last, integrated, big European-Russian project on the books at present. Although there is Russian participation in future missions, notably Luna 27, there is nothing on this scale. ESA once again turned to the Americans for the sequel 26-26-28 sample return mission, despite a long history of joint projects with the Americans being overturned by congressional decisions and re-decisions. Given the financial pressures likely to affect the American return-to-the-Moon programme, it is hard to avoid the possibility that 26-28-28 could also be a casualty of redefined budgets. Shrewd observers say that when the sanctions came down, ExoMars was too far advanced to be stopped, 'but it wouldn't happen now'. Some European scientists believe that turning to the Americans for 26-26-28 was a political decision.

ExoMars was well underway when its progress was interrupted by what could easily pass for a resumption of the cold war, which marked, in some ways, a return to the CoCom era (1949−91) whose role in affecting European-Russian cooperation was described in earlier chapters. As a result of events in Ukraine and Crimea, the United States and the European Union imposed sanctions on Russia, with double force: aerospace companies and critically, the minister responsible for the space industry and later head of the Russian space agency Roskosmos, Dmitry Rogozin.

The sanctions directly affected ExoMars, because the European Union had to grant an exemption to its own rules to permit the ExoMars Trace Gas Orbiter to be filled with European hydrazine fuel. This exemption was by no means unique, because a feature of the sanctions regime, just as during the CoCom days, was its uneven, inconsistent and contradictory application, as examples from financial services, the aluminium industry and even *Baltika* beer illustrated. The exclusion of Rogozin from even travelling to the countries with whom he was formally and legally cooperating brought a sharp edge to the stand-off, one unknown during what was called the original cold war, which was never personalized in this way.

Considering the size and importance of the European space industry, including its scientific community and given the importance of cooperation with Russia, the decision to target Rogozin was a profound one. One would expect that such a decision would have been carefully, delicately, weighed by the Council of the European Union, especially given its substantial diplomatic resources, such as the External Action Service. Anticipation of the consequences would have considered the potential damage to European-Russian cooperation, an assessment of risk, possible retaliation and scientific and industrial gains and losses, so that a finely-balanced decision would have emerged. Instead, the record is clear that no such consideration took place and no background nor information documents were prepared nor circulated, nor criteria discussed. The Council secretariat proposed and the Council of the Union agreed to Rogozin's sanctioning on an unsourced, unreferenced one-line basis that he had supported annexation, which also begs the question of the many others who did so but were untouched. When sanctions against him were renewed in 2019, the justification was a short article in a little-known Russian journal showing that he had given a photograph album of the Kerch Strait Bridge to President Putin, moreover on the strength of a bellicosely misleading machine translation of the text, without even asking for the benefit of the thousands of translators in the employ of the Union. It should be a matter of alarm that European Union foreign policy in an area of conflict should be decided on such an unprofessional basis – in the strict meaning of the word. According to a recent analysis by the German Institute for International and Security Affairs, the past few years have seen the expertise of the European Union in the area of Russia so reduced that important foreign policy decisions are increasingly less well informed [34].

References

1. Fletcher, K: *Mars Express – the scientific investigations*. ESA, Noordwijk, 2009, SP 1291.
2. Kargel, Jeffrey: *Mars – a warmer, wetter planet*. Praxis/Springer, 2004.
3. Wilson, Colin & Ghail, Richard: *Life after Venus Express*; and Wilson Colin *et al: A dynamic atmosphere revealed by the Venus Express mission*. Papers presented 40th COSPAR conference, Moscow, August 2014.
4. For the early history, see *ESA at Mars*. Spaceflight, vol 58, #3, March 2016; Marlow, Jeffrey: *Seeking ET – ExoMars and the continuing search for life on Mars*. Spaceflight, vol 51, #4, April 2009 ; and the research of Anatoly Zak at www.russianspaceweb.com.
5. Lakdawalla, Emily: *The design and engineering of Curiosity – how the Mars rover performs its job*. Cham, Springer-Praxis, 2018.
6. Korablev, Oleg *et al: The Atmospheric Chemistry Suite of three spectrometers for the ExoMars 2016 Trace Gas Orbiter*. Space Science Review, 2018; *How the Sun pumps out water from Mars into space*. Mars Daily, 20 May 2019.
7. Semkova, Jordanka *et al: Radiation environment investigations during ExoMars missions to Mars – objectives, experiments and instrumentation*. Comptes rendus de l'Académie bulgare des Sciences, Tome 68, #4, 2015.
8. Tolker-Nielsen, Toni: *ExoMars 2016 – Schiaparelli anomaly inquiry*. ESA, 2017.
9. ESA: *First results from the ExoMars Trace Gas Orbiter*; Vandaele, Ann Carine & Korablev, Oleg: *Martian dust storm impact on atmosphere H2O observed by ExoMars Trace Gas Orbiter*. Nature, Letters, 10 April 2019.
10. Fedorova, Anna; Montmessin, Frank, Korablev, Oleg *et al: Stormy water on Mars – the distribution and saturation of atmospheric water during the dusty season*. Science, 9 January 2020. For the October 2020 papers, see Institute for Space Research: *The eleventh Moscow solar system symposium, 5–9 October 2020*. Moscow, author, 2020.
11. Manaev, Georgy: *How did Crimea become part of the Russian empire?* TASS, 15 November 2019.
12. Reichardt, Tony: *Might Russia join ESA? Talks pave way for new space partnership*. Nature, 20 May 2004.

13. Arkelyan, Lilian: *EU-Russia security relations – another kind of Europe*, from Roger Kanet (ed): The Russian challenge to the European security environment. Palgrave Macmillian, 2017.

14. Ashford, Emma: *Not so smart sanctions – the failure of western restrictions against Russia.* Foreign Affairs, January/February 2016; Beyer, Andreas & Zogg, Benno: *Time to ease sanctions on Russia.* Policy Perspectives, vol 6, #4, July 2018; Corriero, Davide: *The 'Ukraine crisis' and international unilateral sanctions under international law.* Monograph, accessed academias.edu, 5 January 2020.

15. Jarosłav Ćwiek-Karpowicz & Stanislav Secrieru (eds): Sanctions and Russia. Warsaw, Polish Institute of International Affairs, 2015.

16. Jarosłav Ćwiek-Karpowicz & Stanislav Secrieru (eds): *op cit*; Securieru, Stanislav: *Russia under sanctions – assessing the damage, scrutinizing adaptation and evasion.* Warsaw, Polish Institute of International Affairs, 2015.

17. Jarosłav Ćwiek-Karpowicz & Stanislav Secrieru (eds), *op cit.*

18. Securieru, Stanislav: *op cit.*

19. Information provided by Department of Foreign Affairs, Dublin, following Freedom of Information Act request 25 February 2020.

20. This writer: *Reviving Russia's space programme?* ROOM, spring 2020; see also *Cinquante ans, op cit.*

21. Arkhangelskaya, Svetlana: *Sanctions convince Russia to produce own rocket fuel.* RBTH, 3 March 2016.

22. OHB: *European access to space. Annual report, 2017.* Bremen, author, 2018.

23. Stewart, Jim & Doyle Cillian: *Ireland, global finance and the Russian connection.* Presentation to TASC, Dublin, 27 February 2018; Corcoran, Jason: *Russia's largest bank expands Irish aviation leasing division amid growing losses.* Irish Times, 22 October 2018. Brennan, Joe: *Russian banks deposit €21.5bn in IFSC.* Irish Times, 11 December 2018.

24. Doyle, Dara & Kelly, Fiach: *State considered options to save Rusal's Aughinish Alumina.* Irish Times, 28 April 2018; Lynch, Suzanne: *Aughinish sanctions lifted now but threat still hangs over Limerick plant.* 1 February 2019. Doyle, Dara & Kelly, Fiach: *State considered options to save Rusal's Aughinish Alumina.* Irish Times, 28 April 2018. Paul, Mark: *US lifts sanctions on Aughinish Alumina plan.* Irish Times. 20 December 2018. O'Halloran, Barry: *Deal to lift threat of sanctions against Aughinish Alumina refinery at risk.* Irish Times, 12 November 2018.

25. McLaughlin, Daniel: *Germany reassures Ukraine over Russian sanctions and energy fears.* Irish Times, 19 June 2019.

26. MacCormaic, Ruadhán: *How Europe brought Russia in from the cold.* Irish Times, 29 June 2019.

27. *Airbus DS and Energiya eye new medium-class satellite platform; EU civil satellite purchases unaffected by anti-Russia sanctions.* Space Daily, 26 December 2016; *Airbus' long reach: customers mentioned in the bribery and ITAR allegations.* AW&ST, 10–23 February 2020.

28. Oxenstierna, Suzanne: The Western sanctions against Russia. How do they work? from Steven Rosefielde (ed), *Putin's Russia: economic, political and military Foundations,* World Scientific, August 2018; Jarosłav Ćwiek-Karpowicz & Stanislav Secrieru (eds): Sanctions and Russia Warsaw, Polish Institute of International Affairs, 2015; Hendrickx, Bart: *Burevestnik – a Russian air-launched anti-satellite system.* Space Review, 27 April 2020.

29. Shipenikov, Maxim: *Roskmosmos, ESA, cooperate despite anti-Russian sanctions.* TASS, Moscow, 12 February 2019.

30. Kalinichenko, Paul: *Post-Crimean twister – Russia, the US and the law of sanctions.* Russian Law Journal, vol V, #3, 2017.
31. Baglioni, Pietro: *The ExoMars 2020 rover.* Paper presented to the International Astronautical Congress, Washington DC, October 2019.
32. Pieters, Carle: *Geological evolution of the terrestrial planets – 60 years of exploration and understanding.* Presentation, Sputnik: 60 years along the path of discoveries. Moscow, 4 October 2017.
33. Faust, Jeff: *Mars in limbo.* The Space Review, 16 March 2020; *Mars lander on hold.* Spaceflight, vol 62 #5, May 2020; *ESA Weekly media review*, 14 August 2020; Cozzi, Emilio: *The team leader of the ExoMars mission has revealed to us why Mars can wait.* Forbes, Space Economy, 13 August 2020.
34. Sabine Fischer & Margaret Klein (eds): *Conceivable surprises – eleven possible turns in Russia's foreign policy.* Stiftung Wissenschaft und Politik, Berlin, 2016.

6

Conclusions

Cooperation between Europe and Russia moved through several distinct periods and stages, best guided by decadal landmarks:

At first, in the 1960s and 1970s, cooperation was almost entirely bilateral between France and the Soviet Union. Its start was marked by the visit of de Gaulle and it was refreshed by that of Pompidou. Cooperation began with sounding rockets and balloons, extended into Earth orbiting satellites (e.g. SRET, Signe, Aureole) and incorporated the flights of French equipment to the Moon, Venus and Mars. The second decade ended with the invitation to France to fly an astronaut to the Salyut orbital station, the Premier Vol Habité.

The 1980s saw cooperation broadening into adjacent countries, groups of countries and fields. Cooperation extended to Germany (Foton), Britain and the Netherlands (Kvant) and to the European Space Agency (ESA) countries as a group (Bion). It expanded into astronomy, materials processing and biology. France remained very much the dominant country, most visible through the long-duration flight and spacewalk on the Mir space station and with the French presence in the astronomy 'great observatories' programmes, like Granat. The high moment of cooperation was probably the VEGA project, with representatives from many European nations gathered in Moscow to see the results of their flyby of comet Halley come in.

The 1990s were a period of rapid change. First, the Soviet Union and then Russia opened up its space station Mir to European participation, with France again becoming the lead country. The French were followed by Germany, then ESA, with solo flights by Austria and Britain, between them providing Europe with a much extended, long-duration spaceflight experience, as well as experimental opportunities. Second, the free market in Russia from 1992 opened the door to large-scale industrial cooperation, notably in launcher services, but also in communications

© Springer Nature Switzerland AG 2021
B. Harvey, *European-Russian Space Cooperation*, Springer Praxis Books,
https://doi.org/10.1007/978-3-030-67686-5_6

satellites. Once the Proton rocket established the principle, Russia came to supply launchers (e.g. Rockot, Dnepr , Cosmos 3M) for a broad range of western satellites, large and small, mainly German. Third, cooperation was affected by the financial turmoil in Russia, this being most acute in the frustrating delays affecting the Mars 8 project and ultimately dooming its outcome. Finally, the 1990s were a defining decade for European-Russian cooperation around future human spaceflight. Explorations were made of a joint Hermes programme, Mir 1.5 and later ACTS and Kliper, but they were overtaken by the bilateral American-Russian agreement to set up what became the International Space Station (ISS).

ESA office, Moscow. ESA

The 2000s saw the arrangements settling down, with Europe obtaining regular access via Russia to the ISS and to its own laboratory installed on the station, Columbus. Arrangements were established for European astronauts to train in Moscow for the 18 months or so ahead of each mission and Europeans became a permanent part of Star Town. The most remarkable development of the decade was the construction of a Soyuz launch site at Kourou, making the rocket an integral part of the European launcher fleet, the largest industrial project of cooperation.

The 2010s saw the development of the ExoMars project, probably the most integrated, big project of the two parties, but one threatened by a 'new cold war' through the sanctioning of Russia and the head of its space agency. So far, well-established patterns of working together and the longevity of individual projects (ten years in the case of ExoMars) has enabled such cooperation to survive, but even with ExoMars itself, the politics caused a specific problem around the supply of fuel. In this chapter, to reach conclusions, some specific aspects of this cooperation are explored further.

The France-Russia connection

France has been the dominant country in this narrative, so that relationship is worth exploring in more detail. France remains the country with the most structured, developed and sophisticated approach to its cooperation with Russia. It is seen as a defined theme in the history of French space exploration. Other countries have cooperated with Russia – notably Germany – but they have not seen the need to reflect on or systematize such cooperation. The principles of French space policy have shown remarkable continuity. In 2009, these were enunciated as France being Europe's leading space nation, French leadership of an integrated European space programme, balanced international cooperation and independence – the very principles set down by General de Gaulle and Georges Pompidou, it said **[1]**. Cooperation with Russia had been a consistent theme throughout.

France-Russia: anyone for coffee? Roskosmos

The key role of France arose from its distinct position in Europe in the 1960s and onward. Fifth Republic France saw itself as leader of continental Europe, prepared and able to play a distinct role in foreign policy. Germany might have had the most rapidly developing economy, but for historical reasons could not and did not wish to be as politically assertive. Unlike Britain, which touted its 'special relationship' with and closeness to the United States, France believed that Europe should exercise a certain distance and independence from the US. For de Gaulle, it was important to assert and affirm France's independent line, even if it meant upsetting some of his allies. Although he was a financial, social and political conservative, he did not share the anti-Russia, anti-Communist attitudes of many western leaders, having been in a German prison camp with Russian officers during the first world war and in Moscow with his Russian allies during the second. For de Gaulle, the space programme was an integral part of French post-war recovery and the road to becoming Europe's technological leader. His *dirigiste* policies, so different from the neo-liberalism that later ensorcelled Europe, meant putting state investment into space programmes, setting up CNES and building a technology centre in the underdeveloped rural south-west (Toulouse). It led later to clearing the jungle for a launch site in a remote overseas territory, what the French call the Dominions Outre Mer (Kourou). Between a favourable disposition to the Soviet Union and a belief in spaceflight as a technology leader lies the explanation for the France-Russia connection. Indeed, one could go so far as to say that without de Gaulle's distinct personality, French cooperation with Russia in spaceflight might not have happened until many years later and in much more limited form. It was de Gaulle's achievement and if there is one single, key personality to this story, it is he.

Cooperation with Russia became a distinct, structured element of the narrative of French technological development, in a way unmatched by any other country. It was celebrated thus, as three events illustrate. First, 2010 was declared *Année France-Russie* (France-Russia year), a big event, with exhibitions, conferences, books, films, concerts and themes of student cooperation, culture, education and research. Space cooperation was the highlight of the scientific programme, with events in Biscarosse, Montpelier, Krasnoyarsk and Moscow, with a book to mark space cooperation (*Correspondences Paris Moscou*) and a film *Soyouz sous les tropiques* (Soyuz in the tropics). There was even a literary event, organized by the CNES Observatoire de l'Espace with the Maxim Gorky Literary Institute, in which space-related literature and authors were discussed. There was a conference in Moscow, *Humans in space – how far can we go?* led by Anatoli Grigoriev of IBMP and moderated by Claudie Haigneré. Roskosmos presented an overview of its space programme at a Grand Palais exhibition. It is hard to imagine such a set of events, on such a scale, taking place in any other country. A full-scale model of Mir already had pride of place in Toulouse.

Second, in 2016 at the 50th anniversary celebrations of the 1966 accord, a commemorative publication was issued by the Centre National de la Recherche Scientifique (CNRS, or National Centre for Scientific Research) to convey just how extensive this had become: not just spaceflight but such diverse areas as nuclear and particle physics, big data, prehistoric paintings, 3D modelling, archaeology, molecular design, railways, climate change, linguistics, glaciology, public health, literature, neurobiology, nano-materials, geometry, enzymes and agronomy **[2]**. Third, on 30 January 2019, France and Russia held a round table in Paris, *The challenges of Russian-French cooperation*, organized by the Groupement des Industries Françaises Aéronautiques et Spatiales (GIFAS). The keynote speaker was Jean-Yves Le Gall, president of CNES. The French government's Ministry for Europe and Foreign Affairs even has a webpage called *France and Russia – science technology and space*. No other European country matches this. Even in fiction, the 2020 film *Proxima* about a space mission was set in Star Town and Baikonour: it is hard to imagine Hollywood doing so.

For France, spaceflight opened the door to cooperation in a broader set of areas. The July 1966 space agreement was followed by a bilateral Chamber of Commerce when Alexei Kosygin visited Paris that December. The replacement of the Soviet Union with the Russian Federation meant a revision of the arrangements between the two countries and in 1992, they set up the Conseil Économique, Financier, Industriel et Commercial franco-russe (CEFIC) to extend that model into a broad range of economic, financial and commercial areas.

France and Russia, formal session. Roskosmos.

The CNRS, which dates to 1939 and is the guiding force for all research in France, operates with two joint laboratories, plus 15 internationally associated laboratories and eight international research networks between France and Russia. The two joint units are the French-Russian Mathematics Laboratory, J-V Poncelet and the French-Russian Studies Centre, both in Moscow. The 15 associated laboratories cover several space-related areas, such as low energy particles (Kurchatov Institute and the University of Grenoble Alps), atmospheric sciences and planetology (Zuev Institute of Atmospheric Optics, Tomsk and the universities of Champagne and Grenoble Alps) and plasma (Lomonosov University and the universities of Toulouse and Paris-Saclay). The international research networks, in the space area, cover helio plasmas (the Institute for Space Research (IKI) and the University of Toulouse) and heavy ions (University of Nantes and the Joint Institute for Nuclear Research and Theoretical and Experimental Physics). In 2014, for example, there were 1,178 publications between the two countries, 944 French researchers visiting Russia and 280 the other way around. In starting spaceflight cooperation between France and Russia, de Gaulle had wanted France to have a model for wider fields, an axis outside the bipolar east and west. He may have succeeded more than he might have thought possible.

The connections between France and Russia in the space field were evident, for example, when Pierre Levy was made ambassador to Russia. Within a month of his appointment, on 11 February 2020, he made an official visit to Roskosmos, to be received by Dmitry Rogozin, along with managers and department heads. It would be hard to imagine this being such an early priority for a newly-appointed ambassador from another European country. Cooperation is likewise evident at inter-ministerial level with regard to education. For example, in 2009, the two respective ministers for education, Valérie Pécresse and Andrei Forsnenko, agreed a joint project between students at Bauman Technical University and Montpellier II University to build a micro-satellite, Baumanets 2, which was eventually launched in 2017 with Meteor M2-1, although it was lost when the rocket failed.

France and Russia. Roskosmos.

In August 2019, President Macron spoke to the French ambassadorial corps about the importance of France rebuilding its relationship with Russia, which provoked an angry reaction from the Baltic states. The author of *The Russian reconquest*, Laure Mandeville, quickly pointed out that all he was doing was following in the footsteps of de Gaulle, whose relationships with Russia always transcended ideology [3]. One space engineer spoke of how the US continued to discourage European-Russian cooperation: 'space is a very visible thing to do together. By contrast, a pipeline is not only physically underground, but less visible in the public eye'.

An American assessment of Russian-French cooperation gives us a helpful perspective from further afield. The Office for Technology Assessment (OTA) described it as having primarily political origins, but of acquiring a mixture of political, scientific and economic aims over time, the balance of which fluctuated over the years, but which has remained largely stable [4]. It noted how France sought opportunities to fly instruments to Venus (1970s, 1980s) when they were simply not available on the American side and similarly with piloted flights (1980s) when they were not available on the American side until later. The USSR proved to be a more reliable partner, offering a wider range of opportunities. Whereas in the 1960s, the cooperation project was a flourish of French independence – and a successful one – during the more intensive phase of the cold war in the 1980s, there was a strong French foreign policy view that space cooperation was an important way of keeping channels of cooperation open between the sides. Although such gains might be hard to measure, the alternative of shutting them down offered nothing positive.

Ahead of the 50th anniversary of de Gaulle's visit, at a *table ronde* held in Paris on 20 November 2013, the French reflected, taking a sober, even melancholy view. The events following the fall of the wall had changed everything and, thanks to the ISS, the US had become Russia's primary partner, not Europe, so Franco-Russian cooperation declined. This was accelerated by the collapse of Russian space science, most evident in the failures of Mars 96 and Phobos Grunt. They felt that their partnership with Russia had been based on long historical links dating back to the 18th century and on to the wartime alliance; that it was between two countries of broadly similar technical competence. In contrast, the ISS was less about sharing and more about keeping costs down. There was a sense that Russia was now more interested in cooperation with Germany.

The case could easily be made that by the end of the story, Germany had become the larger player. Germany followed on from France, but not until the 1980s and only wholeheartedly in the 1990s. By 2020, Germany had become the largest contributor to ESA and its piloted flights to the ISS, its experiments (e.g. Plazma Krystall) and Spektr RG gave it higher visibility. In European politics, Germany increasingly came to be seen as the leader of European policy toward Russia. Although German-Russian cooperation came a late second to that of France, it became a substantial field, though this is not reflected in the literature. Whereas French-Russian cooperation had been widely documented and celebrated, it was passed over in the official history of German spaceflight [5].

Germany-Russia meeting. DLR.

Abkommen zwischen der Regierung der Bundesrepublik Deutschland und der
Regierung der Russischen Föderation über die Zusammenarbeit auf dem Gebiet der
Erforschung und Nutzung des Weltraums für friedliche Zwecke, 10.04.2001

Text des Abkommens
Anlage

Abkommen zwischen der Regierung der Bundesrepublik Deutschland und der

Regierung der Russischen Föderation über die Zusammenarbeit auf dem Gebiet der

Erforschung und Nutzung des Weltraums für friedliche Zwecke

Die Regierung der Bundesrepublik Deutschland und die Regierung der Russischen

Föderation im Weltraum als Vertragsparteien bezeichnet

СОГЛАШЕНИЕ МЕЖДУ ПРАВИТЕЛЬСТВОМ
РОССИЙСКОЙ ФЕДЕРАЦИИ И
ПРАВИТЕЛЬСТВОМ ФЕДЕРАТИВНОЙ
РЕСПУБЛИКИ ГЕРМАНИЯ О СОТРУДНИЧЕСТВЕ
В ОБЛАСТИ ИССЛЕДОВАНИЯ И
ИСПОЛЬЗОВАНИЯ КОСМИЧЕСКОГО
ПРОСТРАНСТВА В МИРНЫХ ЦЕЛЯХ
(ЗАКЛЮЧЕНО В Г. САНКТ-ПЕТЕРБУРГЕ 10.04.2001)

РАТИФИЦИРОВАНО РОССИЙСКОЙ ФЕДЕРАЦИЕЙ - ФЗ ОТ 22.05.2004 N 44-ФЗ

СОГЛАШЕНИЕ
МЕЖДУ ПРАВИТЕЛЬСТВОМ РОССИЙСКОЙ ФЕДЕРАЦИИ
И ПРАВИТЕЛЬСТВОМ ФЕДЕРАТИВНОЙ РЕСПУБЛИКИ ГЕРМАНИЯ
О СОТРУДНИЧЕСТВЕ В ОБЛАСТИ ИССЛЕДОВАНИЯ И ИСПОЛЬЗОВАНИЯ
КОСМИЧЕСКОГО ПРОСТРАНСТВА В МИРНЫХ ЦЕЛЯХ

(Санкт-Петербург, 10 апреля 2001 года)

Правительство Российской Федерации и Правительство

Germany-Russia agreement.

Whoever leads, there is at least a consensus among analysts that France and Germany are 'the privileged partners' of European-Russian cooperation **[6]**. Britain presented intermittent examples of cooperation, as did the Netherlands, Belgium and Austria, but lagged far behind. The biggest player to be absent was Italy which, despite being consistently the third largest contributor to the European space effort, never established significant institutional cooperation. Not until 2016 did the Italian Space Agency (ASI), under the leadership of Roberto Battiston, sign an agreement with Roskosmos at the St Petersburg Economic Forum, but there is no readily available information on its subsequent progress.

Drivers of cooperation

Examination of the France-Russia relationship gives us clues as to the drivers of cooperation, notably the politics of western Europe. It would be worth drilling deeper to look at some of the underlying drivers that made cooperation work, or not. On the Soviet side, we know less of the drivers, but we can make some informed speculation. Examination of official publications on spaceflight throughout the Soviet period (e.g. the *Soviet booklets* series) shows that cooperation was a priority theme. The reasons appear to be similar to those which drove cooperation on the American side: a desire to share technical progress with other countries; to spread and share the benefits of science; and to convey the appearance of being a friendly, outward-looking, progressive nation. Such an approach quickly came under western suspicion that the Kremlin was using spaceflight as a means of detaching countries into its political sphere, but such a charge could equally be laid the other way.

One factor which may be underestimated is the degree to which the Soviet Union wished to break out of the isolation in which it found itself after 1949. Whilst critics might say that this was a problem of its own making, the preparedness of the Soviet Union to propose cooperative projects and to respond — often with alacrity — to approaches from others is striking. Much of the flirtation preceding the projects of this story came from the Soviet side. The Soviet Union was undoubtedly isolated — by CoCom, by the range of American bases all around the USSR and by the many American-led alliances — and there was a level of scientific isolation was well. For scientists, though, cooperation with other people and other countries is their life blood, to keep up with new ideas and stay on the cutting edge **[7]**.

From the late 1990s, though, the financial crisis became a strong driver of cooperation from the Russian side. The Russian space programme had to switch from being one of the most state-funded in the world to becoming one of the most commercially-funded in only a few years and the money had to be brought in. This was most evident in industrial cooperation, where western investment was quickly used to keep launch facilities going, as could be seen by the Eurockot investment in Plesetsk and the Starsem investment in Baikonour. For space science, the financial crisis took the form of Russian scientists no longer having access to their own data from their own instruments. Put crudely, the Russians traded their assets for access to projects and scientific experiments. The two striking examples are INTEGRAL, where Russia provided a Proton rocket in exchange for 25 percent observing time; and ExoMars, where Russia offered two Proton rockets and the Kazachok landing platform in exchange for instruments on the Trace Gas Orbiter.

Throughout this period, the Soviet Union and Russia's biggest asset had been its launching power and lower costs compared to other countries. As chapter 4 showed, especially the table of launcher prices, going through CST, Eurockot or bilateral channels offered a lower price and greater speed to get satellites into orbit than any other route, whether that be a cube satellite or a probe to Mars. The Soviet Union and Russia were able to offer access to destinations not available elsewhere. The long-running Venera programme offered repeated opportunities to the French to get instruments to Venus and some opportunities to Mars that were available no other way. The Soviet Union and Russia also offered methods not available elsewhere: Foton and Bion cabins provided flights of a week to several weeks in a zero-gravity environment ideal for materials processing or biological experiments, though it is fair to say that such opportunities could also be negotiated with China. Finally, although there were occasional and spectacular failures, Russian rockets were reliable.

According to Mathieu, cooperation was driven by a strength-weakness model, with Russia trading launchers – and the revenues that they provided – to leverage the state-of-the-art technologies that it lacked, for example with communications satellites or advanced remote sensing technologies that it was denied [8]. Russia valued Europe for its technology, capital and management know-how. For Europe, Russia's big advantage was that it could offer 'full spectrum competence' – in other words, it was able to undertake all aspects of space research and had the full infrastructure to do so. For example, Europe did not and could not do piloted flight on its own. As one engineer put it, 'Europe did not have the skills to build Orion or a piloted spacecraft like it. Europe, though, was good in specialized areas, like manufacturing, instruments and standards'. Another European observer commented that while Russian laboratory conditions and technical equipment were not as good as the Europeans, their scientists were 'equal or even better'. Experts with a good overview always caution against examining cooperation from the perspective of one side being 'better or worse' because 'different' is a much more informative metric.

Such differences can be seen at work in individual sciences. In some areas, the Soviet Union had a specific expertise and for that reason it was globally the most obvious scientific partner. Its near-Earth exploration programme, developed under the Cosmos programme from 1962, emphasized the physics of the northern latitudes, unsurprising given the location of its landmass. Such physics embraced the magnetosphere, the Sun and the Earth-Sun relationship, so the USSR here became an expert through the Cosmos programme, Prognoz and Interball. In working with the Russians on such projects as Aureole, the French were dealing with the world leaders. On the European side, France was a leader in space telescopes (Granat) and Germany in x-ray telescopes (Spektr RG), so it made sense for Russia to get these countries aboard its observatory projects.

In summary, the drivers of cooperation were:

- Political. On the European side, a key element was French foreign policy, whereas on the Soviet side it was both to present a positive image and the desire to reduce isolation.
- Financial. From the point of the financial crisis that affected the space programme since the period of the Russian Federation and that financial shortfalls remain.
- Scientific. The desire by Russia to continue to have access to scientific instruments relaying new information; and on the European side the desire to fly instruments to places it would not otherwise be able to access.
- Trade. Russia had specific tradable assets in terms of rocket lifting power; reliability; access to distant destinations; and spacecraft type.
- Expertise. Both sides have different expertise in different sub-disciplines within space science and looked for economical, efficient ways to share them.

Russia proved to be a reliable partner. Once a project was agreed, it happened. On several occasions, ESA or France chose to partner with the US, only to find the US forced to pull out due to budgets being cancelled (e.g. Netlander, ACRV, original ExoMars, Jupiter mission). Russia did not always find Europe to be so reliable (e.g. Roseau, Éos). One experienced European negotiator spoke of how reaching an agreement with Russia could be difficult, but once done, the Russian side fulfilled all its obligations 'perfectly', with no re-negotiations nor requests for more money.

A final driver of cooperation can be personal and this can be true in a wide range of fields. Studies of cross-border economic, social and political cooperation have long identified the desire and driving force of individuals with an interest in working with the other side as critical [9]. This text has identified many key individuals who have been important to the story: Mstislav Keldysh, Roald Sagdeev,

Lev Zelenyi, Jacques Blamont, Gerry Skinner, Gerry Webb, Sigmund Jähn, to take some examples. On the scientific side, Yuri Galperin was especially important. His colleague, J-J Berthelier wrote later of what he tellingly called 'the human factor':

> 'He [Yuri Galperin] was the driving force. His wide scientific knowledge, his numerous friends scattered across the administration, laboratories and technical institutes were essential in the success of these missions. Above all, his personal ability to understand and interface with colleagues of quite different ways of thinking and working were the basic components which allowed the system to work over more than 30 years, in spite of technical and funding difficulties and political fluctuations [10]'.

Cooperation with Russia may have reduced its isolation, but it was not just a one-way street, as J-J Berthelier referred to how Galperin opened the door to the 'mysterious and unknown world' of Russia's people, culture and history to them. One European scientist, visiting Russia for the first time, admitted that all he knew of Russia was 'what he read in the newspapers', so he took a tourist trip before his professional visits there began in earnest. Visiting scientists remarked on how Russia was so much 'a country on its own, a completely different world, isolated, with its own way of thinking'. In the beginning, no one there spoke the international language of science and engineering (English), although a few spoke German. Gradually, young people learned English, 'but not at the pace of the young Chinese'. It could be hard work, but hugely rewarding.

Comparisons with other cooperation models

The European-Russian experience inevitably invites comparison with the Russian-American and the Russian-Chinese experiences. Despite CoCom, the levels of scientific cooperation between the Russians and the Americans during the early years after Sputnik were both close and cordial, the best example being the reciprocal lecture tours undertaken by James van Allen in Russia and his great rival and colleague Sergei Vernov in the United States [11]. This changed in the early 1960s, to the point that anyone collaborating with Russia was regarded with political suspicion. A cooperation agreement was finally signed in 1965, but limited to what was considered the only non-contentious area, space biology. The Apollo-Soyuz Test Project (1975) was the big exception, the high point of 1970s détente in an otherwise prolonged period of non-cooperation, as evidenced by the bewildered awkwardness of those few Americans who travelled to Moscow for the VEGA mission. A close relationship with Russia did not resume until 1993, when

President Clinton approached Russia to merge the ailing American *Freedom* space station with Mir 2. Cooperation in other areas remained limited, however, leaving us with a contrast between the high level of cooperation in the project of highest visibility (ISS) and rocket engines (e.g. RD-180, Russia supplying no less than 116), but little cooperation at the less visible mezza or lower levels – quite a difference from Europe.

The big casualty of the change in the political atmosphere in the early 1960s was in the sharing of information. Even though the Soviet Union published a substantial body of information regarding its space science outcomes in English (its principal outlet, *Kosmicheskie Issledovaniya* was obligingly published in English as *Space Research*), most of it went unread in the west. Two dissemination worlds developed in space science: a Russian one and an American one (which would include the other English-speaking countries). Perhaps the most extreme example was the deserved award of the Nobel prize to two American scientists for discovering the relict radiation from the big bang – even though this was many years after Prognoz 9 had done so and the results had been published – but there were many others. Russian scientific discoveries were indeed published, but they went unread in two unconnected information worlds whose circles did not overlap.

Cooperation between Russia and China was quite different from the European pattern and was much more formal. There was an original accord in 1954 for Russia to supply R-1 rockets, drawings and technicians (one was Yuri Galperin) to help China start its rocket and missile programme in exchange for the supply of raw materials, part of a number of economic agreements. The accord was not very cordial, with China soon accusing the USSR of supplying outdated designs in a laggardly way and in 1960 the whole agreement ended in tears in a political quarrel between Mao Zedong and Nikita Khrushchev. The two sides did not speak to one another again until 1992, when China went shopping in Moscow for parts that would help them to start a human spaceflight programme, including the training in Star Town of cosmonaut instructors. The Russians were happy to oblige, for a fee. Since then, the relationship has been structured under a series of technical accords, with working groups (10–12 in number) devoted to particular areas, such as space stations and interplanetary probe design. The emphasis has been on information-sharing, with Russia undoubtedly providing advice that has helped China's space station, lunar and interplanetary programmes. Despite Chinese visits to Energomash, Russia has been slow to part with its single greatest area of expertise, rocket engine design. There have been only two joint projects: the Yinghuo sub-satellite on the failed Phobos Grunt mission; and a Chinese participant on the *Mars 520* mission, with China providing important financial support in both cases. Although there are regular meetings, there seems to be a wariness compared to the more intense cooperation with Europe. Documentation on Russian-Chinese cooperation is scarce.

René Pischel giving a briefing in Moscow. ESA.

This brings us to one of the challenges of cooperation, namely documentation. Except for France, which has done so well through articles, books and the reunions, plus Germany with some articles in journals, there is little on record documenting European-Russian cooperation. ESA has commendably issued reports on individual missions and instruments, for example Bion 10 [12]. This was very much on the initiative of René Demets, the leading scientist involved, but it was generally not a task formally built into the mission programme. Otherwise, though, the concept of issuing joint reports on the outcomes of either individual or entire programmes of scientific or other missions is rare. The scientific outcomes are spread over the whole world of scientific publishing, from *Nature* to COSPAR's *Advances in Space Research*. Those wishing to find such information must therefore follow the publications of the individual Principal Investigators (PI) across the literature of the scientific world. Its dispersed nature makes an assessment and weighing of the scientific outcome of this particular axis of endeavour difficult. The lack of a concerted, structured programme of dissemination of the outcomes of cooperative ventures also has a public cost, because to the European mind cooperation with Russia has relatively low visibility. One engineer commented on how 'all the media report is what the Americans do. NASA, NASA, NASA. It's very difficult to interest them in cooperative projects with Russia'.

Winners and losers

To examine cooperation from the perspective of 'winners and losers' can be a superficial approach. According to René Pischel, head of the ESA office in Moscow, 'cooperation is not a matter of gain or loss. If one side is losing and the other is winning, then it's not cooperation. Both sides must get benefits, even if in different ways'. At the same time, it would be useful to unpack the various fields in which there have been gains.

The scientific results from the cooperative record here were substantial and can be seen in the large volume of scientific papers published, from Aureole and the first French satellites onward. They grew with the instruments on Mars probes and the solar missions (e.g. Prognoz) and continue with the great observatories, including such current missions as INTEGRAL and Spektr RG. It is evident that some individual missions generated significant scientific return (e.g. the Bions and Fotons). Europe's greatest single area of gain was probably in human spaceflight, for which it did not otherwise have its own independent means of access. Russia made it possible for Europe to build the third highest level of spaceflight experience, with all the related experimental gains. Europe still has more piloted spaceflight experience than China, although this will change once the Chinese space station begins operation. Scientists were also winners and there is one good example of where they came to pool their knowledge – the European Astrobiology Network Association, in which Russia participated. The tenth meeting, chaired by David Gilichinsky, was held in Puschino and was considered one of the best.

European-Russian astrobiology workshop. René Demets.

It should be remembered that Europe was accessing one of the two most sophisticated space programmes in the world. The French instruments sent to Mars in the 1970s were carried on spacecraft flying with autonomous navigation systems, long before they were developed anywhere else. Europe was cooperating with the country with the longest experience of building, developing and supplying space stations; with the longest expertise in human spaceflight; with a unique knowledge of the planet Venus; with sophisticated computer systems (Cosmos 186, 188); with the most powerful, advanced rocket engines in the world (e.g. RD-253, RD-180); and with one of the world's longest-running scientific programmes (the Cosmos programme, begun in 1962). As a result, Europe will have picked up considerable knowledge gains on the side. A French visitor to Moscow in 1972 who was working on the Aureole programme spoke of how Yuri Galperin and his colleagues gave them 'their first real course in true space research' **[13]**. Forty years later, when German scientists came to examine the results of the flight of the first German in space – from the GDR – they were surprised at the quality of its outcomes and their relevance to new research on the ISS **[14]**. In turn, Russia gained from the type of instruments installed by European countries, from Lunokhod reflectors, Signe, Gémeaux, Echographe and eROSITA. The ISS brought particular examples where they could pool their respective experience of experiments, such as Plazma Krystal.

The principal gains for Russia appear to have been political, financial and scientific. The first, the most difficult to measure, came from reducing its isolation during the cold war by having at least one European country (and others in due course) with whom it could do scientific business. Financially, such cooperation was undertaken at Soviet cost, since the USSR did not charge for western participation, a practice that was reversed in 1992. The financial crisis that hit the Russian space programme made revenue generation an imperative and it is fair to say that without it, the programme would have likely collapsed by the late 1990s. Funding from cooperation and commercialization ultimately saved the programme. There was a perception by Europeans that Russia gained from its experience of western industrial management, especially in the transition after 1991, which would have been quite a culture shock to veterans of the command economy. It is known that Russia brought in, indeed *bought* in, many European methods and approaches to quality standards, certification and quality control.

Cooperation had its costs. Developing missions jointly is inevitably more time-consuming – and therefore money-consuming – than doing so individually. Joint projects are vulnerable to one side coming into financial difficulties and pulling out, meaning that the other partner's time has been wasted (e.g. Hermes). Sometimes there are political difficulties, like the change of government which defeated the Roseau project (Pompidou) and then Éos (Giscard d'Estaing). Lengthy discussions do not always lead to an outcome, with the French balloons to Venus and Mars being the case study. Some negotiations failed to attract the political support necessary (e.g. ACTS, Kliper) or were overtaken by events (e.g. Mir 1.5). All told, though, the list of failed cooperative projects over this period is surprisingly short.

Even successful projects have costs and challenges. They can be resource and time-consuming and depend on each party being able to define priorities, themes and projects. Matthieu and other commentators described the European decision-making processes as complicated, slow and inefficient, compounded by the twists-and-turns of European policies toward Russia. Russia found itself dealing with multiple organizations, institutions and people, often uncoordinated and disconnected and would have preferred fewer, clearer priorities and more efficient decision-making. The Russian system was highly structured and coordinated, but Europe tended to respond with projects or programmes independent of each other. Europe drew attention to the practical difficulties such as language; different business, technical and engineering practices and standards; and issues of export control, technology transfer and intellectual property. Such issues do not stop projects but they inhibit long-term cooperation, incentivizing individual, one-off agreements. The individual agreements over Luna 27, then 25 and 26 were a case in point, when a longer-term arrangement could have been made, as with ExoMars.

Missed opportunities?

Analyses of European-Russian cooperation are scarce, most coming unsurprisingly from a francophone perspective [15]. Their insightful commentaries have a recurrent theme: cooperation has fallen far short of its potential, as seen in the title of one, *La coopération spatiale Russie-Europe, une entreprise inachevée?* (*Russian-European cooperation: an uncompleted enterprise?*). Europe was slow on the uptake – Russia suggested *Soyouz à Kourou* in 2001, but it was not signed until 2005 – and missed opportunities, like Kliper and ACTS. When Europe announced the Galileo satellite navigation system in 1999, Russia proposed a cooperative effort with its GLONASS system. Instead, Europe made an interoperability agreement with the United States in 2004, so Russia rebuilt GLONASS on its own, making an arrangement with India instead. Russian overtures met with a tepid response from a Europe unable to establish a consensus on how to form a potential partnership. Europe must be a difficult entity with which to negotiate, considering the complexity and heterogeneity of its national, multilateral, ESA and EU programmes, let alone a political context that was at times quite hostile.

The problem was very much on the European side, as it lacked the building blocks necessary for a positive dynamic for cooperation, such as a common, coherent, coordinated policy or efficient decision-making and all within the framework of an overall policy toward Russia. For example, Russia could be welcomed institutionally as an associate or full member of ESA, as considered twice over between 1992–2005. Cooperation could formally be extended into the areas of navigation and Earth observations. Much could be done to achieve the harmonization of parts and procedures, an area where the European Union had expertise in the completion of the single market. Europe seemed unable to develop a portfolio of long-term partnerships with Russia and as a result ended up with a series of one-off, short-term agreements.

Perhaps the biggest missed opportunity in recent years was Ural and its related programmes. Europe began this programme, which made rapid initial progress, only to abandon it in favour of a conservative but cheaper successor, Ariane 6. The study outcomes were not even published. This was a unique opportunity to work with the Russians, the world's best rocket engine makers, on the next level of technological development. At the *table ronde*, Daniel Sacotte argued this very point, but 'it was a taboo subject' and gathered no traction. Russia would be a reliable, stable partner, he said. In the end, the Russians went their own way with the Angara rocket and the Soyuz 5 and between them they left innovation to SpaceX and the Chinese.

These judgements may be harsh. Whatever their merits, at this stage, European-Russian cooperation may be a story in search of new projects. ExoMars will finish in the 2020s, although if all goes well the science arising from it should continue for years. Launcher cooperation dating to the 1990s (e.g. Eurockot) may have run its course, but Russia is planning a new set of rockets (Angara fleet, Soyuz 5, Yenisei, Amur and even a nuclear rocket, the Nuklon), piloted spacecraft (Orel) and scientific opportunities which may offer openings for Europe. Luna 25, 26 and 27, where initial cooperation has been agreed, are part of a programme for a return to the Moon. Time will tell.

Final remarks

Cooperation between Europe and Russia has come under most scrutiny for its political aspects. Such examinations may have been looking in the wrong place and may overlook the efficiency gains, in integrating the world space industry and in achieving economies of scale. Although the term 'world space industry' may be a cliché, cooperative experiences may be one of the key ways through which it is achieved. Cooperation with Russia enabled European countries to fly experiments, instruments, satellites and people to places where they could not otherwise have gone, or which would otherwise have been considerably more costly. Cooperation with Europe enabled Russia to attract vital financial resources and specific areas of expertise which it had not itself developed. For some time, the economic experts who built the European Union, like Paolo Cecchini, pointed out that the cost of *not* cooperating is ultimately more expensive and wasteful than doing so **[16]**. The principal gains of European-Russian space cooperation may be a more integrated world space industry (e.g. launchers), efficiencies in the gathering of scientific knowledge and applications (e.g. the great observatories), shared learning (e.g. ExoMars) and accrued and consolidated expertise (e.g. in human spaceflight). These may be bland, unspectacular gains, but are nonetheless important and real ones. Despite the shortcomings and difficulties of European-Russian cooperation, who could have imagined that so much would be achieved when the general's Caravelle flew into Moscow on that hot summer day in June so many years ago?

References

1. *Rendezvous cannois.* CNESMag 53, April 2012.
2. CNRS: *France-Russia 50 years of cooperation in science and technology.* Paris, author, undated.
3. Marlowe, Lara: *Concern over French-Russian connection.* Irish Times, 14 November 2019.
4. Office of Technology Assessment, 1985, *op cit.*
5. Reinke, Niklas: *Geschichte der deutschen Raumfahrt.* Bonn, DLR, 2010 (in German and English).
6. Facon and Sourbes-Verger, *op cit.*
7. Walsh, TC: *Communicating science in the Sputnik era.* MA in communication studies, Dublin City University, 2002.

8. Mathieu, Charlotte: *Assessing Russia's space cooperation with China and India – opportunities and challenges for Europe*. Vienna, European Space Policy Institute, 2008.
9. Triskele: *The emerald curtain*. Triskele, Carrickmacross, 2006.
10. Berthelier, J-J: *Yuri Galperin, a scientist and friend*, from LM Zelenyi, MA Geller & JH Allen (eds): <u>Auroral phenomenon and solar terrestrial relations. Proceedings of a conference in memory of Yuri Galperin</u>. Moscow, Institute of Space Research, 2003.
11. Foerstner, Abigail: James *Van Allen – the first eight billion miles*. Iowa City, University of Iowa Press, 2007.
12. Demetz, R; Jansen, WH; Simeone, E: *Biological experiments on the Bion 10 satellite*. ESA, 2002, SP-1208.
13. Sauvaud, J-A: *Yuri Galperin, a teacher and friend*, from LM Zelenyi, MA Geller & JH Allen (eds): <u>Auroral phenomenon and solar terrestrial relations. Proceedings of a conference in memory of Yuri Galperin</u>. Moscow, Institute of Space Research, 2003.
14. Kowalski, Gerhard: *40th anniversary of the first German space flight*. <u>DLR Magazine</u>, #157, April 2018.
15. Mathieu, *op cit*; Facon & Sourbès-Verger, *op cit*; *Cinquante ans, op cit*.
16. Cecchini, Paolo: *The cost of non-Europe*. Brussels, European Commission, 1988.

Correction to: European-Russian Space Cooperation

Brian Harvey

Correction to:
B. Harvey, *European-Russian Space Cooperation*, **Springer Praxis Books,**
https://doi.org/10.1007/978-3-030-67686-5

The book was inadvertently published without updating the following corrections:

Abbreviations:

p.   = page

l.   = line

fb = from bottom

The updated versions of these chapters can be found at
https://doi.org/10.1007/978-3-030-67686-5_1
https://doi.org/10.1007/978-3-030-67686-5_2
https://doi.org/10.1007/978-3-030-67686-5_3
https://doi.org/10.1007/978-3-030-67686-5_4
https://doi.org/10.1007/978-3-030-67686-5_5
https://doi.org/10.1007/978-3-030-67686-5

Corrections:

p.121: The photo of Rashid Sunyaev was replaced with this new photo:

Rashid Sunyaev.

p.298, l.8: "RD-0110" was replaced with "RD-0162".

p.298, l.13, fb: The sentence "Methane-powered engines had long been a theme of Russian rocket engine development – a methane engine, the RD-301 had even been built…" was replaced with "Exotic-powered engines had long been a theme of Russian rocket engine development – a flourine ammonia engine, the RD-301 had even been built…".

p.299: The figure caption was replaced to read "RD-301 - the original exotic fuel engine".

p.3, l.4: The phrase "the next day" was deleted, so the sentence reads "His delegation…".

p.57, l. 10, fb: "M-2" was replaced with "M-3".

p. 57, l. 6, fb: "Bioxbox" was replaced with "Biobox".

p.58: The figure caption was replaced to read "Foton M-3. Marcel van Slogteren, ESA".

p.58, l.1, fb: The text in bracket was deleted: "(outside experiments had not been possible in the shuttle)".

p.59: "Bion" was replaced with "Biopan" in figure caption.

p.63, l.22: The word "agency" before "plane" was deleted.

p.69, l.11, fb: The phrase "Orenburg in the Urals" was replaced with "Orenburg to the south-west of the Urals".

p.145, l.7, fb: The text "M-1" was deleted from list, to read as "(Bion 9, 10)".

p.243, l.4, fb: The word "built" was replaced with "overseen" so as to read "…were overseen by…".

p.244, l.5 and l.6: The sentence "Some adopted a dormant or dessicated state to survive" was deleted, but retaining reference "[45]".

p.263, l.13: The word "Altas" has been replaced with "Atlas".

p.329, In Table 5.3, the word "impact" was replaced with "touchdown" in the last line of the table so as to read "intended touchdown".

p.333, l.16: "2000" was replaced with "2020" so as to read "October 2020".

Acronyms and abbreviations

ABRIXAS	A BRoadband Imaging X-ray All-sky Survey
ARAKS	Artificial Radiation and Aurora between Kerguelen and the Soviet [Union]
ACTS	Advanced Crew Transportation Spacecraft
ARCAD	ARC Aurorale et Densité
AUREOLE	AURora and EOLus
BNSC	British National Space Centre
CESR	Centre d'Etudes Spatiales des Rayonnements
CMEA	Council for Mutual Economic Assistance (COMECON)
CNES	Centre National d'Études Spatiales (French Space Agency)
CNRS	Centre National de la Recherche Scientifique
CoCom	Coordinating Committee for Multilateral Export Controls
COMPASS	Complex Orbital Magneto Plasma Autonomous Small Satellite
COSPAR	Committee on Space Research
DGA	Direction Général des Armaments
DLR	Deutsches Zentrum für Luft- und Raumfahrt (German space agency)
DPRK	Democratic Peoples' Republic of Korea
ERA	European Robotic Arm
ELDO	European Launcher Development Organization
eROSITA	extended Röntgen Survey with Imaging Telescope Array
ESA	European Space Agency
ESRO	European Space Research Organization
ESTEC	European Space and Technology Centre
GDR	German Democratic Republic

© Springer Nature Switzerland AG 2021
B. Harvey, *European-Russian Space Cooperation*, Springer Praxis Books,
https://doi.org/10.1007/978-3-030-67686-5

IBMP	Institute for Bio Medical Problems
IKI	Institute of Space Research, Moscow (*Institut Kosmicheskie Issledovaniya*)
InSIGHT	Interior exploration using Seismic Investigations, Geodesy and Heat Transport
INTEGRAL	INTernational Gamma Ray Astrophysical Laboratory
ISS	International Space Station
ITAR	International Traffic in Arms Regulations
IZMIRAN	Pushkov Institute of Terrestrial Magnetism, Ionosphere and Radio Wave Propagation of the Russian Academy of Sciences
LAS	Laboratoire d'Astronomie Spatiale
LEND	Lunar Exploration Neutron Detector
LRO	Lunar Reconnaissance Orbiter
MEX	Mars Express
MRO	Mars Reconnaissance Orbiter
NATO	North Atlantic Treaty Organization
ONERA	Office National d'Études et de Recherches Aérospatiales
PI	Principal Investigator
PRC	People's Republic of China
PROSPECT	PROSPecting for Exploration, Commercial exploitation and Transportation
PVH	Premier Vol Habité
ROSEAU	Radio Observation par Satellite Excentrique à Automatisme Unique
SARSAT	Search And Rescue SATellite
SCARAB	SCAnner for RAdiation Budget
SIGMA	Satellite d'Imagerie Gamma Monté sur Ariane
SIGNE	Solar International Gamma Ray and Neutron Experiment
SRET	Satellites de Recherche et sur l'Environnement et la Technologie
SRON	Stichting Ruimte Onderzoek Nederland
SSC	Swedish Space Corporation
TsKB	*Tsentralnoye Spetsializorovannoye Konstruktorskoye Buro* (Samara, Russia)
TsUP	*Tsentr Upravleniye Polyotom* (Mission Control, Moscow)
VEGA	VEnus HAlley ('H' is 'G' in Russian)
VEX	Venus Express

Appendix 1: Timeline

1961	Visit by Alla Masevich to Jodrell Bank to track lost Venus probe
1963	Visit by Bernard Lovell to USSR
1964	Visit by Gaston Palewski, Minister of State for Scientific Research and Atomic and Space Questions, to the Academy of Sciences in Moscow
1965	Discussion of space cooperation between General de Gaulle and Andrei Gromyko
	Visit by French scientist Jacques Blamont to Mstislav Keldysh, Vice President, Academy of Sciences
1966	Visit by President de Gaulle to Soviet Union, including Baikonour cosmodrome
	Signing of cooperation agreement
1967	First Soviet-French sounding rocket campaign
1968	Roseau mission cancelled
	Soviet invitation to fly instruments on next Venus probes, Venera 5, 6
1970	Visit by President Georges Pompidou, including Baikonour
	French laser carried to Moon by Soviet Moon rover, Lunokhod
1971	French instrumentation on Mars 3 probe to Mars
	Aureole 1, French-Russian space probe, launched by USSR
1972	SRET 1 launched by USSR
	Start of Prognoz programme, with French instruments on Prognoz 2, 6, 7, 9, M1, M2
1973	French instrumentation on Mars 5, 6 and 7 to Mars
1974	Éos balloon to Venus cancelled by France
	First proposal by France for USSR to fly French cosmonaut
1975	Soviet-French ARAKS programme
	Start of French experiments on Bion programme (Cosmos 782)
	Founding of European Space Agency (ESA)
1976	Swedish instrument flies on Interkosmos 16
1978	Sweden joins Prognoz programme (Prognoz 7)
1979	Invitation to France to participate in human spaceflight
	Offer to fly ESA Marecs satellite on Proton
1982	PVH (Premier Vol Habité): French astronaut Jean-Loup Chrétien to Salyut 7 station
1983	Astron observatory, with French spectrometer
1984	Start of VEGA mission to Venus with substantial European participation
1986	Start of study of Hermes spaceplane programme

© Springer Nature Switzerland AG 2021

B. Harvey, *European-Russian Space Cooperation*, Springer Praxis Books,
https://doi.org/10.1007/978-3-030-67686-5

1987	Start of ESA experiments on Bion programme (Bion 8)
	Kvant observatory, with equipment from Britain, the Netherlands
1988	Phobos mission to Mars with European experiments
	President François Mitterrand flies to Baikonour on Concorde
	Second French piloted spaceflight: Jean-Loup Chrétien to Mir space station
1989	Amendment of 1966 agreement for four French flights to Mir space station
	Start of participation in Foton programme (Foton 2)
	Granat observatory, with French telescope, French and Danish instrumentation
1990	Agreement between USSR and Germany on piloted missions
	Gamma observatory, with French telescope
	ESA begins contracting Russian equipment for Hermes programme
1991	Flight by Helen Sharman to Mir space station (Britain)
	Flight by Franz Viehböck to Mir space station (Austria)
1992	Flight by Klaus Dietrich Flade to Mir space station (Mir 92, Germany)
	Agreement for ESA flights to Mir
	Cancellation of Hermes programme
	Start of ESA-Russia cooperation on Mir 2 programme
	First western satellite launched by Proton (Inmarsat, for Britain)
1993	Start of International Space Station (ISS) partnership
	US lifts sanctions on Proton rocket flying western payloads
1994	First ESA flight to Mir, Ulf Merbold, EuroMir 94
1995	ESA opens office in Moscow
	Start of launches arranged by Commercial Space Technologies
	Formation of Eurockot consortium, with launches from 2002
1996	Replacement of 1966 agreement between France and USSR, now Russia
	Failure of Mars 8 mission
	Formation of joint company Starsem for launches by Soyuz rocket
1999	First launch of Soyuz Ikar
2000	Launch of Europe's Cluster satellites by Soyuz
	Netlander project (France-USA)
2001	Flight by Claudie Haigneré to ISS (France/ESA), first of taxi missions
	ESA announces Aurora programme
2002	ESA observatory INTEGRAL, launch on Proton rocket, 25% Russian observing time
	Start of *Soyouz à Kourou* project
2003	Russia sends Mars Express to Mars
2005	Period of cooperation on Kliper project
	Russia sends Venus Express to Venus
	France-Russia Ural launcher development studies
	Ground-breaking of *Soyouz à Kourou*
2008	Period of cooperation on ACTS project
	ExoMars project developed in cooperation with United States
2011	US withdraws from ExoMars programme
	Russia joins ExoMars programme, offering two Proton rockets
	First launch of Soyuz from Kourou, French Guyana
2014	United States and European Union sanction Russia
2016	Commemoration in Astana of visit by President de Gaulle
	Launch of first ExoMars mission; second part delayed from 2018 to 2020.
	ExoMars Trace Gas Orbiter enters Mars orbit, Schiaparelli crashes
2018	Dmitry Rogozin appointed head of Russian space agency, Roskosmos
2019	Spektr RG, with German eROSITA telescope, launched by Russia
2020	Second part of ExoMars project delayed to 2022, now ExoMars 2022

Appendix 2: Lists of heads of governments and space agencies

France (5th Republic onward)

Presidents

1959–69	Charles de Gaulle
1969–74	Georges Pompidou
1974–81	Valéry Giscard d'Estaing
1981–95	François Mitterrand
1995–2007	Jacques Chirac
2007–12	Nicholas Sarkozy
2012–17	François Hollande
2017–	Emmanuel Macron

Prime ministers

1959–62	Michel Debré
1962–8	Georges Pompidou
1968–9	Maurice Couve de Murville
1969–72	Jacques Chaban-Delmas
1972–4	Pierre Messmer
1974–6	Jacques Chirac
1976–81	Raymond Barre
1981–4	Pierre Mauroy
1984–6	Laurent Fabius
1986–8	Jacques Chirac
1988–91	Michel Rocard
1991–2	Édith Cresson
1992–3	Pierre Bérégovoy
1993–5	Édouard Balladur
1995–7	Alan Juppé

© Springer Nature Switzerland AG 2021
B. Harvey, *European-Russian Space Cooperation*, Springer Praxis Books,
https://doi.org/10.1007/978-3-030-67686-5

1997–2002	Lionel Jospin
2002–5	Jean-Pierre Raffarin
2005–7	Dominique de Villepin
2007–12	François Fillon
2012–14	Jean-Marc Ayrault
2014–16	Manuel Valls
2016–17	Bernard Cazeneuve
2017–20	Édouard Philippe
2020	Jean Castex

Chancellors of Germany

1949–63	Konrad Adenauer
1963–6	Ludwig Erhard
1966–9	Kurt Georg Kiesenger
1969–74	Willi Brandt
1974–82	Helmut Schmidt
1982–98	Helmut Kohl
1998–2005	Gerhard Schröder
2005–	Angela Merkel

Soviet Union/Russia

1953–64	Nikita Khrushchev
1964–82	Leonid Brezhnev
1982–4	Yuri Andropov
1984–5	Konstantin Chernenko
1985–91	Mikhail Gorbachev
1992–99	Boris Yeltsin
2000–	Vladimir Putin

Their leadership took the form of General Secretary of the Communist Party of the Soviet Union, but also President and President of the Praesidium.

Presidents of CNES

1961–2	Pierre Auger
1962–7	Jean Coulomb
1967–73	Jean-François Denisse
1973–6	Maurice Lévy
1976–84	Hubert Curien
1984–92	Jacques-Louis Lions
1992–5	René Pellat
1995–6	André Lebau
1996–2003	Alain Bensoussan
2003–13	Yannick d'Escatha
2013–	Jean-Yves Le Gall

Director Generals of the European Space Agency

1975–80	Roy Gibson
1980–4	Erik Quistgaard
1984–90	Reimar Lüst
1990–7	Jean-Marie Luton
1997–2003	Antonio Radotà
2003–15	Jean-Jacques Dordain
2015–21	Johann-Dietrich (Jan) Wörner

Directors of Roskosmos (and predecessor agencies)

1992–2004	Yuri Koptev
2004–11	Anatoli Perminov
2011–13	Vladimir Popovkin
2013–15	Oleg Ostapenko
2015–18	Igor Komarov
2018–	Dmitry Rogozin

Bibliography

Several excellent histories have already explored European-Russian cooperation, the most comprehensive being the 50th anniversary history by CNES of French-Russian cooperation *Cinquante ans de coopération France-URSS/Russia* (Paris, Institut Française d'Histoire de l'Espace, Éditions Tessier et Ashpool, 2014). This had a wealth of detail covering all aspects of bilateral Franco-Russian cooperation and benefits from the long standing tradition of French space journalism of Christian Lardier and his colleagues. Jacques Blamont wrote *Vénus dévoilée – voyage autour d'une planète (*Paris, Éditions Odile Jacob, 1987), both a personal account and a remarkable insider view of Soviet-French cooperation. There have been some specialized works, two of which should be especially highlighted: Facon, Isabelle & Sourbès-Verger, Isabelle: *La coopération spatiale Russie-Europe, une entreprise inachavée.* Revue Géoéconomie, 2007, #43; and Mathieu, Charlotte: *Assessing Russia's space cooperation with China and India – opportunities and challenges for Europe*. Vienna, European Space Policy Institute, 2008.

Two texts in the Praxis/Springer series address important aspects of this story and are recommended:

Lardier, Christian & Barensky, Stefan: T*he Soyuz launch vehicle – the two lives of an engineering triumph.* (Praxis/Springer, 2010); and

O'Sullivan, John: *In the footsteps of Columbus – European missions to the International Space Station* (Praxis/Springer, 2016).

The writer has made as much use as possible of original scientific papers that capture the collaborative efforts and projects of the two sides. They come principally from the International Astronautical Federation (IAF) annual conferences, called the International Astronautical Congress (IAC); those of COSPAR; and conferences convened by the Institute for Space Research (IKI), Moscow.

For the political background, Philip Short's biography *Mitterrand – a study in ambiguity (*London, Bodley Head, 2013) is recommended. For a contemporary context, readers should go to Marco Aliberti and Ksenia Lisitsyna's *Russia's*

© Springer Nature Switzerland AG 2021

B. Harvey, *European-Russian Space Cooperation*, Springer Praxis Books, https://doi.org/10.1007/978-3-030-67686-5

posture in space – prospects for Europe (Vienna, European Space Policy Institute with Springer, 2019). Two earlier histories are also recommended:

Rebrov, M; Kozyrev, V & Denissenko, V: *URSS-France – exploration de l'espace*. Moscow, Progress editions, 1983; and

Carlier, Claude & Gilli, Marcel: *The first thirty years at CNES, 1962–92*. CNES, Paris, 1994.

Magazines
Independent and in-house magazines and journals have provided reports and commentary on Russian and European cooperation. Those most prominent are:

Air & Cosmos
Air & Space
Astronomy Ireland
Aviation Week and Space Technology
Flight International
Spaceflight (British Interplanetary Society)
Space:UK
DLR Magazine
La Lettre d'information du CNES (–1997)
CNES Magazine (1998–2005); CNESMag (2005–)
Wallonie Espace

Websites
Several websites provide quality documentation that includes European-Russian cooperation. Foremost of these are the European Space Agency (ESA); the Russian space agency, Roskosmos; the French space agency, CNES; and the German space agency, DLR. The best narrative, commentaries and analyses are provided by Anatoli Zak's Russianspaceweb.

CNES: www.cnes.fr
DLR: www.dlr.de
ESA: www.esa.int
Roskosmos: www.roskosmos.ru
Russianspaceweb: www.russianspaceweb.com
Space.com: www.space.com
Space Daily: www.spacedaily.com

Index

Printed in the United States
by Baker & Taylor Publisher Services